ENGINEERED PLUMBING DESIGN II

Original book by
Alfred Steele, P.E., CIPE

Revised by
A. Calvin Laws, P.E., CPD

Published by
American Society of Plumbing Engineers

Engineered Plumbing Design II is designed to provide accurate and authoritative information for the design and specification of plumbing systems. The publisher makes no guarantees or warranties, expressed or implied, regarding the data and information contained in this publication. All data and information are provided with the understanding that the publisher is not engaged in rendering legal, consulting, engineering, or other professional services. If legal, consulting, or engineering advice or other expert assistance is required, the services of a competent professional should be engaged.

American Society of Plumbing Engineers
2980 S. River Rd.
Des Plaines, IL 60018
(847) 296-0002
E-mail: aspehq@aspe.org • Internet: www.aspe.org

Copyright © 2004 by American Society of Plumbing Engineers

Originally Printed 2004, Reprinted with Corrections 2006

All rights reserved. No part of this book may be reproduced or transmitted in any form or by any means, electronic or mechanical, including photocopying and recording, or by any information storage or retrieval system, without permission in writing from the publisher.

ISBN 978-1-891255-20-5
Printed in the United States of America

10 9 8 7 6 5 4 3

Acknowledgments

Numerous people have contributed to, reviewed, and otherwise helped to make possible this book, which has been many years in the making. While it would be impossible to list the names of all these individuals, The American Society of Plumbing Engineers appreciates the work they have done. Special acknowledgment goes to the following:

A. Calvin Laws, P.E., CPD, for his effort in rewriting and revising the original;

Frank Teebagy, P.E., C.I.P.E., and **Harold Olsen, P.E.,** for their peer review and contributions;

Jill Dirksen, ASPE Technical Director, for her editorial work and contributions.

The Society also would like to thank Coordinating Editor **Mina Freehill** who had the ultimate challenge of bringing together the many editorial comments and author and committee variations.

Page design and layout by David Ropinski

About ASPE

The American Society of Plumbing Engineers (ASPE) is the international organization for professionals skilled in the design and specification of plumbing systems. ASPE is dedicated to the advancement of the science of plumbing engineering, the professional growth and advancement of its members, and the health, welfare, and safety of the public.

The Society disseminates technical data and information, sponsors activities that facilitate interaction with fellow professionals, and, through research and education programs, expands the base of knowledge of the plumbing-engineering industry. ASPE members are leaders in innovative plumbing design, effective materials and energy use, and the application of advanced techniques from around the world.

Worldwide Membership — ASPE was founded in 1964 and currently has 7,500 members. Spanning the globe, members are located in the United States, Canada, Asia, Mexico, South America, the South Pacific, Australia, and Europe. They represent an extensive network of experienced engineers, designers, code officials, educators, contractors, and manufacturers interested in furthering their careers, their profession, and the industry. ASPE is at the forefront of technology. In addition, ASPE represents members and promotes the profession among all segments of the construction industry.

ASPE Membership Communication — All members belong to ASPE worldwide and have the opportunity to belong and participate in one of the 61 state, provincial, or local chapters throughout the U.S. and Canada. ASPE chapters provide the major communication links and the first line of services and programs for the individual member. Communications with the membership is enhanced through the Society's bimonthly magazine, *Plumbing Systems and Design*, and the bimonthly newsletter, ASPE Report, which is incorporated into of the magazine.

Technical Publications — The Society maintains a comprehensive publishing program, spearheaded by the profession's basic reference text, the ASPE *Plumbing Engineering Design Handbook* (PEDH). The PEDH encompassing 47 chapters in four volumes, provides comprehensive details of the accepted practices and design criteria used in the field of plumbing engineering. New additions that will shortly join ASPE's published library of professional technical manuals and handbooks include: *Pharmaceutical Facilities Design Manual, High-Technology Electronic Facilities Design Manual, Health Care Facilities and Hospitals Design Manual*, and *Water Reuse Design Manual*.

Convention and Technical Symposium — The Society hosts biennial Conventions in even-numbered years and Technical Symposia in odd-numbered years so professional plumbing engineers and designers can improve their skills, learn original concepts, and make important networking contacts to help them stay abreast of current trends and technologies. In conjunction with each Convention there is an Engineered Plumbing Exposition, the largest gathering of plumbing engineering and design products, equipment, and services. Products on display include pipes to pumps to fixtures, compressors to computers to consulting services, giving engineers and specifiers the opportunity to view the newest and most innovative materials and equipment available.

Certified in Plumbing Design — ASPE sponsors a national certification program for engineers and designers of plumbing systems, which carries the designation Certified in Plumbing Design or CPD. The certification program provides the profession, the plumbing industry, and the general public with a single, comprehensive qualification of professional competence for engineers and designers of plumbing systems. The CPD, designed exclusively by and for plumbing engineers, tests hundreds of engineers and designers at centers throughout the United States biennially. Created to provide a single, uniform national credential in the field of engineered plumbing systems, the CPD program is not in any way connected to state-regulated Professional Engineer (P.E.) registration.

ASPE Research Foundation — The ASPE Research Foundation, established in 1976, is the only independent, impartial organization involved in plumbing engineering and design research. The science of plumbing engineering affects everything… from the quality of our drinking water, to the conservation of our water resources, to the building codes for plumbing systems. Our lives are impacted daily by the advances made in plumbing engineering technology through the Foundation's research and development.

Preface

Alfred Steele, P.E., CIPE, authored the first Engineered Plumbing Design book in 1977 and revised it in a 1982 second edition. Al passed away, at age 82, in February, 1998, leaving behind an enduring work in the field of plumbing engineering. With his passing, the American Society of Plumbing Engineers had the foresight to purchase all rights to the original book.

With this book, I have tried to carry on where he left off. Engineered Plumbing Design is a unique book in that the mix between hands-on practical information and the scientific underpinnings of plumbing design is perfectly matched for the designer's needs. In contrast, many books have been published on plumbing that are filled with numerous illustrations of tools, fittings, tables, etc., which do little to assist in the design of engineered plumbing systems and, in particular, they leave out the scientific basis needed to make sound decisions in designing plumbing systems.

This edition was a collaborative effort to update all parts of the text, where appropriate, correct some minor errors, and improve the graphics and the general presentation of the book — to make it relevant and available for the next generation of plumbing engineers. I thank Mina Freehill and the Society Staff for coordinating the many editorial comments. The largest share of gratitude goes to my fellow ASPE members and educators, Frank G. Teebagy, P.E., CIPE and Harold L. Olson, P.E. for their input in editing of the second edition of Engineered Plumbing Design and to Jill J. Dirksen, Director of Technical Services with the American Society of Plumbing Engineers, for all her help in editing and in bringing this edition to publication.

To the American Society of Plumbing Engineers, a special "Thank You" for your wisdom and foresight to continue the tradition of the profession with the publication of this book; and for offering me the honor and privilege of following in a great man's footsteps.

A. Calvin Laws, P.E., CPD

Table of Contents

Chapter 1: Systems and Fixtures 1
   Fixture Selection 2
   Quality of Fixtures 2
   Fixture Classification 4
   Water Closets 4
   Shape and Size 7
   Water Closet Seat 8
   Flushing Performance 8
   Installation Requirements 9
   Flushing Systems 10
   Flush Tank Requirements 11
   Urinals 12
   Urinal Styles 13
   Flushing Performance 13
   Installation Requirements 14
   Flushing Requirements 15
   Lavatories 15
   Size and Shape 15
   Sinks 17
   Service Sinks 17
   Bathtubs 18
   Showers 18
   Drinking Fountains 19
   Bidets 19

Chapter 2: Fixture Traps 20
   Prohibited Traps 20
   Trap Seal 21
   Siphonage 21
   Induced Siphonage 21
   Self-Siphonage 22
   Limitation of Trap Seal Loss 24
   Trap Installation 24
   Intercepting Fixture Traps 25
   Grease Interceptors 26
   Questions 26

Chapter 3: Flow in Horizontal Drainage Piping 27
   Uniform Flow 27
   Scouring Action 30

   Surcharging 30
   Sewer Shapes 31
   Questions 32

Chapter 4: Soil and Waste Stacks 33
   Stack Connections 33
   Flow in Stacks 33
   Terminal Velocity Length 34
   Stack Capacities 35
   Hydraulic Jump 35
   Questions 37

Chapter 5: Drainage Systems 38
   Storm Water Disposal 38
   Combined Systems 38
   High-Temperature Wastes 39
   Drainage Systems Below Sewer Level 39
   Backwater Valves 39
   House Traps and Fresh Air Inlets . 39
   Connections to Sanitary House Drains 41
   Branch Connections to Stack Offsets 41
   Piping Installation 42
   Cleanouts 44
   Indirect Wastes 47
   Special Wastes 47
   Questions 49

Chapter 6: Drainage System Sizing 50
   Rate of Flow in Branches 50
   Fixture Drain Size 50
   Sanitary Drainage Fixture Unit 51
   Stack Sizing 52
   Procedure for Sizing Stacks 55
   House Drains 58
   Questions 62

Chapter 7: Storm Water Systems 63
Collection Areas 64
Vertical Walls 65
Sizing 66
Roof Gutters 66
Roof Drains 66
Material 67
Flow Velocity 68
Controlled-Flow Roof Drainage 68
Roof Loading 68
Storm Intensity 69
Drain-Down Time 69
Design Configuration is the Key ... 69
Suggested Code for Controlled-Flow Roof Drainage 70
Subsoil Drainage 70
Combined Storm and Sanitary System 70
Questions 73

Chapter 8: Vent Systems 74
Static Head 75
Pneumatic Effects in Sanitary Systems 75
Rate of Flow from Outlets 75
Static Pressure of Air 76
Friction Head Loss 76
Air Flow in Stacks 77
Air Flow in Horizontal Drains 78
Permissible Length of Vent Pipe ... 78
Gravity Circulation 78
Vent Stacks 79
Vent Terminals 79
Fixture Trap Vents 79
Distance of Vent from Trap 81
Various Methods of Fixture Trap Venting 83
Relief Vents 83
Continuous Venting 84
Wet Venting 84
Combination Waste and Vent Venting 86
Circuit and Loop Venting 87
Common Vents 88
Suds Pressure 88
Vapor Vents (Local Vents) 90
Ejector and Sump Vents 90
Frost Closure 91
Tests of Plumbing Systems 92
Questions 93

Chapter 9: Vent Sizing 94
Sizing Vent Extensions and Terminals 94
Sizing Vent Headers 94
Sizing Individual Vents and Branch Vents 96
Sizing Relief Vents 97
Sizing Circuit Vents 97

Chapter 10: Sumps and Ejectors 100
Terminology 100
Ejector Basin 100
Basin Materials 100
Lifting Devices 101
Operation of a Pneumatic Ejector 101
Ejector Pump Sizing (Centrifugal) 103
Ejector Basin Sizing 106
Controls 107
Installation 107
Sump Basin Sizing 108
Sump Pump Sizing 108
Controls 108
A General Rule for a Subdrainage System 108

Chapter 11: Flow in Water Piping 109
Physical Properties of Water 109
Types of Flow 110
Velocity of Flow 110
Potential Energy 111
Kinetic Energy 111
Static Head 112
Velocity Head 113
Bernoulli's Theorem 114
Friction 115
Flow from Outlets 115
Flow in Piping 115
Friction in Piping 117

Chapter 12: Velocity Effects in Piping 120
Erosion, Noise, and Cavitation ... 124

Chapter 13: Water System Design 125
Flow Pressure 125
Flow at an Outlet 126
Constant Flow 126
Material Selection 128

Parallel Circuits 128
Inadequate Pressure..................... 130
Flow Definitions 130
Demand Types 131
Estimating Demand..................... 131
Design Loads 131
Water Supply Fixture Units 132

Chapter 14: Water System Sizing...............................138
Friction Head Loss....................... 138
Maximum Velocity...................... 139
Minimum Sizes 140
Procedure for Sizing.................... 140
A Hydropneumatic or Booster Pump System 141

Chapter 15: Water System Components 151
Protection of the Potable Water Supply 151
   Protective Methods for Below-the-Rim Supply.............................. 152
      Rules Relative to Submerged Inlets 152
Water Meters 154
   Displacement Meters 155
   Current (or Velocity) Meters... 156
   Proportional Meters 156
   Compound Meters 156
Water Meter Rules 157
Piping Installation 157
Valve Types................................... 159

Chapter 16: Hot Water System Design 160
Objectives 160
Safety Devices 161
Water Heaters 162
   Directly Heated Automatic Storage Heaters 162
   Instantaneous Heaters 163
   Booster Heaters........................ 164
   Semi-Instantaneous Heaters ... 164
   Storage Water Heaters 165
      Sizing Storage-Type Heaters....... 167
      Installation 176
      Hot-Water Temperature 178
      Safety and Health Concerns 178

Chapter 17: Sizing the Hot Water Circulating System........... 181
System Types................................ 181
Sizing 185
   Procedure 185
   Rules of Thumb....................... 191

Chapter 18: Pipe Expansion and Contraction 192

Chapter 19: Water Piping Tests 196
Disinfection 196
   Disinfection of Water Systems 196

Chapter 20: Chilled Drinking Water Systems................... 198
Drinking Water Coolers.............. 198
Refrigeration Components 199
Stream Regulators 200
Central and Unitary Systems...... 200
Central Chilled Drinking Water Design... 202
   Circulating Pump Capacity 204
   Makeup Water Mixture 204
   Storage Tank............................ 204
   Piping 205

Chapter 21: Private Sewage Disposal Systems 207
Sewage System Criteria 207
Cesspools...................................... 208
Septic Tanks 208
   Removal of Solids.................... 208
   Sludge and Scum Storage 209
   Septic Tank Location 209
   Tank Capacity 210
   Tank Material 210
   Tank Access 210
   Tank Inlet 210
   Tank Outlet 210
   Tank Shape 211
   Scum Storage Space 211
   Compartments 211
   Cleaning of Tanks.................... 211
   Chemical Additives 212
   Septic Tanks for Nonresidential Buildings 213
Subsurface Soil Absorption System 214
   Criteria for Design 214

Percolation Tests 214
    Procedure for Percolation Tests .. 215
Absorption Area 216
Absorption Trenches 216
    Construction 218
Seepage Beds 218
    Design Criteria for Seepage Beds 219
    Distribution Boxes 220
Seepage Pits 220
    Effective Area of Seepage Pit 220
    Construction of Seepage Pit 221

Chapter 22: Valves 222
Valve Selection 222
Gate Valves 222
    Valve Stems 222
    Bonnets 225
    Discs 226
    Materials 229
    Trim 229
    Packing 230
    End Connections 230
    Application 231
    Operation and Maintenance 232
    Trends 232
Globe Valves 232
    Globe Valve Seating 232
    Globe Valve Structure 232
    Materials and End
    Connections 233
    Bonnets 233
    Stems 233
    Discs 233
        Tapered Plug Disc 233
        Conventional Disc 233
        Composition Disc 233
    Seals 234
    Angle Valves 234
    Installation 236
Check Valves 236
    Swing-Check Valves 237
    Double-Disc Check Valves 237
    Slanting-Disc Check Valves 238
    Lift Check Valves 239
    Silent Check Valves 239
    Installation 241
    Sizing 241
Pressure-Regulating Valves
(PRV) 241
    PRV Characteristics 243
    Outlet Pressure 246
    PRV Sizing 247
    Cavitation 247
    Series Hookup 247
    Parallel Hookup 248
    Installation 249
Quarter-Turn Valves 249
    Plug Valves 250
    Ball Valves 250
    Butterfly Valves 251

Figures

Figure 1-1 The older styles of water closets were identified as (A) reverse trap, (B) blowout, and (C) siphon jet, to name a few. Though still used in the industry, these terms are no longer used in the standards. 4

Figure 1-2 Water closets are identified as (A) close coupled, (B) one piece, and (C) flushometer types. 5

Figure 1-3 A wall hung water closet attaches to the back wall; the water closet does not contact the floor. 6

Figure 1-4 Carrier for a Water Closet 6

Figure 1-5 A floor-mounted, back outlet water closet is supported on the floor with the piping connection through the back wall. 7

Figure 1-6 The standard rough-in dimension is 12 in. from the centerline of the water closet outlet to the back wall. The floor flange must be permanently secured to the building structure. 7

Figure 1-7 (A) A Gravity Tank and (B) a Flushometer Tank 11

Figure 1-8 Flush Tank (Gravity) 12

Figure 1-9 Urinal spacing must be adequate to allow adjacent users to access the urinals without interference. 14

Figure 1-10 Slab-Type Lavatory 16
Figure 1-11 Splash-Back Lavatory ... 16
Figure 1-12 Shelf-Back Lavatory 16
Figure 1-13 Ledge-Back Lavatory 16
Figure 1-14 Surgeon's Sink 18
Figure 1-15 Double-Compartment Sink 18
Figure 2-1 Typical Fixture Trap 20
Figure 2-2 Crown Vented Trap 20
Figure 2-3 Trap Action Under Excess Positive Pressure 21
Figure 2-4 Trap Action Under Negative Pressure 22
Figure 2-5 Maximum Permissible Positive and Negative Pressure 22
Figure 2-6 Location of Traps 25

Figure 2-7 Grease Interceptor 25
Figure 3-1 Half-Full Flow 28
Figure 3-2 Illustration of Sewer with Surcharge 31
Figure 3-3 Sewer Shapes 32
Figure 4-1 Cross-Section of Stack Flowing at Design Capacity 34
Figure 4-2 Hydraulic Jump at Offset 36
Figure 5-1 Backwater Valve and Combination BWV with Manually Operated Gate Valve 40
Figure 5-2 Installation of a House Trap with Fresh Air Inlet 42
Figure 5-3 Piping for Fixtures Directly Above Offset 43
Figure 5-4 Underground Drainage Piping 42
Figure 5-5 Hangers and Supports 45
Figure 5-6 Maximum Vent Connection Distance 47
Figure 5-7 Designing Cleanouts 48
Figure 6-1 Diagrammatic Representation of Interference of Flows in Stack Fitting 53
Figure 6-2 Branch Interval 54
Figure 6-3 Procedure for Sizing a Stack 55
Figure 6-4 Procedure for Sizing an Offset Stack 56
Figure 6-5 Sizing Example 57
Figure 6-6 Sanitary System with an Ejector 59
Figure 6-7 A Simple Drainage System 60
Figure 7-1 Connection of Leader of Combined Building Drain 64
Figure 7-2 A Simple Drainage System 65
Figure 7-3 Typical Roof Drain 67
Figure 7-4 Plan of a Combined Storm and Sanitary System 72
Figure 8-1 Various Vent Stack Connections 80
Figure 8-2 Vent Pipe Opening 81
Figure 8-3 Various Fixture Trap Vents 83

Figure 8-4 Horizontal Run of Vent .. 84
Figure 8-5 Venting for Stacks Having More Than 10 Branch Intervals 85
Figure 8-6 Venting at Stack Offsets . 86
Figure 8-7 Wet Venting at Top Floor 87
Figure 8-8 Wet Venting Below Top Floor 87
Figure 8-9 Stack Vented Unit........... 88
Figure 8-10 Circuit and Loop Venting............. 89
Figure 8-11 Suds Pressure Zones 91
Figure 9-1 Developed Length of a Vent Stack 95
Figure 9-2 Vent Stack Sizing............ 97
Figure 9-3 Sizing Vent Headers 99
Figure 10-1 Submersible Motor and Pump............ 102
Figure 10-2 Top-Mounted Motor with Submersible Pump......... 103
Figure 10-3 Top-Mounted Motor and Pump............ 104
Figure 10-4 Dry Pit......... 105
Figure 10-5 Operation of a Pheumatic Ejector......... 107
  A) Filling Position............. 107
  B) Rising Sewage 107
  C) Completing Cycle 107
Figure 11-1 Bernoulli's Theorem (Disregarding Friction).................. 114
Figure 11-2 Toricelli's Theorem..... 116
Figure 12-1 Illustrations of a Shock Wave 121
Figure 12-2 Shock Absorber 123
Figure 12-3 Pressure Surge Control Curves............. 123
Figure 13-1 Flow Control 126
Figure 13-2 Flow Control Device Curve (Dole Valve) 127
Figure 13-3 Typical Parallel Pipe Circuit............. 129
Figure 13-4 Example of Division of Flow in a Parallel Pipe Circuit 129
Figure 13-5 Conversion of Fixture Units to gpm.......... 135

Figure 13-6 Conversion of Fixture Units to gpm (enlarged scale) 135
Figure 13-7 Conversion of Fixture Units to gpm (Mixed System) 137
Figure 14-1 Sizing of Distribution System 149
Figure 15-1 Typical Meter Setting Including Bypass.......... 154
Figure 15-2 Typical Internal View of A Disc Meter 155
Figure 16-1 Conventional U-Tube Instantaneous Water Heater......... 163
Figure 16-2 Straight Tube,Floating Head Instantaneous Water Heater 163
Figure 16-3 Shell and Coil Limited Storage Type Instantaneous Heater 165
Figure 16-4 Conventional Storage Water Heaters—Note Temperature Regulator Location 166
Figure 16-5 Apartments................ 169
Figure 16-6 Office Buildings.......... 169
Figure 16-7 Nursing Homes 169
Figure 16-8 Motels........ 169
Figure 16-9 Food Service 170
Figure 16-10 Elementary Schools .. 170
Figure 16-11 High Schools 170
Figure 16-12 Dormitories.............. 170
Figure 16-13 Typical Hot Water Storage Tank Heater................... 177
Figure 16-14 Dual Temperature Hot Water Heating.............. 178
Figure 17-1 Upfeed System (Heater Located at Bottom of System) 182
Figure 17-2 Downfeed System (Heater Located at Bottom of System) 182
Figure 17-3 Combination Upfeed and Downfeed System (Heater Located at Bottom of System) 183
Figure 17-4 Downfeed System (Heater Located at Top of System) 183
Figure 17-5 Combination Upfeed and Downfeed System(Heater Located at Top of System) 184
Figure 17-6 Upfeed System (Heater Located at Top of System) 184

Figure 17-7 Riser Diagram for 24-Story Building........................... 189
Figure 18-1 Piping to Absorb Movement...................................... 193
Figure 18-2 Pipe Deformity........... 194
Figure 18-3 Riser Expansion 195
Figure 20-1 Schematic of Refrigeration Cycle 200
Figure 20-2 Downfeed Return Loop 201
Figure 20-3 Upfeed Return Loop... 202
Figure 20-4 Multi-Riser Downfeed System 204
Figure 20-5 Multi-Riser Upfeed System 205
Figure 21-1 Precast Septic Tank..... 212
Figure 21-2 Section through Typical Absorption Trench 219
Figure 22-1 Rising Stem — Outside Screw and Yoke 223
Figure 22-2 Rising Stem — Inside Screw 224
Figure 22-3 Nonrising Stem — Inside Screw 224
Figure 22-4 Bronze Gate Valve Rising Stem — Wedge Disc 225
Figure 22-5 Bronze Gate Valve Rising Stem — Double Disc 226
Figure 22-6 Non-Rising Stem Valve Open Position 227
Figure 22-7 Bronze Gate Valve Non-Rising Stem — Wedge Disc 227
Figure 22-8 Solid Wedge Disc 228
Figure 22-9 Split-Wedge Design 228
Figure 22-10 Union Bonnet, Plug Disc, Renewable Seat........................... 234
Figure 22-11 Union Bonnet, Conventional Disc 235
Figure 22-12 Screwed Bonnet, Conventional Disc 235

Figure 22-13 Union Bonnet, Composition Disc 236
Figure 22-14 Conventional-Swing Check 237
Figure 22-15 Swing Check with Outside Lever & Weight............... 238
Figure 22-16 Double-Disc Check Valve 238
Figure 22-17 Slanting-Disc Check Valve 239
Figure 22-18 Horizontal-Lift Check Valve 239
Figure 22-19 Horizontal-Ball Lift Check Valve 240
Figure 22-20 Silent Check Valve.... 240
Figure 22-21 Operation of a Pressure-Regulating Valve 242
Figure 22-22 Direct-Operated, Spring Loaded PRV 243
Figure 22-23 Direct-Operated Diaphragm PRV 244
Figure 22-24 Pilot-Operated Pressure-Regulating Valve 245
Figure 22-25 Cavitation Chart....... 248
Figure 22-26 Typical PRV Assembly 249
Figure 22-27 Plug Valve................. 250
Figure 22-28 Ball Valve.................. 250
Figure 22-29 Full and Reduced Port Ball Valves...................................... 251
Figure 22-30 Dynamic Forces Acting on Disc of Butterfly Valve 252
Figure 22-31 Butterfly Valve Actuators 252
Figure 22-32 Butterfly Valve Disc Types... 253
Figure 22-33 Butterfly Valve Types... 253

Tables

Table 1-1   Standards That Affect Plumbing Design and Engineering 3
Table 3-1   Values of R, $R^{2/3}$ and A for Full Flow 28
Table 3-2   Values of R, $R^{2/3}$ and A for Half-full Flow 29
Table 3-3   Values of S and $S^{1/2}$ 29
Table 3-4   Uniform Flow Velocity and Capacity of Sanitary Drains at ¼" Slope 29
Table 3-5   Uniform Flow Velocity and Capacity of Storm Drains at ¼" Slope and Full Flow (n = 0.0145) 30
Table 4-1   Maximum Capacities Stacks 36
Table 6-1   Minimum Size of Trap for Plumbing Fixtures 51
Table 6-2   Fixture Units Per Fixture or Group 52
Table 6-3   Maximum Permissible F.U. Loads for Sanitary Stacks 53
Table 6-4   Maximum Permissible F.U. Loads for Sanitary Branches 54
Table 6-5   Maximum Permissible Fixture Unit Loads for Sanitary Building Drains and Runouts from Stacks 58
Table 6-6   Approximate Flow Velocity of Sewage 59
Table 7-1   Maximum Permissible Loads for Storm Drainage Piping 66
Table 7-2   Maximum Permissible Loads for Semicircular Gutters 66
Table 7-3   Rectangular Leaders Equivalent to Round Leaders 66
Table 8-1   Discharge Rates of Air (1 Inch Water Pressure) 76
Table 8-2   Air Required by Attendant Vent Stacks (Drainage Stack Flowing 7/24 Full) 77
Table 8-3   Rate of Air In Horizontal Drains 78
Table 8-4   Distance of Vent from Fixture Traps 81
Table 9-1   Size and Length of Vent 96
Table 9-2   Maximum Permissible Lengths of Vents for Horizontal Branches 98
Table 9-3   Horizontal Circuit and Loop Vent Sizing Table 98
Table 11-1   Density of Pure Water at Various Temperatures 109
Table 11-2   Heads of Water in Feet Corresponding to Pressure in Pounds per Square Inch 112
Table 11-3   Average Values for Coefficient of Friction, f 118
Table 11-4   Values of $d^{2½}$ 119
Table 11-5   Equivalent Pipe Length for Valves and Fittings 119
Table 13-1   Actual Inside Diameter of Piping, in Inches 128
Table 13-2   Demand at Individual Fixtures and Required Pressure 131
Table 13-3   Demand Weight of Fixtures, in Fixture Units 133
Table 13-4   Conversion of Fixture Units to Equivalent gpm 134
Table 14-1   ½ inch 142
Table 14-2   ¾ inch 142
Table 14-3   1 inch 143
Table 14-4   1¼ inch 143
Table 14-5   1½ inch 144
Table 14-6   2 inch 144
Table 14-7   2½ inch 145
Table 14-8   3 inch 145
Table 14-9   4 inch 146
Table 14-10   5 inch 146
Table 14-11   6 inch 147
Table 14-12   8 inch 147
Table 14-13   Minimum Size of Fixture Supply Pipes 140
Table 14-14   Cold Water 148
Table 14-15   Hot Water 149
Table 15-1   Disc Meter 155
Table 15-2   Maximum Support Distance 158
Table 16-1   Hot Water Demands and Use for Various Types of Buildings 167
Table 16-2   Estimated Hot Water Demand Characteristics for Various Types of Buildings 168
Table 16-3   Hot Water Demand per Fixture for Various Types of Buildings (Gallons of water per hour per fixture, calculated at a final temperature of 140°F) 175
Table 16-4   Mixing of Hot and Cold Water (Ratio of Gallons) 176
Table 16-5   Mixing of Hot and Cold Water (Percentage Basis) 177

Table 16-6 Typical Hot-Water Temperatures for Plumbing Fixtures and Equipment 179
Table 16-7 Time/Water Temperature Combinations Producing Skin Damage 179
Table 17-1 Piping Heat Loss (Btu/hr. Per Lineal Ft. For 140°F. Water Temp and 70°F. Room Temp.) 186
Table 17–2 Heat Loss and Circulation Rate 190
Table 17-3 Sizing The Hot Water Circulating System 191
Table 18-1 Pipe Expansion For 80° F Temperature Change 192
Table 18-2 Developed Length of Pipe to Absorb ½-In. Movement 194
Table 20-1 Summer Tap Water Temperatures 198
Table 20-2 Drinking Water Requirements 203
Table 20-3 Refrigeration Load 203
Table 20-4 Circulating System Line Loss (Heat Gain) Approx. 1-In. Insulation 203
Table 20-5 Circulating Pump Heat Input 203
Table 20-6 Circulating Pump Capacity 204
Table 21-1 Liquid Capacity of Tank (gal) 210
Table 21-2 Quantities of Sewage Flows 213
Table 21-3 Sewage Flow from Country Clubs 214
Table 21-4 Sewage Flow at Public Parks 214
Table 21-5 Absorption Area Requirements for Individual Residences[a] 214
Table 21-6 Allowable Rate of Sewage Application to a Soil Absorption System 215
Table 21-7 Minimum Distance Between Components of Sewage Disposal System 217
Table 21-8 Vertical Wall Areas of Seepage Effective Strata Depth Below Flow Line (below inlet) 221
Table 22-2 Valve Standards Agencies 230
Table 22-1 Valve Material Specifications 230
Table 22-3 Pressure-Regulating Valve Glossary 246

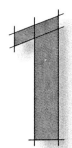

Systems and Fixtures

The American Society of Plumbing Engineers defines plumbing systems as all potable water supply and distribution pipes, plumbing fixtures and traps, drainage and vent pipes, and building (house) drains, including their respective joints and connections, devices, receptacles, and appurtenances within the property lines of the premises and including potable water piping, potable water treating or using equipment, fuel gas piping, water heaters, and vents for same.

A model code defines plumbing systems as "all potable water building supply and distribution pipes, all plumbing fixtures and traps, all drainage and vent pipe(s), and all building drains and building sewers, including their respective joints and connection devices, receptors, and appurtenances within the property lines of the premises and shall include potable water piping, potable water treating or using equipment, medical gas and medical vacuum systems, fuel gas piping, water heaters and vents for same."

Plumbing engineers are responsible for systems that serve all types of buildings, including commercial, residential, and institutional buildings, such as hospitals, laboratories, industrial plants, jails, schools, shopping centers, housing developments, power plants, research centers, and sports complexes.

The plumbing engineer is now responsible for design of the following systems:

1. Sanitary drainage
2. Sanitary sewage disposal
3. Storm water drainage
4. Site drainage
5. Storm water disposal
6. Venting
7. Domestic water
   A. Cold water
   B. Hot water
   C. Hot water circulation
   D. Tempered water
   E. Tepid water for emergency eyewash and showers
8. Fire protection
   A. Standpipe
   B. Sprinkler
   C. CO2
   D. Clean agent
9. Acid and industrial waste
10. Chilled drinking water
11. Gas
    A. Natural and manufactured
    B. Liquefied petroleum (LP)
12. Compressed air
13. Vacuum
    A. Clinical and surgical
    B. Laboratory
    C. Cleaning
14. Argon
15. Oxygen
16. Carbon dioxide
17. Nitrogen
18. Nitrous oxide
19. Helium
20. Deionized water
21. Distilled water

22. Water treatment
23. Liquid soap dispensing
24. Disinfectant
25. Food waste disposal and solid waste handling
26. Radioactive waste
27. Pools and decorative fountains
28. Lawn sprinkler and irrigation.

Although this list may seem extensive, there are many additional specialized and exotic systems for which the plumbing engineer is called upon to furnish his or her professional expertise.

Fixture Selection

The type, quantity, and arrangement of plumbing fixtures is usually the prerogative of the architect, but the engineer must evaluate and advise the architect as to type and arrangement and, particularly, space requirements. The type and quantity of fixtures to be installed in a building is predicated upon the number of people served and the type of building occupancy. These requirements are clearly delineated in every building code. Separate facilities must be provided for male and female personnel and these facilities must be within easy access from any floor of the building. "Easy access" has been interpreted to mean within one floor distance so that a person never has to walk more than one floor up or down.

Quality of Fixtures

Manufacturers have accepted certain standards for the manufacture of plumbing fixtures. Most manufacturers adhere to these standards so that, at the present time, fixture quality is a minor problem. A list of standards affecting plumbing fixtures, plumbing design, and engineering is included in Table 1-1. These standards include some that apply to important recent trends: The limitation of water consumption in water closets to 1.6 gallons per flush (gpf) is required in most jurisdictions and it is required that fixtures and designs ensure accessibility for the handicapped in public and private buildings. Engineers should be familiar with these standards and must consult applicable codes for the jurisdiction in which the design is being done.

When evaluating fixtures, the following characteristics should be carefully checked:
1. Strength
2. Durability
3. Corrosion resistance (acid resisting)
4. Abrasion resistance
5. Absence of defects
6. Adequate performance for the service intended
7. Concealed fouling surfaces.

Materials most commonly used in the manufacture of fixtures are enameled cast iron, enameled pressed steel, vitreous china, vitrified earthenware, and stainless steel. Additionally, plastics, aluminum, and stone compositions have been used.

Table 1-1 Standards That Affect Plumbing Design and Engineering

The following list contains the significant industrial standards that affect plumbing and related products. Most standards are not free of charge. Contact appropriate organizations for costs. Bold face listings are the name and address of the organization where the specific standards can be obtained. Beneath each organization is the name and title of the plumbing related standard.

American National Standards Institute (ANSI)
25 West 43rd Street, 4th Floor, New York, NY 10036
Phone: (212) 642-4900, Toll-free: (888) 267-4783
Fax: (212) 398-0023
Email: info@ansi.org
www.ansi.org
Dr. Mark W. Hurwitz, CAE, President/CEO

ANSI A112.1.2: Air gaps in Plumbing Systems. ANSI A112.6.1M Supports for Off-the-Floor Plumbing Fixtures for Public Use. ANSI A112.18.1M Finished and Rough Brass Plumbing Fixtures. ANSI A112.19.1M Enameled Cast Iron Plumbing Fixtures. ANSI A112.19.2 Vitreous China Plumbing Fixtures. ANSI A112.19.3 Stainless Steel Plumbing Fixtures. ANSI A112.19.4 Porcelain Enameled Formed Steel Plumbing Fixtures. ANSI A112.19.5 Trim for Water Closet Bowls, Tanks and Urinals. ANSI A117.1 Specifications for Making Buildings and Facilities Accessible to and Usable by Physically Handicapped People.

ANSI Z124.1 Plastic Bathtub Units. ANSI Z124.2 Plastic Shower Receptors and Shower Stalls. ANSI Z124.3 Plastic Lavatory Units. ANSI Z124.4 Plastic Water Closet Tanks and Bowls Proposed. ANSI Z124.5 Water Closet Seats Proposed.

American Society of Heating, Refrigeration, & Air Conditioning Engineers
1791 Tullie Circle N.E., Atlanta, GA 30329-2305
Phone: (404) 636-8400, Toll-free: (800) 527-4723
Fax: (404) 321-5478
Email: ashrae@ashrae.org
www.ashrae.org
Frank M. Coda, Secretary & Executive VP

ASHRAE 90.1 Energy Efficient Design of New Buildings Except Low-Rise Residential Buildings. ASHRAE 90.2 Energy Efficient Design of New Low-Rise Residential Buildings. ASHRAE 100 Energy Conservation in Existing Buildings – Residential.

American Society of Sanitary Engineering (ASSE)
901 Canterbury, Suite A, Westlake, OH 44145
Phone: (440) 835-3040, Fax: (440) 835-3488
Email: asse@ix.netcom.com
www.asse-plumbing.org
Shannon Corcoran, Executive Director

ASSE 1001 Pipe Allied Vacuum Breakers. ASSE 1002 Water Closet Flush Tank Ball Cocks. ASSE 1008 Household Food Waster Disposer. ASSE 1009 Commercial Food Water Disposer. ASSE 1010 Water Hammer Arrestors. ASSE 1014 Handheld Showers. ASSE 1016 Individual Shower Control Valves, Anti-Scald Type. ASSE 1017 Thermostatic Mixing Valves, Self-Activated for Primary Domestic Use. ASSE 1018 Trap Seal Primer Valves. ASSE 1019 Wall Hydrants, Frostproof Automatic Draining Anti-Backflow Type. ASSE 1020 Vacuum Breakers, Anti-Siphon Pressure Type. ASSE 1021 Air Gap Drains for Domestic Dishwashers. ASSE 1025 Diverters for Plumbing Faucets with Hose Spray, Anti-Siphon Type, Residential Applications. ASSE 1028 Automatic Flow Controller for Faucets and Showerheads. ASSE 1029 Water Supply Valves: Mixing Valves for Single Control Mixing Valves. ASSE 1034 Fixed Flow Restrictors. ASSE 1035 Laboratory Faucet Vacuum Breakers. ASSE 1037 Proposed Flushometers.

American Society for Testing and Materials (ASTM)
100 Barr Harbor Drive, P.O. Box C700,
West Conshohocken, PA 19428-295
Phone: (610) 832-9585, Fax: (610) 832-9555
Email: service@astm.org
www.astm.org
James Thomas, President

ASTM F444 Consumer Safety Specification for Scald-Preventing Devices and Systems in Bathing Areas. ASTM F445 Consumer Safety Specification for Thermal-Shock Preventing Devices and Systems in Showering Areas. ASTM F446 Consumer Safety Specification for Grab Bars and Accessories Installed in the Bathing Areas. ASTM F462 Consumer Safety Specification for Slip-Resistant Bathing Facilities.

Fixture Classification

Fixtures may be divided into the following classes:
1. Water closets
2. Urinals
3. Lavatories
4. Sinks
5. Service sinks
6. Bathtubs
7. Showers
8. Drinking fountains
9. Bidets.

Water Closets

Water closets are manufactured in a number of styles and with various features that make them distinct from each other. These include siphon jet reverse trap, wash down, blowout, siphon vortex, siphon wash, flush valve (flushometer valve), gravity tank, flushometer tank, dual flush, wall-hung tank, corner tank, prison, handicapped design, pneumatic assist flush, wall mounted, floor mounted, back outlet, one-piece tank type, two-piece tank type, round front bowl, and elongated bowl. Traditionally water closets have been made of vitreous china; however, water closets are now also made of plastics, cultured marble, or—for institutional installations—of stainless steel or aluminum.

Certain types of water closets are unacceptable. They are those that have:
1. An invisible water seal
2. Unventilated spaces
3. Surfaces that are not thoroughly cleansed with each flushing action
4. A design that permits contamination of the domestic water supply.

Quiet operation and economical use of water are important qualities of a water closet. Water closets must be emptied of waste after each use without using any moving parts within the trapway, and the flushing action must cleanse the walls of the bowl and then refill the bowl and trap.

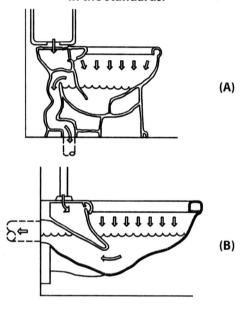
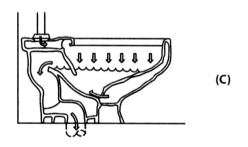

Figure 1-1 The older styles of water closets were identified as (A) reverse trap, (B) blowout, and (C) siphon jet, to name a few. Though still used in the industry, these terms are no longer used in the standards.

Water closets may be floor-outlet mounted on special closet flange connections in the floor or wall-hung closets mounted on a combination chair carrier and fitting that supports the water closet without placing any stress on the wall. The wall-hung water closet permits greater ease in cleaning the floor around and below the closet. Manufacturers supply bowls for a variety of flushing actions.

Passage of the Energy Policy Act of 1992 by the US government changed the design of a water closet. It imposed a maximum flushing rate of 1.6 gallons per flush (gpf) (6 L per flush). This was a significant drop in the quantity of water used, previously 3.5 gal per flush, and was considered to be a water savings. Prior to the first enactment of water conservation in the late 1970s, water closets typically flushed between 5 and 7 gal of water. The greatest water use, 7 gal per flush, was by blowout water closets.

With the modification in water flush volume, the style of each manufacturer's water closet changed. The former terminology for identifying water closets no longer fit. Water closets were previously categorized as blowout, siphon jet, washout, reverse trap, and wash down. (See Figure 1-1.) The new style of 1.6 gpf water closets fit between the cracks of these old categories. The standards have since changed, no longer identifying a water closet by these designations.

Water closets are currently placed into one of three categories:
- A *close-coupled water closet* is one with a two-piece tank and bowl fixture.
- A *one-piece water closet* is, as it suggests, one with the tank and bowl as one piece.
- A *flushometer style water closet* is a bowl with a spud connection that receives the connection from a flushometer valve. Flushometer type water closets are also referred to as "top spud" or "back spud bowls." The "spud" is the name for the connection for the flushometer valve and the top or rear identifies the location of the spud. (See Figure 1-2.)

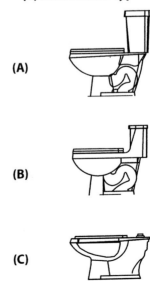

Figure 1-2 Water closets are identified as (A) close coupled, (B) one piece, and (C) flushometer types.

There are also three distinct means for identifying the flushing of a water closet:
- In a *gravity flush*, used with tank type water closets, the water is not under pressure and flushes by gravity.
- With a *flushometer tank*, also for tank type water closets, however, the water is stored in a pressurized vessel and flushed under a pressure ranging between 25 and 35 psi.

Engineered Plumbing Design

- A *flushometer valve* type of flush uses the water supply line pressure to flush the water closet. Because of the demand for a flush of a large volume of water in a short period of time, the water supply pipe must be larger in diameter than that for a gravity or flushometer tank type of flush.

Another distinction used to identify a water closet is the manner of mounting and connection. The common designations for water closets are the following:

- A *floor-mounted water closet* is supported by the floor and connected directly to the piping through the floor. (See Figure 1-2.)
- A *wall hung water closet* is supported by a wall hanger and never comes in contact with the floor. Wall hung water closets are considered superior for maintaining a clean floor in the toilet room since the water closet doesn't interfere with the cleaning of the floor. (See Figure 1-3 and 1-4.)
- *Floor-mounted, back outlet water closets* are supported by the floor yet connect to the piping through the wall. The advantage of the floor-mounted, back outlet water closet is that the penetrations of the floor are reduced for the plumbing. It should be noted that with the change to 1.6 gal per

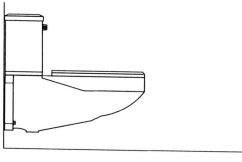

Figure 1-3 A wall hung water closet attaches to the back wall; the water closet does not contact the floor.

Figure 1-4   Carrier for a Water Closet
Source: Courtesy of Jay R. Smith Company.

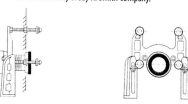

Side View   Front View

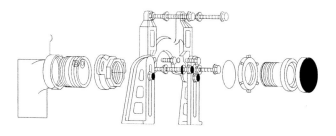

flush it is more difficult for manufacturers to produce a floor-mounted, back outlet water closet that meets all of the flushing performance requirements in the standard. (See Figure 1-5.)

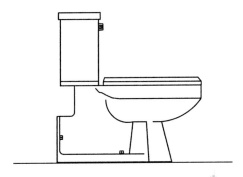

Figure 1-5   A floor-mounted, back outlet water closet is supported on the floor with the piping connection through the back

Shape and Size

A water closet bowl is classified as either a round front or elongated. An elongated bowl has an opening that extends 2 in. farther to the front of the bowl. Most plumbing codes require elongated bowls for public and employee use. The additional 2 in. provides a larger opening, often called a "target area." With the larger opening, there is a greater likelihood of maintaining a cleaner water closet for each user.

For floor-mounted water closets, the outlet is identified based on the rough-in dimension. The rough-in is the distance from the back wall to the center of the outlet when the water closet is installed. A standard rough-in bowl outlet is 12 in. Most manufacturers also make water closets with a 10-in. or 14-in. rough-in. (See Figure 1-6.)

The size of the bowl is also based on the height of the bowl rim measured from the floor:
- A *standard water closet* has a rim height of 14 to 15 in. This is the most common water closet to install.

Figure 1-6   The standard rough-in dimension is 12 in. from the centerline of the water closet outlet to the back wall. The floor flange must be permanently secured to the building structure.

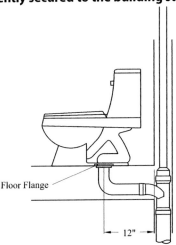

- A *child's water closet* has a rim height of 10 in. above the floor. Many plumbing codes require child's water closets in day-care centers and kindergarten toilet rooms for use by small children.
- A *water closet for juvenile* use has a rim height of 13 in.
- A *water closet for the physically challenged* has a rim height of 18 in. With the addition of the water closet seat, the fixture is designed to conform to the accessibility requirements.

Water Closet Seat

A water closet seat must be designed for the shape of the bowl to which it connects. There are two styles of water closet seat: solid and split rim. Plumbing codes typically require a split rim seat for public and employee use water closets. The split rim seat is designed to facilitate easy wiping by females, and to prevent contact between the seat and the penis with males. This is to maintain a high level of hygiene in public facilities.

A new style of water closet seat has a plastic wrap around the seat. The intent of this seat is to allow a clean surface for each use. The seat is intended to replace the split rim seat in public and employee locations.

Flushing Performance

The flushing performance requirements for a water closet are found in a separate standard, ANSI/American Society of Mechanical Engineers (ASME) A112.19.6. This standard identifies the test protocol that must be followed to certify a water closet. The tests include a ball removal test, granule test, ink test, dye test, water consumption test, trap seal restoration test, water rise test, back pressure test, rim top and seat fouling test, a drain line carry test, and a bulk media test.

The ball removal test utilizes 100 polypropylene balls that are ¾ in. in diameter. The water closet must flush at least an average of 75 balls on the initial flush of three different flushes. The polypropylene balls are intended to replicate the density of human feces.

The granule test utilizes approximately 2500 disc shaped granules of polyethylene. The initial flush of three different flushes must result in no more than 125 granules on average remaining in the bowl. The granule test is intended to simulate a flush of watery feces (diarrhea).

The ink test is performed on the inside wall of the water closet bowl. A felt tip marker is used to draw a line around the inside of the bowl. After flushing, no individual segment of line can exceed ½ in. The total length of the remaining ink line must not exceed 2 in. This test determines that the water flushes all interior surfaces of the bowl.

The dye test uses a color dye to add to the water closet trap seal. The concentration of the dye is determined both before and after flushing the water closet. The dilution ratio of 100:1 must be obtained for each flush. This test determines the evacuation of urine in the trap seal.

The water consumption test determines that the water closet meets the federal mandate of 1.6 gal per flush.

The trap seal restoration test determines that the water closet refills the trap of the bowl after each flush. The remaining trap seal must be a minimum of 2 in. in depth.

The water rise test evaluates the rise of water in the bowl when the water closet is flushed. The water cannot rise above a point 3 in. below the top of the bowl.

The back pressure test is used to determine that the water seal remains in place when exposed to a back pressure (from the outlet side of the bowl) of 2½ in. of water column (wc). This test determines that no sewer gas will escape through the fixture when high pressure occurs in the drainage system piping.

The rim top and seat fouling test determines if the water splashes onto the top of the rim or seat of the water closet. This test ensures that the user will not encounter a wet seat when using the water closet.

The drain line carry test determines the performance of the water closet flush. The water closet is connected to a 4-in. drain 60 ft in length pitched ¼ in./ft. The same 100 polypropylene balls used in the flush test are used in the drain line carry test. The average carry distance of all the polypropylene balls must be 40 ft in length. This test determines the ability of the water closet to flush the contents in such a manner that they properly flow down the drainage piping.

The bulk media test is a test of a large quantity of items placed in the bowl. The bowl cannot be stopped up by the bulk media during the flush, and a certain flushing performance of the bulk media is required. The debate over this test is the repeatability of the test. In Canada, water closets must conform to Canadian Standards Association (CSA) B45.1, CSA B45.4, or CSA B45.5. While Canada does not have a federal mandate requiring 1.6-gal-per-flush water closets, many areas require these water closets. It should also be noted that Canada requires a bulk media test for water closet flush performance.

Installation Requirements

The water closet must be properly connected to the drainage piping system. For floor-mounted water closets, a water closet flange is attached to the piping and permanently secured to the building. For wood framed buildings, the flange is screwed to the floor. For concrete floors, the flange sits on the floor.

Noncorrosive closet bolts connect the water closet to the floor flange. The seal between the floor flange and the water closet is made with either a wax ring or an elastomeric sealing connection. The connection formed between the water closet and the floor must be sealed with caulking or tile grout.

For wall hung water closets, the fixture must connect to a wall carrier. The carrier must transfer the loading of the water closet to the floor. A wall hung water closet must be capable of supporting a load of 500 lb at the end of the water closet. When the water closet is connected to the carrier, none of this load can be transferred to the piping system. Water closet carriers must conform to ANSI/ASME A112.6.1.

The minimum spacing required for a water closet is 15 in. from the centerline of the bowl to the side wall, and 21 in. from the front of the water closet to any obstruction in front of the water closet. The standard dimension for a water closet compartment is 30 in. wide by 60 in. in length. The water closet must be installed in the center of the standard compartment. The minimum distance required between water closets is 30 in.

The change in the flushing performance of the 1.6-gal-per-flush water closet has affected the piping connection for back-to-back water closet installations. With a 3.5-gal-per-flush water closet, the common fitting used to connect back-to-back water closets was either a 3-in. double sanitary tee or a 3-in. double fixture fitting. With the superior flushing of the 1.6-gpf water closet, the plumbing codes have prohibited the installation of a double sanitary tee or double fixture fitting for back-to-back water closets. The only acceptable fitting is the double combination wye and eighth bend. The fitting, however, increases the spacing required between the floor and the ceiling.

The minimum spacing required to use a double sanitary tee fitting is 30 in. from the centerline of the water closet outlet to the entrance of the fitting. This spacing rules out a back-to-back water closet connection.

One of the problems associated with the short pattern fittings is the siphon action created in the initial flush of the water closets. This siphon action can draw the water out of the trap of the water closet connected to the other side of the fitting. Another potential problem is the interruption of flow when flushing a water closet. The flow from one water closet can propel water across the fitting, interfering with the other water closet.

Flushing Systems

Gravity flush The most common means of flushing a water closet is a gravity flush. This is the flush with a tank type water closet, described above, wherein the water is not pressurized in the tank. The tank stores a quantity of water to establish the initial flush of the bowl. A trip lever raises either a flapper or a ball, allowing the the flush is at the maximum siphon in the bowl, the flapper or ball reseals, closing off the tank from the bowl.

The ballcock, located inside the tank, controls the flow of water into the tank. A float mechanism opens and closes the ballcock. The ballcock directs the majority of the water into the tank and a smaller portion of water into the bowl to refill the trap seal. The ballcock must be an antisiphon ballcock conforming to American Society of Sanitary Engineers (ASSE) 1002. This prevents the contents of the tank from being siphoned back into the potable water supply. (See Figure 1-7.)

Flushometer tank A flushometer tank has the same outside appearance as a gravity tank. However, inside the tank is a pressure vessel that stores the water for flushing. The water in the pressure vessel must be a minimum of 25 psi* to operate properly. Thus, the line pressure on the connection to the flushometer tank must be a minimum of 25 psi.* A pressure regulator prevents the pressure in the vessel from rising above 35 psi (typical of most manufacturers).

The higher pressure from the flushometer tank results in a flush similar to a flushometer valve. One of the differences between the flushometer tank

* Check Manufacturers Requirements

Figure 1-7 (A) A Gravity Tank and (B) a Flushometer Tank

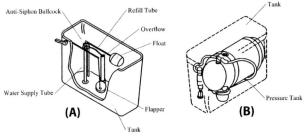

and the flushometer valve is the sizing of the water distribution system. The water piping to a flushometer tank is sized the same way the water piping to a gravity flush tank is sized. Typically, the individual water connection is ½ in. in diameter. For a flushometer valve, there is a high flow rate demand, resulting in a large piping connection. A typical flushometer valve for a water closet has a connection of 1 in. in diameter. (See Figure 1-7.)

Flushometer valve A flushometer valve is also referred to as a "flush valve." The valve is designed with upper and lower chambers separated by a diaphragm. The water pressure in the upper chamber keeps the valve in the closed position. When the trip lever is activated, the water in the upper chamber escapes to the lower chamber, starting the flush. The flush of 1.6 gal passes through the flush valve. The valve is closed by line pressure as water reenters the upper chamber, closing off the valve.

For 1.6-gpf water closets, flushometer valves are set to flow 25 gpm at peak to flush the water closet. The flushing cycle is very short, lasting 4 to 5 s. The water distribution system must be properly designed to allow the peak flow during a heavy use period for the plumbing system.

Flushometer valves have either a manual or an automatic means of flushing. The most popular manual means of flushing is a handle mounted on the side of the flush valve. Automatic flushometer valves are available in a variety of styles. The automatic can be battery operated or directly connected to the power supply of the building.

Flush Tank Requirements

There are certain essential requirements which must be satisfied when a flush tank is employed:

1. There must be an overflow to prevent tank flooding. The overflow should discharge into the water closet. (See Figure 1-8).
2. The ballcock, which controls the flow of water into the tank, should be equipped with a means of replenishing the trap seal after each flushing action.
3. The ballcock should be equipped with an adequate means of protection against back siphonage into the domestic water supply. A vacuum breaker is satisfactory for this purpose.

Urinals

A urinal was developed as a fixture to expedite the use of a toilet room. It is designed for the removal of urine and the quick exchange of users. The Energy Policy Act of 1992 included requirements for the water consumption

Engineered Plumbing Design

Figure 1-8 Flush Tank (Gravity)

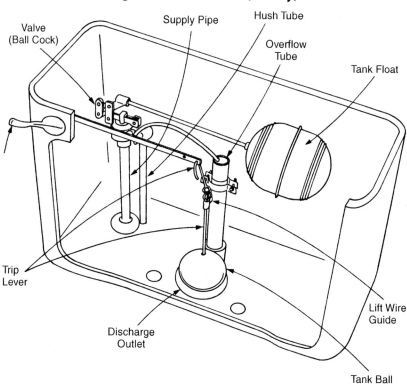

of urinals. A urinal is now restricted to a maximum water use of 1.0 gal per flush. This change in water consumption resulted in a modified design of the fixture.

One of the main concerns in the design of a urinal is the maintenance of a sanitary fixture. The fixture must contain the urine, flush it down the drain, and wash the exposed surfaces. Prior to the passage of the Energy Policy Act of 1992, urinals were developed using larger quantities of water to flush the contents. This included a blowout model that could readily remove any of the contents thrown into the urinal in addition to urine. Blowout urinals were popular in high-traffic areas such as assembly buildings. However, the older blowout urinals require more than 1 gal of water to flush. The newer urinals identified as blowout urinals do not have the same forceful flush.

Urinals have been considered a fixture for the male population. However, that has not always been the case. Various attempts have been made to introduce a female urinal. The female population has never embraced the concept of a female urinal. Problems that have been encountered include a lack of understanding of the use of the urinal. (The first female urinals required the woman to approach the urinal in the opposite way a man would. She would be facing away from the urinal slightly bent over.) Another continuing concern is privacy during use. Finally, there have been concerns regarding cleanliness with its use

Flushing Requirements

With the federal requirements for water consumption, urinals must be flushed with a flushometer valve. The valve can be either manually or automatically actuated.

A urinal flushometer valve has a lower flush volume and flow rate than a water closet flushometer valve. The total volume is 1 gal per flush and the peak flow rate is 15 gpm. The water distribution system must be properly sized for the peak flow rate for the urinal.

Urinal flushometer valves operate the same as water closet flushometer valves. For additional information see the discussion of flushing systems under "Water Closets" earlier in this chapter.

A modern version of the century-old waterless urinal is available where water savings are paramount. The waterless urinal has a special trap that is filled with a liquid that is lighter than water and urine. Urine travels down the interior sides of the urinal, through the liquid, and safely into the waste piping. The liquid must be replenished periodically, thus scheduled maintenance is required, the schedule depending on the frequency of use of the urinal.

Lavatories

A Lavatory is a washbasin used for personal hygiene. In public locations, a lavatory is used for washing one's hands and face. Residential lavatories are intended for hand and face washing, shaving, applying makeup, cleaning contact lenses, and similar hygienic activities.

Lavatory faucet flow rates are regulated as a part of the Energy Policy Act of 1992. The original flow rate established by the government was 2.5 gpm at 80 psi for private use lavatories and 0.5 gpm, or a cycle discharging 0.25 gal, for public use lavatories. Since the initial regulations, there has been a change to 2.2 gpm at 60 psi for private (and residential) lavatories, and 0.5 gpm at 60 psi, or a cycle of 0.25 gal, for public lavatories.

Size and Shape

Manufacturers produce lavatories in every conceivable size and shape, providing an unlimited selection. Lavatories are square, round, oblong, rectangular, shaped for corners, with or without ledges, decorative bowls, and molded into countertops. They can be classified into five different types: slab, splashback, shelf back, ledge back, and countertop. Several are illustrated. Special purpose lavatories can usually be placed within these five categories.

- The slab type lavatory comes in vitreous china and is supported by concealed or exposed arms, wall brackets, and chrome legs. The back of the fixture is usually installed 2 in. from the wall to facilitate cleaning of the wall behind the lavatory. (See Figure 1-10.)
- The splashback lavatory has an integral back and is recommended for sanitary purposes. Splashing, which can run down the back of the slab type, stays on the fixture. The fixture is made of vitreous china and is supported by wall hangers or by concealed or exposed arms. (See Figure 1-11.)

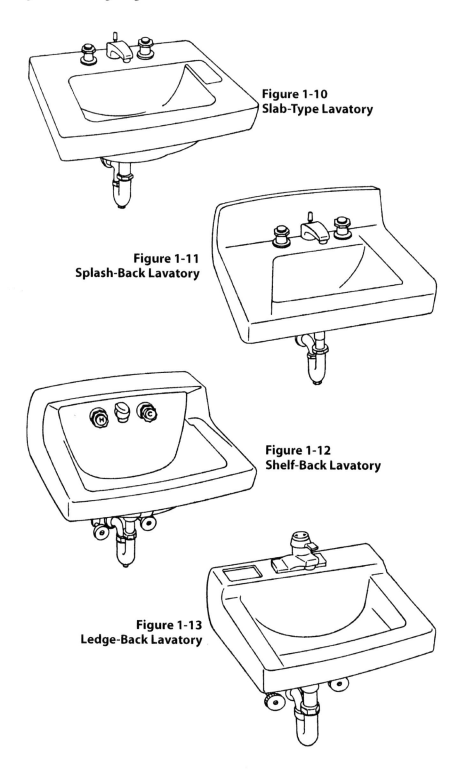

Figure 1-10
Slab-Type Lavatory

Figure 1-11
Splash-Back Lavatory

Figure 1-12
Shelf-Back Lavatory

Figure 1-13
Ledge-Back Lavatory

- The shelf-back lavatory reduces splashing and in addition provides a shelf for the storage of toiletries. The fixture is made of either vitreous china or enameled iron. (See Figure 1-12.)
- The ledge-back lavatory offers some splash reduction and some shelf area. This type as well as the others has depressions molded into the fixture for holding bar soap. Manufacturers offer modifications to provide or delete additional holes or depressions. An additional hole may be provided for a liquid soap dispenser. (See Figure 1-13.)
- Counter self-rimming and undercounter mounted lavatories are the most recent and probably the most diversified of any category. They come in various materials, including vitreous china, enameled cast iron, stainless steel, plastics, fiberglass, and precast artificial marble. A development that has contributed to the popularity and acceptance of countertop lavatories is the self-rimming feature that does not require the use of a stainless steel rim.

The standard outlet for a lavatory is 1¼ in. in diameter. The standard lavatory has three holes on the ledge for the faucet. A normal faucet hole pattern spaces the two outside holes 4 in. apart. The faucets installed in these lavatories are called 4-in. center sets. When spread faucets are to be installed, the spacing between the two outer holes is 8 in.

For many years, the fixture standards required lavatories to have an overflow. This requirement was based on the use of the fixture whereby the basin was filled prior to cleaning. If a user left the room while the lavatory was being filled, the water would not overflow on the floor.

The engineer is warned to be especially aware of possible problems created when specifying the newer materials and to carefully analyze their application to a given installation. Special care must be exercised with these products regarding their abrasion-resistance characteristics. Abrasive cleaners tend to destroy the luster of the surface of these materials much more quickly than they do traditional materials. In addition, some of the newer materials are fire resistant while some are not.

Sinks

There is a wide selection of sink types available. They come in single, double, and triple-compartment models. Two-compartment sinks, with both compartments the same size, are the most widely used models (see Figure 1-15). It is recommended that one compartment be at least 15 in. by 18 in. in size for residential use to allow the acceptance of a roasting pan. Faucet spouts should be high enough to place a large pot beneath without any difficulty. Specialty sinks, such as the surgeon's sink shown in Figure 1-14, are available.

Service Sinks

The most popular service sinks are those that have a high back and are wall mounted and supported on a trap standard or low type mop basins that are mounted on, or recessed into, the floor. Protective rim guards are recommended for both.

Engineered Plumbing Design

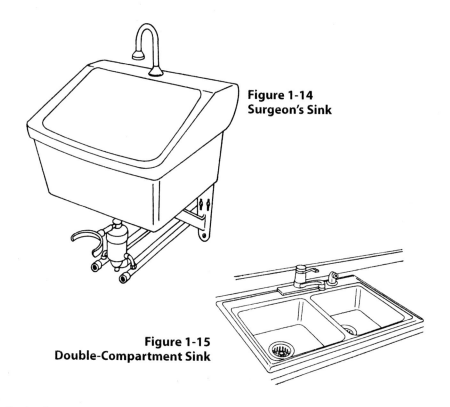

**Figure 1-14
Surgeon's Sink**

**Figure 1-15
Double-Compartment Sink**

Bathtubs

Bathtubs are available in various sizes and shapes. The 5 ft bathtub has become practically a standard, but the public has recently indicated a preference for a much longer model. Perhaps there will be a return to 5 ft 6 in. and 6 ft 0 in. tubs, which were more readily available in the not too distant past.

Fiberglass and plastic models have recently entered the market. The plumbing engineer is strongly advised to check thoroughly with the manufacturer as to the hardness of the surface and the resistance to abrasive cleaners.

Enameled cast iron tubs have been preferred because of their ability to resist chipping and rusting, which frequently happens with enameled steel tubs. The thickness of the enamel coating on cast iron is two to three times heavier than that on pressed steel and has superior adherence to the base.

Showers

Shower receptors are available in various sizes and shapes. They are available in standard precast sizes (minimum 30" × 30") but may be obtained in custom built models to fit practically any application.

Where precast receptors are not employed, the built-up type is used. The pan for a built-up shower can be fabricated from lead, copper, or various compositions presently available. The pan should turn up at least 6 in. and turn over the threshold to provide a watertight installation. An exception to this is shower enclosures made to be wheelchair accessible.

Drinking Fountains

Drinking fountains (nonrefrigerated) are available as free-standing, surface mounted, semirecessed, fully recessed, bi-level (regular height and handicapped accessible height), pedestal, or deck type for counter tops. When selecting a semirecessed or fully recessed model, the plumbing engineer should ascertain that the wall or pipe space is deep enough to accommodate the fountain and necessary piping.

Electric water coolers are available in as many variations as drinking fountains. It is extremely important to provide adequate wall thickness to accommodate the chiller unit and piping. Location of the chiller unit and grill finish should be coordinated with the architect.

Bidets

The bidet is about the same size and shape as a water closet and could be classified as a small bath. It is used primarily for washing the anal regions after using the water closet.

The hot and cold water supply and the drain fitting are very similar to those used for lavatories. Instead of the water entering the bowl from a spout, however, it is introduced through a flushing rim. The tepid water flows through the rim and while filling the bowl it warms the china hollow rim which serves as a seat.

A spray rinse is optional, and recommended, for external rinsing. Although this rinse is often called a "douche" it should not be construed as being designed or intended for internal use.

The plumbing engineer should recommend that a soap dispenser and towel rack be provided within easy reach for the convenience of the user.

The foregoing has been a very brief discussion of the salient features of some of the most common fixtures. The reader is referred to the catalogs of various fixture manufacturers for a complete presentation of fixtures and trim. The catalogs are an excellent source of information and give all the detailed data required; it is unnecessary to fill these pages with that information, they are better devoted to design criteria.

Fixture Traps

A fixture trap is a fitting or device that provides a liquid seal that will prevent the back passage of air without materially affecting the flow of sewage or waste water through it. Acceptable fixture traps must have the following characteristics:
1. They must be self-cleaning.
2. They must have smooth interiors.
3. They must have no movable parts or partitions.
4. They must have a minimum seal of 2 in.

Figure 2-1 shows a typical fixture trap.

Prohibited Traps

Crown-vented traps and traps that depend on movable parts for their seal are prohibited. A trap is considered to be crown vented when the vent is within two pipe diameters of the weir. This proximity of the vent to the surface of the water in the trap tends to foster evaporation of the seal as the air circulates in the venting system. If a fixture is not used over an extended period of time, it is possible to lose the water seal. (See Figure 2-2.)

Another danger is the probability of closing the vent opening with foreign matter present in the waste water as it flows out of the trap and is centrifuged to the top of the pipe.

When a trap depends upon movable parts for its seal, it is obvious that any malfunction in the moving parts would result in the loss of the seal.

Figure 2-1 Typical Fixture Trap

Figure 2-2 Crown Vented Trap

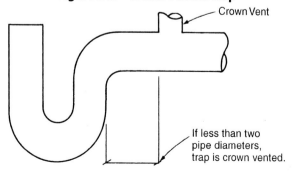

Trap Seal

Trap seal is the maximum vertical depth of liquid that a trap will retain, measured between the crown weir and the dip of the trap. A trap must provide a minimum seal of 2 in. All drainage and venting systems are designed so that pneumatic pressure variations in the system are limited to ±1 in. of water column.

A clear concept of positive and negative pressure is necessary to understand the dangers to maintenance of the trap seal. Atmospheric pressure at sea level exerts a pressure of 14.7 pounds per square inch (psi). This is *absolute* pressure. If 14.7 psi is equated to zero, then any pressure greater than the absolute pressure of 14.7 psi is positive and anything less is negative. Negative pressure is also called "vacuum." The pressures that are read on gauges are the pressures above and below atmospheric pressure. If a gauge reading –10 lb or 10 lb vacuum, it is the same as an absolute pressure of 4.7 psia (14.7 - 10 = 4.7 psi absolute). The pressures dealt with in plumbing systems are always expressed as gauge pressures.

Siphonage

The possible siphoning of water out of the trap is of major importance. Two kinds of siphonage are possible:
1. Induced siphonage
2. Self-siphonage.

Induced Siphonage

In a plumbing system at rest (no flow occurring) it can be stated that atmospheric conditions exist throughout the system. Stated another way, zero pressure exists at the inlet and outlet of all traps in the system. When flow occurs in any part of the system, fluctuations in pressure occur. These pressure variations may possibly induce the siphonage of a fixture trap with the consequent loss of the protective water seal. Figure 2-3(A) illustrates a trap with zero pressure at the inlet and outlet. Figure 2-3(B) illustrates the condition when an excess pressure develops in the drain. Let ΔP represent the excess pressure, then the

Figure 2-3 Trap Action Under Excess Positive Pressure

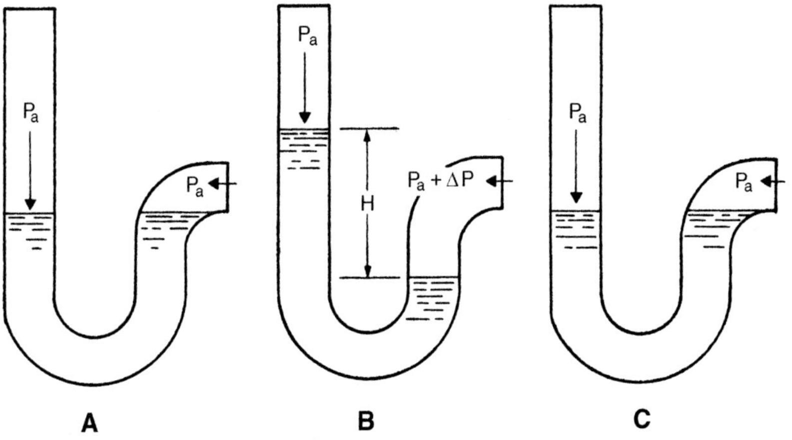

total pressure at the outlet would be atmospheric pressure (P_a) plus the excess pressure (ΔP). The pressure at the inlet would still be atmospheric (P_a). The water would rise in the inlet leg of the trap to a height (ΔH) to balance ΔP. When the pressure at the outlet returns to zero (atmospheric), the water in the two legs of the trap will return to the initial level as shown in Fig. 2-3(C), with a very minor loss of trap seal due to the overshooting of the water column as it returns to equilibrium.

Figure 2-4 represents the condition when a negative pressure ($P_a - \Delta P$) develops at the outlet of a trap. This condition will cause some of the water in the trap seal to flow over the weir and down the fixture drain. When the outlet pressure again returns to zero, the water columns in both legs will stabilize as shown in Figure 2-4 and only part of the original trap seal will remain.

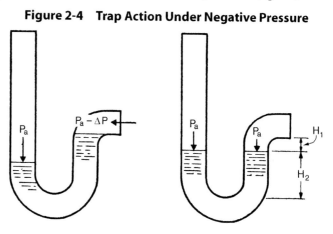

Figure 2-4 Trap Action Under Negative Pressure

The trap seal loss is equal to H_1, and the final trap seal is H_2.

Figure 2-5 represents the case when the positive pressure at the outlet is great enough to force the water in the downstream leg down to the dip of the trap. This is the maximum positive pressure differential for which the seal of the trap will prevent the flow of sewer air through the trap into the room. This excess pressure is equal to twice the initial depth of the trap seal. Thus, when there is a trap seal of H_2 as shown in Figure 2-4, the positive pressure at the outlet will not force sewer air through the trap seal until the excess pressure exceeds $2H_2$.

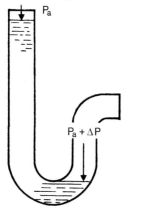

Figure 2-5 Maximum Permissible Positive and Negative Pressure

Self-Siphonage

The phenomenon of self-siphonage is not as easily analyzed as induced siphonage. It can best be illustrated by examining the discharge of a lavatory. A lavatory is prone to self-siphonage to a greater degree than any other fixture. Most lavatories are constructed with rounded bottoms. The discharge rate is high at first, decreasing as the depth of water in the basin decreases, then suddenly falling off to nearly zero, with an accompanying formation of a vortex, which sucks air down into the drain.

This does not occur in a flat- bottomed fixture such as a sink or bathtub. In these fixtures there is a trail flow of the film of water on the surface of the fixture that slowly drains off, tending to replenish the seal of the trap. Trail flow is very brief with a lavatory, so there is very little tendency to replenish the seal.

A typical case of flow in a drain as a result of a discharging lavatory is described in the National Bureau of Standards Report BMS 126 (1951):

"At first thought, it might seem that the flow taking place through a trap ordinarily consists of water only. However, this is not the case, as has already been pointed out. In most instances, air is carried out with the water also. We have observed this phenomenon in a transparent trap and fixture drain connected to a lavatory, and these observations indicated that air entered the trap in three ways. First, air was entrained by the water as it passed the overflow outlet just below the lavatory, and this air was carried through the trap in the form of bubbles.

"Second, as the water surface in the lavatory continued to recede, a vortex formed in the lavatory, and in this way additional air was carried through the trap into the drain with the water.

"Third, near the end of the discharge of the lavatory, with its attendant rapid decrease in the rate of flow through the lavatory outlet orifice, water was flowing out of the trap more rapidly than it entered, owing primarily to the inertia of the water in the trap and drain. Hence the water surface in the inlet leg of the trap receded to such an extent that in many instances air bubbled past the dip of the trap and entered the outlet leg of the trap. This latter manner in which air enters or passes through a trap is especially noticeable when a large pressure reduction occurs in the fixture drain near the end of the discharge period.

"Now if the rate of flow in the drain and the diameter of the drain are not sufficient to close off the passageway in the inlet branch of the stack fitting, this air can pass off to the vent or stack and exert no particular effect on the nature of flow in the drain.

"However, if the flow is sufficiently great to close off the passageway referred to, then the air in the drain becomes trapped between the solid mass of water in the vent fitting and the water in the trap and causes changes in the nature of the flow in the drain. This in turn affects the pressures in the drain and hence on the outlet end of the trap when discharge from the fixture ceases.

"The following is typical of the phenomena that were observed when air entered the drain in the manner described above. The entrained air from the overflow outlet entered the drain in the form of bubbles, which rose to the top of the drain at some definite point along it. If the quantity of air that collected in this way was sufficiently large, the drain, which was flowing full from the trap weir to this point, flowed only part full from this point on to the vent fitting. As the volume of entrained air diminished near the end of the lavatory discharge, the water frequently rose to the top of the drain along one or more portions of its length, filling the entire cross-section of the drain. Thus, one or more plugs of water formed in the drain near the end of the period of discharge from the lavatory.

"As the flow from the lavatory trailed off, the velocity of these plugs of water toward the vent fitting decreased, owing to the adverse head of water in the trap, which slowed them down; and at the same time the length of the plugs of water diminished, owing to the sloughing off of water at their upstream and downstream ends. If one of these plugs was sufficiently near the trap when this sloughing off occurred, some of the water that sloughed off from the upstream end of the plug flowed backward up the drain and partly or entirely refilled the trap. When the plug was further down the drain, this temporary backflow of water in the drain did not reach the trap, and hence no refill from this cause occurred."

From the foregoing it can be seen that there are two possible ways of replenishing the trap seal:
1. By trail flow from the fixture
2. By backflow in the drain.

It can also be seen that the presence of an adequate supply of air carried along with the water helps prevent loss of trap seal.

Limitation of Trap Seal Loss

Reduction of the trap seal is the basis for establishing limitations of drain lengths and slopes. Protection of the building against the passage of sewer gas through the trap depends upon the depth of trap seal that can be maintained under adverse conditions, or on the magnitude of the effects of induced siphonage and self-siphonage. The *remaining* trap seal is of paramount importance. Authorities have established this value at 1 in. of water and this depth of seal will then protect against the passage of sewer air so long as the excess pressure in the drain does not exceed twice the depth of the seal. This would permit a pressure of ±2 in. of water in the drain. As a factor of safety, *all sanitary systems are designed to limit the pressure variations in the sanitary system to ±1 in. of water column.*

Trap Installation

Traps should be set level with respect to their water seal. At first glance, this would appear to be impossible, since the trap is connected to piping that has a pitch of ¼ in./ft and thus the water surface in the trap would have the same slope. To overcome this, all trap outlets are tapped at a ¼ in./ft pitch and thus when screwed onto the pitched pipe they are automatically level.

Traps should always be located in an area that is not subjected to freezing temperatures. They should be located as close to the fixture as possible and in no case more than 2 ft vertically from the fixture outlet to the weir of the

Chapter 2—Fixture Traps

Figure 2-6   Location of Traps

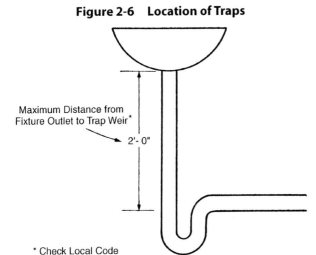

Maximum Distance from Fixture Outlet to Trap Weir*

2'- 0"

* Check Local Code

trap. This will minimize the loss of trap seal due to the momentum of the fixture discharge as well as minimize the odors from the fouled interior of the tailpiece from the fixture waste outlet to the trap inlet. (See Figure 2-6.)

Minimum size of the trap should be the size of the fixture outlet and maximum size should be the size of the drain connection.

Intercepting Fixture Traps

Intercepting traps should be used only when it is deemed the best method of removing objectionable wastes. The objection to their use is that they all require maintenance. The various types and their applications are as follows:

1. **Grease Interceptors.** Pot sinks, scullery sinks, baking sinks, food scrap sinks, the scraper section of commercial dishwashers, kettle trench drains and floor drains that receive spillage for local grease interceptors. (See Figure 2-7.) (*Note*: The American Society of Plumbing Engineers defines a grease interceptor as an automatic or manual device used to separate and retain grease, with a capacity greater than 50 gal [227.3 L] and generally located outside a building. A grease trap is defined as an automatic or manual device used to separate and retain grease, with a capacity of 50 gal [227.3 L] or less and generally located inside a building. See the discussion below.)
2. **Plaster or Barium Trap.** Orthopedic and dental areas.
3. **Sand Trap.** Planter drains and subsoil drainage.

Figure 2-7   Grease Interceptor

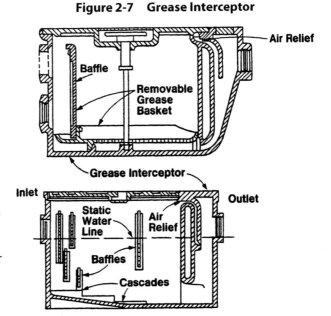

25

4. **Hair and Lint Trap.** Beauty parlors and barber shops.

Note: Some municipalities may require an exterior grease trap to collect all fixtures (except for water closets) associated with large kitchens such as hotels or large restaurants.

Grease Interceptors

In the drainage from commercial kitchens, grease, fats, and oils must be separated from sewage. This function is performed by grease interceptors installed in drain lines where the presence of grease in the sewage is expected.

It is sometimes practical to discharge the waste from two or more sinks into a single interceptor. This practice is recommended only when all the fixtures are close together to avoid installing long piping runs to the interceptor. The closer the interceptor can be installed to the fixture(s) the better. The longer the run of pipe, the cooler the waste water is. As the waste water cools, the grease congeals, coating and clogging the interior of the pipe.

The procedures for sizing grease interceptors are as follows:
1. Determine the cubic content of the fixtures by multiplying length by width by depth.
2. Determine the capacity in gallons (1 gal = 231 in.^{3}) (liters [1 L = 1000 cm^3]).
3. Determine the actual drainage load. The fixture is usually filled to about 75% of capacity with waste water. The items being washed displace about 25% of the fixture content. Therefore, actual drainage load = 75% of fixture capacity.
4. Determine the flow rate and the drainage period. In general, good practice dictates a 1-min drainage period; however, where conditions permit, a 2-min period is acceptable. The drainage period is the actual time required to completely empty the fixture.
5. Flow rate $= \dfrac{\text{Actual drainage load}}{\text{Drainage period}}$
6. Select the interceptor that corresponds to the flow rate calculated. It is recommended to provide the automatic removal of grease from the interceptor to a storage tank that can be cleaned regularly.

Questions

1. *Why are movable parts or partitions prohibited in fixture traps?*
2. *Why are crown-vented traps prohibited?*
3. *Why is an accessible cleanout required on a fixture trap?*
4. *If there is a remaining seal of 1½ in. in a fixture trap and a pressure of +1 in. of water column develops at the outlet, will the trap perform its function?*
5. *If a fixture trap has a remaining seal of 1½ in., how much pressure could be tolerated at the outlet before sewer gas would pass through into the room?*
6. *Why would a trap never be located in an area subjected to freezing temperatures?*
7. *What is the minimum distance a vent can be located from the weir of a 2-in. fixture trap? From a 1½-in. fixture trap?*

Flow in Horizontal Drainage Piping

When flow occurs in drain piping, it does not entirely fill the pipe under normal conditions of flow. If the pipe were to flow full, pressure fluctuations would occur that could possibly destroy the seal of traps within the building. Gravity flow in sloping (horizontal) drain lines of a plumbing system is similar to the flow of water in open channels. Flow in open channels is not dependent upon a pressure applied to the water but is caused by the gravitational force induced by the slope of the drain and the height of the water in that drain.

Uniform Flow

Uniform flow is that flow that is achieved in an open channel of constant shape and size and uniform slope. The slope of the water surface then matches the slope of the channel. Many formulas have been developed for determining the velocity of uniform flow in sloping drains. The one that has become most popular in the solving of plumbing problems is the Manning formula, first proposed by Robert Manning in 1890. Its form in the metric system is simpler, but converting to the English system, the formula takes the following form:

(3-1)
$$V = \frac{1.486}{n} \times R^{2/3} \times S^{1/2}$$

where V = Velocity of flow, feet per second (fps)
        n = A coefficient representing roughness of pipe surface, degree of fouling and pipe diameter
        R = Hydraulic radius (hydraulic mean depth of flow), ft
        S = Hydraulic slope of surface of flow, ft/ft

The quantity rate of flow is equal to the cross-sectional area of flow times the velocity of flow. This can be expressed as

(3-2)
$$Q = AV$$

where Q = Quantity rate of flow in cubic feet per second, cfs
        A = Cross-sectional area of flow, ft^2
        V = Velocity of flow, fps

By substituting the value of V from Manning's formula, we obtain

(3-3)
$$Q = A \times \frac{1.486}{n} \times R^{2/3} \times S^{1/2}$$

Particular note should be made of the units in the above equations. It is of extreme importance to convert all values to the proper units whenever utilizing any formula.

Engineered Plumbing Design

The hydraulic mean depth of flow (R), called the "hydraulic radius," is the ratio of the cross-sectional area of flow to the wetted perimeter of the pipe surface.

$$R = \frac{\text{area of flow}}{\text{wetted perimeter}}$$

For conditions of half-full flow (See Figure 3-1), the hydraulic radius is

(3-4)
$$R = \frac{\pi D^2}{8} \div \frac{\pi D}{2} = \frac{D}{4}$$

Note: The area of a circle is

(3-5)
$$\frac{\pi D^2}{4}$$

For full flow conditions

(3-6)
$$R = \frac{\pi D^2}{4} \div \pi D = \frac{D}{4}$$

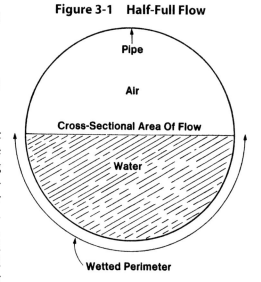

Figure 3-1 Half-Full Flow

It can be seen that the hydraulic radius is exactly the same when the pipe is flowing full as it is flowing half full. It follows that the velocity of flow is identical for full or half-full flow for the same slope. The quantity of flow, however, varies because the cross-sectional area of flow varies. Tables 3-1 and 3-2 give values of R, $R^{2/3}$, and A for full and half-full flow. Note that values are the same in both tables except for the cross-sectional area of flow. Tables for other depths of flow are available in various books on hydraulics and in catalogs.

Values of S, the slope of the surface of flow, and $S^{1/2}$ are tabulated in Table 3-3 for the most common slopes of drainage piping employed in plumbing systems.

Manning established values for n experimentally. He found that n varied with the roughness of the pipe surface and the diameter of the pipe. Research has been pursued throughout the years to obtain more precise values of n and to understand the factors that affect this coefficient. Experiments were performed by Bloodgood and Bell and reported in the February 1961 issue of the *Journal Water Pollution Control Federation*. Their conclusions

Table 3-1 Values of R, $R^{2/3}$ and A for Full Flow

Pipe Size inches	R=D/4 feet	$R^{2/3}$	Cross-sectional area (A) of flow, sq. ft.
1½	0.0335	0.1040	0.01412
2	0.0417	0.1200	0.02180
2½	0.0521	0.1396	0.03408
3	0.0625	0.1570	0.04910
4	0.0833	0.1910	0.08730
5	0.1040	0.2210	0.13640
6	0.1250	0.2500	0.19640
8	0.1670	0.3030	0.34920
10	0.2080	0.3510	0.54540
12	0.2500	0.3970	0.78540
15	0.3125	0.4610	1.22700

Table 3-2 Values of R, $R^{2/3}$ and A for Half-full Flow

Pipe Size inches	R=D/4 feet	$R^{2/3}$	Cross-sectional area (A) of flow, sq. ft.
1½	0.0335	0.1040	0.00706
2	0.0417	0.1200	0.01090
2½	0.0521	0.1396	0.01704
3	0.0625	0.1570	0.02455
4	0.0833	0.1910	0.04365
5	0.1040	0.2210	0.06820
6	0.1250	0.2500	0.09820
8	0.1670	0.3030	0.17460
10	0.2080	0.3510	0.27270
12	0.2500	0.3970	0.39270
15	0.3125	0.4610	0.61350

Table 3-3 Values of S and $S^{1/2}$

Slope inches per ft.	S ft. per ft.	$S^{1/2}$
⅛	0.0104	0.102
¼	0.0208	0.144
½	0.0416	0.204

were that the slope, flow rate, and size of pipe are all significant in the determination of n. They found that the type of pipe material was not significant.

Recommended values of n for use in the design of plumbing sanitary drains are as follows:

Pipe Size	n
1½"	0.012
2" through 3"	0.013
4"	0.014
5" and 6"	0.015
8" and larger	0.016

For storm drains, a value of 0.0145 can be safely employed for all sizes of pipe.

Using Manning's formula and employing the recommended values of n, the uniform flow velocity and capacity of various size sanitary drains installed at a slope of ¼ in./ft are tabulated in Table 3-4. Cubic feet per second has been converted to gallons per minute by use of proper conversion factors.

Values of velocity and capacity of a storm drain flowing full at a slope of ¼ in./ft are tabulated in Table 3-5.

For slopes of ⅛ in. and ½ in. the values given in the tables can be multiplied by 0.707 and 1.414, respectively, to obtain correct figures.

All of the foregoing values are for the uniform flow condition, which is achieved when the flow has had sufficient time to adjust itself and arrive at the state of equilibrium where the water surface slope is equal to the pipe slope. These velocities and capacities are the minimum that will occur, regardless of the velocity of the water entering the drain from stacks or leaders. High entrance velocity produces surges in the drain and thus velocities and capacities are increased for a relatively short distance downstream. The increase does not create any hydrostatic head in the drain, but does provide an added factor of safety when the design is based upon uniform flow conditions.

Table 3-4 Uniform Flow Velocity and Capacity of Sanitary Drains at ¼" Slope

Pipe Size inches	Full or Half-Full Flow Velocity V, fps	Half-Full Flow Capacity q, gpm	Full Flow Capacity q, gpm
1½	1.85	5.85	11.7
2	1.98	9.70	19.4
2½	2.30	17.60	35.2
3	2.59	28.60	57.2
4	2.91	57.00	114.0
5	3.15	96.50	193.0
6	3.58	157.50	315.0
8	4.07	318.50	637.0
10	4.69	574.00	1148.0
12	5.31	936.00	1872.0
15	6.15	1690.00	3380.0

Scouring Action

The minimum velocity of flow to achieve scouring action in piping is 2 *fps*. Sand, grit, pebbles, and other foreign matter that are held in suspension in the waste water will begin to deposit in the pipe when velocities fall below 2 fps. A minimum velocity of 4 fps is required to maintain greasy wastes in suspension. Minimum velocities are obtained by proper sizing and slope of pipe. An examination of Tables 3-4 and 3-5 reveals that several of the pipe sizes yield velocities less than 2 fps. It is for this reason that runs of 1½-in. and 2-in. pipe must be held to a minimum length so that entrance velocities will maintain the minimum required scouring velocity for the short distance involved.

Table 3-5 Uniform Flow Velocity and Capacity of Storm Drains at ¼" Slope and Full Flow (n = 0.0145)

Pipe Size, inches	Velocity V, fps	Capacity q, gpm
2	1.72	17.4
2½	1.99	31.5
3	2.25	51.3
4	2.74	111.0
5	3.16	201.0
6	3.58	327.0
8	4.35	705.0
10	5.04	1268.0
12	5.67	2070.0
15	6.58	3730.0

Surcharging

There are conditions when it is desirable to increase the capacity of a sewer line beyond that possible under conditions of gravity flow. Such a situation could occur where an existing sewer line was not designed for future conditions which in actuality far exceed original estimates, thus resulting in a sewer that can not handle peak demands. Another situation might be one of economics. It is possible to impose a pressure on a sewer line by means of *surcharging* and thus increase the capacities of the sewers in relation to the amount of surcharge. It is a very simple concept, as illustrated in Figure 3-2. The discharge into the sewer is increased beyond its capacity of open-channel flow at full flow. The water then rises in the manholes and, depending upon the height of the water in the manholes, imposes a pressure in excess of the gravitational force and thus increases the velocity of flow and capacity. Surcharge is usually expressed as *feet of surcharge*, which is the vertical distance measured above the invert elevation of the pipe. Smaller size pipe can be used and less slope utilized by surcharging to obtain the same capacity as exists under gravity flow.

When a sewer is pressurized, the flow is no longer "open-channel" flow as defined by Manning's formula. This pressurized flow is now defined as closed-conduit flow. To analyze closed-conduit flow we use Bernoulli's theorem. Bernoulli's theorem is explained in detail in Chapter 11, "Flow in Water Piping," but to conclude the explanation of surcharge we need to understand that Bernoulli's theorem is based on the idea of conservation of energy. This theorem states that the sum of static head, kinetic energy, and potential energy is a constant, if losses are ignored, for points in the same closed-conduit system. This is stated as follows:

Chapter 3—Flow in Horizonal Drainage Piping

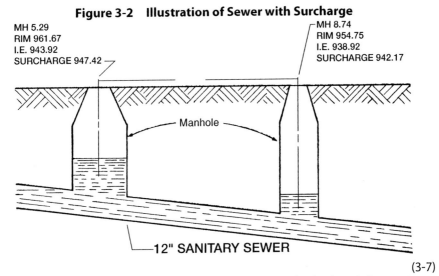

Figure 3-2   Illustration of Sewer with Surcharge

$$P/\delta + V^2/2g + z = \text{a constant and total energy at a point in the piping system} \tag{3-7}$$

where P/δ = Static head, ft
  $V^2/2g$ = Kinetic head, or velocity head, ft
  z = Potential energy, or head, due to elevation, ft

The concept of hydraulic and energy grade lines is helpful in analyzing these flow problems. If, at each point along a pipe system, the term P/δ is determined and plotted as a vertical distance above the centerline of the pipe, the locus of end points is the hydraulic grade line. More generally, the plot of the two terms $P/\delta + z$ for the flow, as ordinates, against length along the pipe as abscissas, produces the hydraulic grade line. The hydraulic grade line is the locus of heights to which water would rise in the manholes.

When the pressure intensity in the pipe is less than atmosphere, P/δ is negative and the hydraulic grade line is below the pipe. This obviously is then "open channel" and the water is not rising up in the manholes.

If in a similar way, the term $V^2/2g$ is added to the terms $P/\delta + V^2/2g$ and if, at each point along a pipe system, these terms are plotted as a vertical distance above the centerline of the pipe, the locus of end points is the energy grade line.

Sewer Shapes

It is important to maintain a minimum of 2 fps velocity in public sewers as well as in house drains. It can be seen that this presents a difficult problem in situations where there are extremely high peak demands and long periods of relatively low flow. This is particularly true of combined sewers where large sizes are required to accommodate storm water flow and much smaller sizes would be adequate to maintain the 2 fps velocity of the decreased sanitary flow. Engineers have overcome this problem by designing sewers of egg or oval shape. Figure 3-3 illustrates these various sewer shapes capable of maintaining the required minimum velocities at low-flow conditions. Examine Manning's formula and observe how the hydraulic radius affects the velocity. Now compare the hydraulic radius for circular, egg, and oval sewers. By decreasing the wetted perimeter at low-flow conditions, we have increased the hydraulic radius and thus increased the velocity.

Questions

1. What causes flow to occur in a sloping drain?
2. A sanitary house drain is installed at ¼ in./ft slope and is to carry a flow of 50 gpm. What size pipe is required to maintain a minimum velocity of 2 fps when flowing full? What size pipe is required for half-full flow?
3. For a velocity of 2 fps what must the slope of a 5-in. pipe be when flowing full? What is the slope for half-full flow?
4. What is the capacity of the pipe in Problem 3 for full flow?
5. A 3-in. sanitary drain line is installed at a pitch of ¼ in./ft. What is the velocity of flow when flowing full? When flowing half full? What is the quantity of flow when flowing full and half full?
6. What is the capacity of a 15-in. sanitary drain flowing full at a pitch of ¼ in./ft? What is the capacity and velocity at half-full flow?
7. What is the velocity of flow at full and half-full flow for 1½-in., 2-in., 2½-in., and 3-in. pipe installed at ⅛ in./ft pitch? Utilize Table 3-4 and explain the significance of the velocities.

Figure 3-3   Sewer Shapes

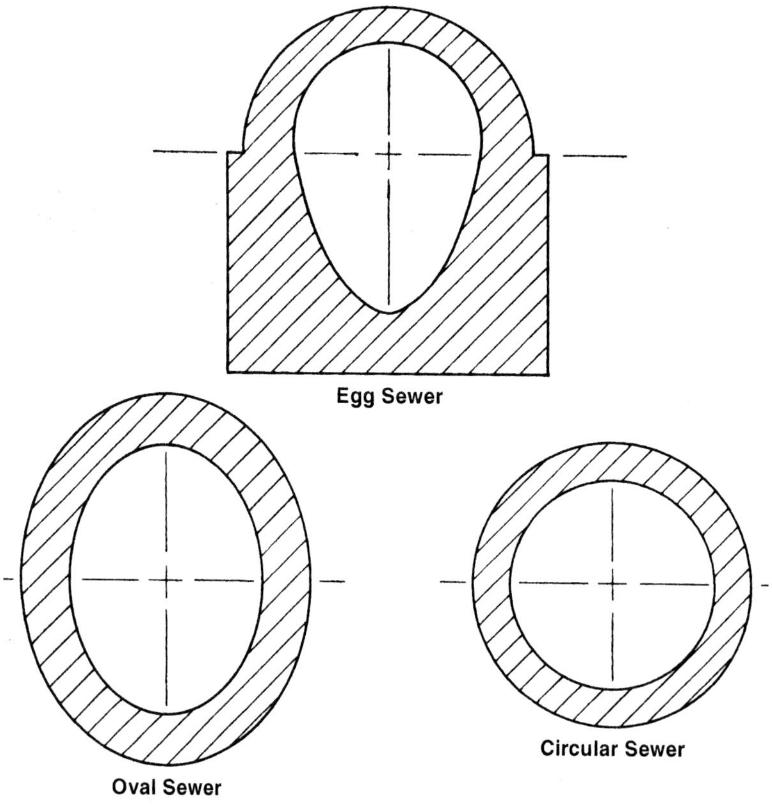

Egg Sewer

Oval Sewer

Circular Sewer

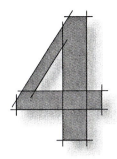

Soil and Waste Stacks

"Stack" is a general term for any vertical line of soil, waste, or vent piping that collects water and wastes from fixture drains and horizontal branch drains from two or more floors of a building. This does not include vertical fixture drains and branch vents that do not pass through more than two stories before being reconnected to the soil or waste stack or vent stack. The waste water comes from water closets, urinals, lavatories, bathtubs, showers, sinks, and various other fixtures. A soil stack collects wastes that contain fecal matter, such as those from water closets and urinals. A waste stack collects wastes that do not contain fecal matter, such as those from sinks, lavatories, and bathtubs.

Stack Connections

The horizontal branch connections to the stack are made with a sanitary tee, which is a short-radius tee wye, or with a tee wye that has a long radius. These fittings direct the waste into the stack with an initial downward velocity that permits the stack to accept an increased rate of flow at any one level. The tee wye gives the water a greater downward component than the sanitary tee and hence is more advantageous for greater stack capacities; but it is less favorable because of a tendency to create self-siphonage of the fixture traps connected to the branch. The tee wye is often called a combination wye and ⅛ bend. The sanitary tee can be used only in the vertical position, whereas the tee wye may be used in the vertical or horizontal position.

Flow in Stacks

The characteristics of flow down a stack can best be described in the words of Dr. Roy B. Hunter in the *Report of the Subcommittee on Plumbing of the Building Code Committee,* U.S. Department of Commerce, Bureau of Standards, BH 13 (1932):

> "The character of the flow of water in partially filled vertical pipes varies with the extent to which the pipe is filled… For small volumes of flow, amounting to little more than a trickle, the flow is entirely on the inner wall of the stack. With the increase in volume, this adherence to the wall continues up to a point where the frictional resistance of the air causes it to diaphragm across the pipe temporarily, forming a short slug of water which descends as a slug filling the stack until the increased air pressure breaks through, the water forming the slug either being thrown against the wall or falling a short distance as separate streamlets in the center of the pipe. This diaphragming and forming slugs probably first appears in a 3" stack when the stack is from one-fourth to one-third full. This intermittent rate partially accounts for the rapid erratic oscillations of pressure in a plumbing system."

Engineered Plumbing Design

From this it can be seen that a stack should never be designed for a capacity greater than one-third full (see Figure 4-1). Otherwise, the pressure fluctuations in the system could greatly exceed the ±1-in. column of water pressure criterion and traps could possibly lose their seal by siphonage or blowout.

Figure 4-1   Cross-Section of Stack Flowing at Design Capacity

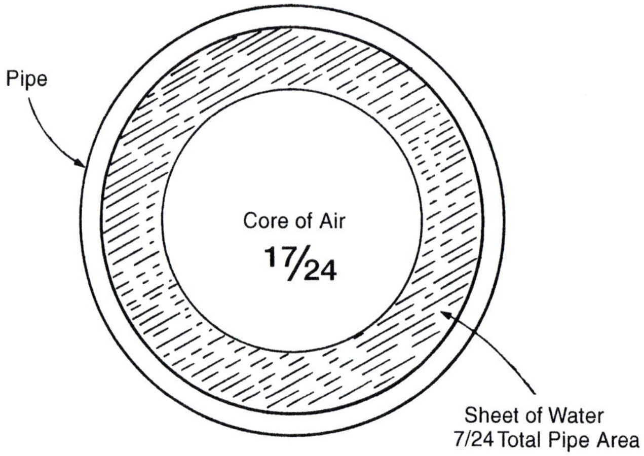

Pipe

Core of Air
17/24

Sheet of Water
7/24 Total Pipe Area

Terminal Velocity Length

When the plumbing engineer is called upon to design the systems for a very tall building, he/she is invariably asked how to accommodate the extremely high velocities developed at the base of the stacks. How is he/she going to prevent the base fitting from being blown out or broken? This scenario is one of the oldest and most persistent myths in the plumbing profession and, unfortunately, is still believed by too many uninformed designers.

Depending upon the rate of flow from the branch drain into the stack, the type of stack fitting, the diameter of the stack, and the flow down the stack from upper levels, the discharge from the branch may or may not entirely fill the cross section of the stack at the point of entry. As soon as the water enters the stack, it is immediately accelerated by the force of gravity and in a very short distance it forms a sheet around the inner wall of the pipe. This sheet of water, with a core of air in the center, continues to accelerate until the frictional force exerted by the pipe wall on the falling sheet of water equals the gravitational force. From this point downward, provided no flow enters the stack as the sheet passes a fitting, the sheet of water will fall at a velocity that remains practically unchanged. This ultimate vertical velocity is called "terminal velocity" and the distance in which this maximum velocity is achieved is called the "terminal length."

F. M. Dawson and A. A. Kalinske in *Report on Hydraulics and Pneumatics of Plumbing Drainage Systems*, State University of Iowa Studies in Engineering, Bulletin 10 (1937); and R. S. Wyly and H. N. Eaton in *Capacities of Plumbing Stacks in Buildings*, National Bureau of Standards, Building Materials and Structures Report BMS 132 (1952) investigated terminal velocity and derived a workable formula by treating the sheet of water as a solid hollow cylinder sliding down the inside wall of the pipe. The formulas developed for terminal velocity and terminal length, without going through the complicated calculus involved, are as follows:

(4-1)
$$V_T = 3.0 \left(\frac{q}{d}\right)^{2/5}$$

(4-2)
$$L_T = 0.052 \, V_T^2$$

where V_T = Terminal velocity in stack, fps
L_T = Terminal length below point of flow entry, ft
q = Quantity rate of flow, gpm
d = Diameter of stack, in.

Applying the formulas for various size pipes, it is found that terminal velocity is achieved at approximately 10 to 15 fps and this velocity is achieved within 10 ft to 15 ft of fall from the point of entry. The importance of this research is that it conclusively destroys the myth that water falling in a stack from a great height will destroy the fitting at the base of the stack. *The velocity at the base of a 100-story stack is only slightly and insignificantly greater than the velocity at the base of a three-story stack!*

Stack Capacities

The flow capacity of a stack can be expressed in terms of the ratio of the cross-sectional area of the sheet of water to the cross-sectional area of the pipe when the flow down the stack is at terminal velocity. Both Dawson and Hunter, in independent investigations, found that slugs of water and the resultant violent pressure fluctuations did not occur until the stack flowed one-quarter to one-third full. The maximum permissible flow rates in the stack can be expressed by the following formula:

(4-3)
$$q = 27.8 \, r^{5/3} \, d^{8/3}$$

where q = Capacity, gpm
r = Ratio of cross-sectional area of the sheet of water to cross-sectional area of the stack
d = Diameter of the stack, in.

Values of flow rates when $r = 5/24$, $7/24$, and $8/24$ are tabulated in Table 4-1.

Most code authorities have based their stack loading tables on a value of $r = 5/24$ or $7/24$. The upper limit of $r = 8/24$ is very rarely used due to the real probability that diaphragming will occur with resultant problems. There are other factors that must be considered and that govern the maximum permissible loading of stacks; these are discussed in detail in another portion of this book.

Hydraulic Jump

At the base of a stack, flow enters the horizontal drain at a relatively high velocity when compared to the velocity of flow in a horizontal drain under uniform flow conditions. For a 3-in. stack flowing at capacity, the terminal velocity is 10.2 fps. For a 3-in. drain installed at a slope of ¼ in./ft, the velocity under uniform flow conditions, at full or half-full flow, is 2.59 fps. When the water reaches the bend at the bottom of the stack, it is turned at right angles to its original flow and for a few pipe diameters downstream it will continue to flow at relatively high velocity along the lower part of the horizontal pipe. Since the slope of the horizontal drain is not adequate to maintain the velocity of the water that existed when it reached the bottom of the stack, the velocity of the water flowing in the horizontal drain slowly decreases with a corresponding increase in the depth of flow until a critical velocity is reached where the depth of flow suddenly increases. This increase in depth is often great enough to completely fill the cross-sectional area of the pipe. This phenomenon of sudden rise in depth is called the "hydraulic jump." The critical distance at which the hydraulic jump may occur varies. It is dependent upon the entrance velocity, depth of water that may already exist in the horizontal drain when the new flow is introduced, roughness of the pipe, diameter of the pipe, and the slope. The distance varies from immediately at the stack fitting up to ten times the diameter of the stack downstream. Less jump occurs if the horizontal drain is larger in size than the stack. Increasing the slope of the horizontal drain will also minimize the jump. After the hydraulic jump occurs and fills the drain, the

Table 4-1 Maximum Capacities Stacks

Pipe Size inches	Flow in GPM		
	r = ¼	r = 7/24	r = ⅓
2	18.5	23.5	—
3	54	70	85
4	112	145	180
5	205	270	324
6	330	435	530
8	710	920	1145
10	1300	1650	2055
12	2050	2650	3365

Figure 4-2 Hydraulic Jump at Offset

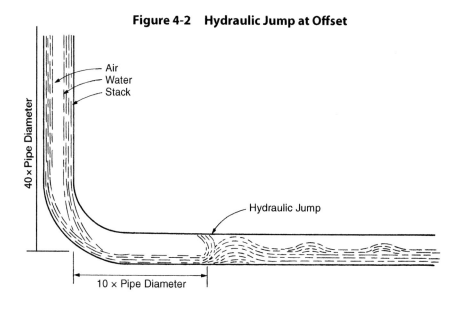

pipe tends to flow full, large bubbles of air moving along the top of the pipe with the water. Surging flow conditions will exist until the frictional resistance of the pipe retards the velocity to that of uniform flow conditions.

Any offset of the stack greater than 45° can cause a hydraulic jump. See Figure 4-2 for an illustration of the hydraulic jump.

Questions
1. What is the difference between a soil stack and a waste stack?
2. When a stack is flowing at maximum capacity, is it flowing full? If not, what is the recommended ratio of water to total pipe area?
3. Why should stacks never be designed for a capacity greater than ⅓ of its total area?
4. What is meant by terminal velocity?
5. What is meant by terminal length?
6. Discuss the implications of terminal velocity and terminal length.
7. Why do codes differ in their stack loading tables?
8. What is the hydraulic jump? Explain why it occurs.
9. A stack is offset 45°. Will a hydraulic jump occur? Substantiate your answer.

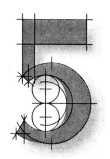# Drainage Systems

Wherever public sanitary sewers or combined sewers are available for disposal of sewage from a building, the sanitary drainage system of the building must be connected to the public system. Availability of public sewers is usually defined by the municipality and can vary from 100 ft to 500 ft and greater. Whenever it is practical to do so, it is recommended that connection be made to the public sanitary or combined sewer as it is the most economical method of sewage disposal. Where public sewers are not available, an approved private sewage disposal system must be provided. This private system must conform to all the rules and regulations of all authorities having jurisdiction in that locality. The sanitary drainage system should never discharge onto the ground and it should never discharge into a waterway until it is treated and rendered harmless.

The sanitary drainage system should never discharge into the public storm sewer. Storm water is not treated and any discharge of sanitary wastes into this system could contaminate other water sources.

The American Society of Plumbing Engineers defines a building (house) drain as that part of the lowest piping of a drainage system that receives the discharge from soil, waste, and other drainage pipes inside the walls of the building (house) and conveys it to the building (house) sewer, which begins outside the building (house) walls. A building (house) sewer is defined as that part of the horizontal piping of a drainage system that extends from the end of the building (house) drain and receives the discharge from the building (house) drain and conveys it to a public sewer, private sewer, individual sewage-disposal system, or other approved point of disposal.

Storm Water Disposal

Every building must have adequate provisions for draining water from roofs, paved areas, courts, and yards. Storm water drainage systems should connect to the public storm sewer or combined sewer. The storm drain should never be connected to the public sanitary sewer. When public storm or combined sewers are not available, storm water may be run to an existing stream or into an adequate system of dry wells.

Combined Systems

A combined plumbing drainage system or a combined sewer is one that conveys both storm and sanitary sewage in the same conduit. More and more communities are eliminating the combined sewers and providing separate public sanitary and storm sewers. The storm water in a combined system imposes too great a load upon the sewage treatment plants. This load can be great enough to cause a municipality to bypass the overload around the treatment facility during heavy storms and thus dump great quantities of raw untreated sewage into the public waterways.

Sanitary and storm water drainage systems within a building should be independent of each other unless they discharge into a combined public sewer.

High-Temperature Wastes

Excessively high-temperature wastes (above 140° F) should never discharge directly into the drainage system. High temperatures can cause excessive expansion and contraction of the piping with resulting harmful effects. Joints may be pulled apart or loosened and solidly bedded pipe may be broken. The discharges from boiler blowoffs, steam exhaust, condensate, etc., must be cooled down to at least 140°F before connecting to the drainage system. This may be accomplished by piping the high-temperature discharge to a water supplied sump or a cooling tank. The high-temperature waste can also be used to preheat the cold water supply to hot water heaters—thus accomplishing two purposes at once and at the same time conserving energy.

Drainage Systems Below Sewer Level

Where a drainage system is below the elevation of the house drain (which drains by gravity to the sewer) or the public sewer, the discharge should be conveyed to a sump or ejector and pumped or automatically lifted up into the gravity drainage system. Sumps are used to handle clear water and need not be airtight and vented. Ejectors are used for sewage and must be airtight and vented.

Backwater Valves

There is often the danger of backflow of sewage into a building when the public sewer becomes overloaded or is surcharged. To prevent backflow and possible flooding of the building, a backwater valve should be installed in the drainage piping from all fixtures that are at an elevation below the surcharge level of the public sewer.

Rather than use a multitude of backwater valves (at each fixture), it is feasible to install a backwater valve, a manually operated gate valve, or a combination backwater and gate valve in the house drain at the point of exit inside the building and downstream from the house trap if one is installed. The latter installation has the added advantage of not interfering with the circulation of air throughout the entire drainage system. The gate valve is recommended where there is a history of backflow as a positive means of protecting the building from flooding in case of a malfunction of the backwater valve. The backwater valve is nothing more than a swing check valve. (See Figure 5-1.)

House Traps and Fresh Air Inlets

There has been constant debate on the subject of the use of house traps. Some codes insist upon their installation, some allow an option and others prohibit them. None of the arguments, pro and con, appear to be conclusive and the plumbing engineer must adhere to the applicable code, or where there is no code he/she is free to follow his/her preference or prejudice.

Engineered Plumbing Design

Figure 5-1　Backwater Valve and Combination BWV with Manually Operated Gate Valve

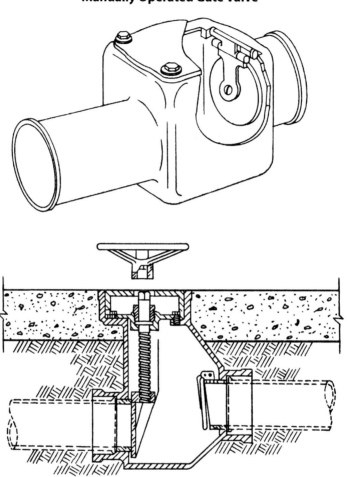

Storm water systems were developed long before sanitary drainage systems. The sanitary system developed from the original storm system. Before plumbing fixtures were installed inside buildings, drainage systems were provided to convey the storm water to storm water disposal systems. When fixtures were finally brought indoors, they were connected to the existing storm water system. This combined system soon became fouled and malodorous from sewage and became infested with rats and other vermin which traveled freely from one building to another through the sewer piping. Originally, house traps were installed in the house drain to eliminate this health hazard.

Fixture traps provide the primary safeguard against the flow of sewer gases into the building. This protection was not adequate during the years before the application of the basic principles of venting to protect the seal of fixture traps. The installation of house traps provided a secondary safeguard against sewer gas.

Proponents of the house trap claim that systems inside buildings are frequently opened for repairs and during this work the house trap prevents sewer gases from entering the building.

The anti–house trap forces counter with the valid statement that a house trap tends to create undesirable flow conditions in the house drain by adding resistance to flow. They also claim that when house traps are not installed, every stack in the building acts as a means of venting the public sewer. Recent research, however, has found that this action is negligible and is also of questionable value.

These, then, are the history and arguments and you are free to make your choice.

When house traps are installed, it is mandatory that fresh air inlets be provided to ventilate the building drainage system adequately. The term "fresh air inlet" is a misnomer. As much, if not more, air is discharged from the fresh air inlet as is ever drawn in. As water flows down the house drain, the air in the downstream piping is pushed along and escapes up available stacks or out the fresh air inlet. There have been constant complaints about the malodorous conditions existing in the area adjacent to fresh air inlets.

House traps should be of the two–hand hole type. When the piping is underground, the cleanouts can be extended to the floor level or a pit with a removable cover for accessibility must be provided. The maximum height for the cleanout extensions is limited to 2 ft to permit ease of rodding the piping.

The fresh air inlet should be a minimum of one-half the size of the house drain it serves, but never less than 3 in. The terminal should be a minimum of 6 in. above grade. The connection to the house drain should be upstream, as close as possible but no farther than 4 ft from the house trap. Fresh air plates, to protect the pipe opening from the entry of foreign matter, should be provided and the open area of the perforations in the plate must equal the pipe area. Figure 5-2 illustrates the installation of a house trap with a fresh air inlet. The fresh air inlet may be terminated with a return bend, or, if located inside, may be terminated in the outside wall with a fresh air plate.

Connections to Sanitary House Drains

House drains are generally designed to flow half full to two-thirds full at peak load conditions. Branch connections to the house drain should be made to the upper half of the pipe (the air space portion) wherever possible for the following reasons:
1. There is less of a chance of stoppages in the branch.
2. There is less flow interference at the point of connection.
3. When there is no flow in the branch, the full area of the pipe is available to relieve pneumatic pressure fluctuations in the house drain.

Branch Connections to Stack Offsets

Connections should not be made to the horizontal offset of a stack if at all possible. If a connection must be made, it should be made at least 10 diameters downstream to avoid the hydraulic jump area where there is danger of excessive pressures. Connections should be made at a minimum of 2 ft above

Engineered Plumbing Design

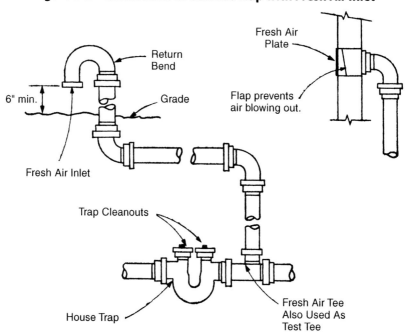

Figure 5-2 Installation of a House Trap with Fresh Air Inlet

the offset and 2 ft below the offset to avoid areas that are subject to extreme pressure fluctuations. (See Figure 5-3.)

Piping Installation

Horizontal piping must be installed in alignment, parallel to walls and at sufficient pitch to ensure a minimum velocity of flow of 2 fps. To attain this minimum velocity for scouring action a minimum pitch of ¼ in./ft should be employed for all pipe 3 in. and smaller in diameter. A pitch of ⅛ in./ft for larger pipe is satisfactory to maintain an adequate velocity in the piping. Drainage 90° elbows, traps, sanitary tees, and tee wyes are all tapped at a pitch of ¼ in./ft to provide proper pitch to the horizontal pipe when transferring from the horizontal to the vertical position.

Horizontal drainage piping should be so routed as not to pass over any equipment or fixtures where leakage from the line could possibly cause contamination. Drainage piping must never pass over switchgear or other electrical equipment. Route piping around electrical closets. If it is impossible to avoid these areas and piping must be run in these locations, then a pan must be installed below the pipe to collect any water from leaks or condensation. A drain line is run from this pan to a convenient floor drain or service sink.

Underground drainage piping should always be laid on a firm bed for its entire length with the earth scooped out at the bells to make this possible. (See Figure 5-4.) Clean earth or screened

Figure 5-4 Underground Drainage Piping

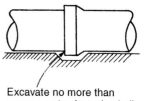

Excavate no more than necessary to clear pipe bell.

Figure 5-3 Piping for Fixtures Directly Above Offset

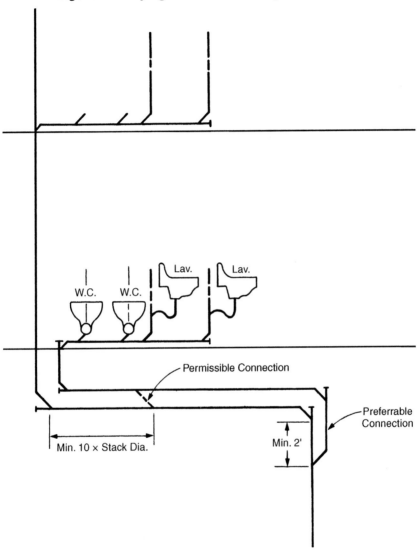

gravel should be firmly tamped under, around, and above the pipe to a level of 1 ft above, and thereafter backfilling can be completed to grade, compacting the fill every 2 ft.

Piping aboveground in the building should be securely supported from the building construction. Piping should never be hung from ductwork or other piping. Inserts, rods, hangers, piers, and anchors should be of durable material having adequate strength to perform their function. Maximum spacing between supports for various piping materials are

- Cast iron soil pipe
   Horizontal: at every fitting and every joint
   Vertical: at every story

- Screwed piping
  Horizontal: every 12 ft
  Vertical: alternate floors
- DWV copper:
  Horizontal: every 10 ft for 2 in. and larger, every 6 ft for smaller sizes
  Vertical: every story.
  Figure 5-5 illustrates various hangers and supports.

Changes in the direction of piping must be made with fittings that will not cause excessive reduction in the velocity of flow or create any other adverse effects. Short turn fittings should never be used in horizontal piping but may be used in the vertical to transfer from the horizontal. A sanitary tee (short turn fitting) can be used in the vertical, but a double sanitary tee should never be installed. The possibility of flow crossover and the buildup of excessive pressures in the opposite inlet when one branch is discharging is an ever-present danger.

Any method of installation or use of fittings that retards flow to a greater degree than normal should not be used—e.g., double hub, hub facing downstream, and tee inlet.

There can be no greater length than 2 ft from the vent connection for any future drain outlets. If this maximum distance is not observed, a dead end is created in which fungi, slime, and sludge can collect (See Figure 5-6). The only exception to this rule is where it is necessary to extend the piping for a cleanout to an accessible location.

Cleanouts

Cleanouts are provided in piping to permit access to piping for clearing of stoppages without the necessity of dismantling or breaking the piping. The size of the cleanout should be the same size as the piping up to 4 in. For larger size piping, 4 in. cleanouts are adequate for their intended purpose. Cleanouts should be provided at the following locations:
1. At the upper end of the building (house) drain
2. At the point where the building drain joins the building sewer. Use a wye branch or a house trap.
3. At every change of direction greater than 45°.
4. A maximum distance of 50 ft should be maintained between cleanouts for piping 4 in. and less, and 100 ft for larger piping. Underground piping larger than 10 in. in diameter should be provided with manholes at every change of direction and every 150 ft.
5. At the base of all stacks.

All cleanouts must be accessible and where necessary should be extended to the floor or wall. Adequate clearances must be provided around the cleanout for the manipulation of equipment in rodding out stoppages. (See Figure 5-7.)

Some cleanouts are designed with a neoprene seal plug, which prevents "freezing" or binding to the ferrule. All plugs are machined with a straight or running thread and a flared shoulder for the neoprene gasket, permitting quick and certain removal when necessary. A maximum opening is provided for tool access. Recessed covers are available to accommodate carpet, tile, terrazzo, and

Chapter 5—Drainage Systems

Figure 5-5 Hangers and Supports

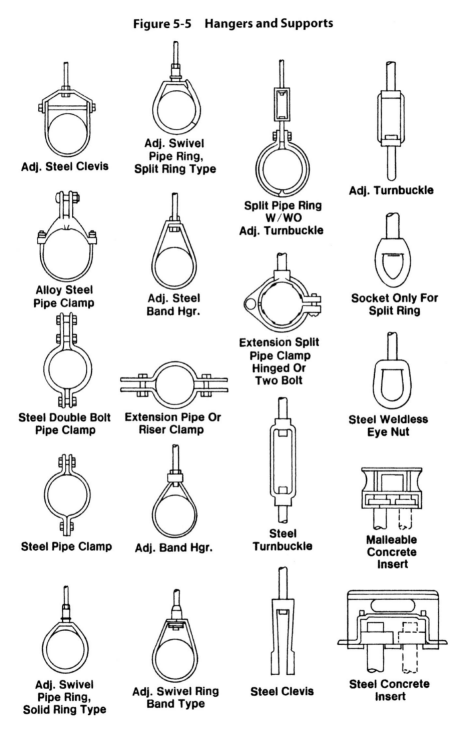

Figure 5-5 (con't)   Hangers and Supports

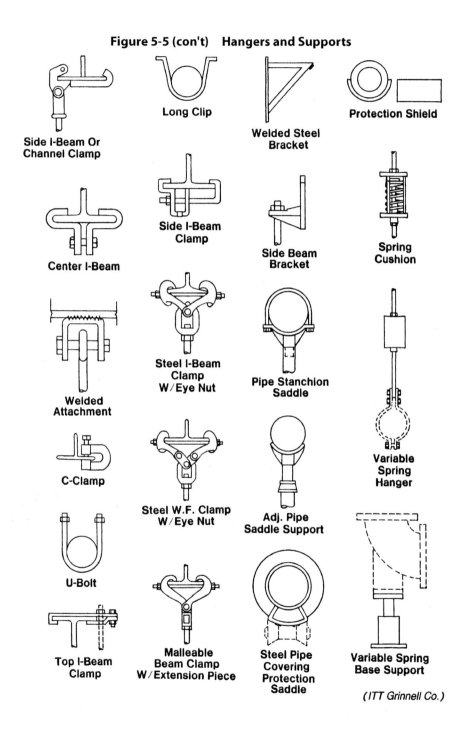

(ITT Grinnell Co.)

Figure 5-6   Maximum Vent Connection Distance

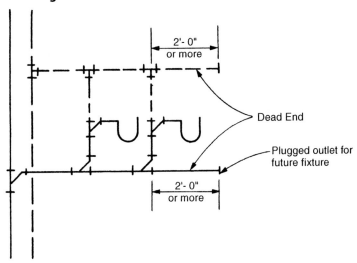

other surface finishes, and are adjustable to the exact floor level established by the adjustable housing or by the set screws.

Indirect Wastes

Waste piping that does not connect directly to the sanitary system is termed an "indirect waste." The discharge from an indirect waste should be conveyed into a water supplied, trapped, and vented receptacle or fixture. The discharge outlet should be a minimum of 1 in. above the flood level of the receptacle. The installation and sizing of indirect wastes is in accordance with all the rules of direct connected waste piping with one exception. Due to the low velocities usually present in indirect waste piping, stoppages are more prevalent and cleanouts should be provided at every possible location.

Fixtures that connect to indirect waste piping must be trapped but do not require vents. There are no severe pneumatic effects in this piping because of the extremely low rates of flow.

Where the piping exceeds 100 ft in developed length it should be extended to the atmosphere. Preferably it should be run independently through the roof. This ventilation is required to prevent the rapid fouling of the pipe due to the growth of slime and fungi in the absence of air circulation.

Special Wastes

Tank overflows, tank emptying lines and relief valve discharges should not connect directly to the drainage system due to the danger of contamination of the water supply. The discharge should be through an air break to an acceptable receptacle or floor drain or onto a roof. The same method should be applied for the drains from sprinkler systems, cooling jackets, drip pans, etc. Steam expansion tank drains and overflows, boiler blowoff, condensate and cooling tank drains may be treated in a similar manner or by a direct connection to the house sewer on the street side of the house trap where permitted by code.

Figure 5-7 Designing Cleanouts

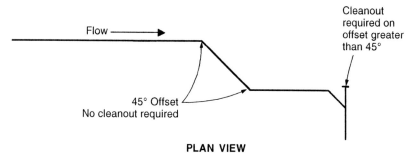
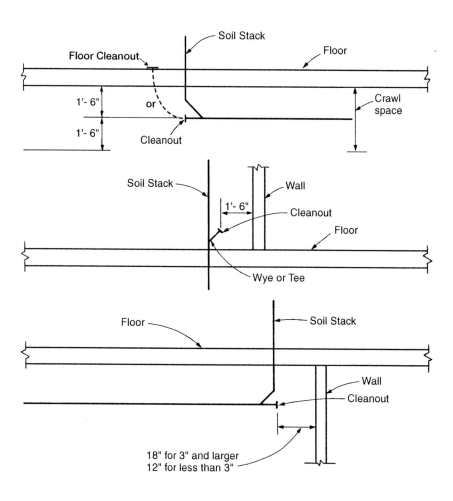

Questions

1. Must house drains and storm water drains always connect to public sewers?
2. What is a combined sewer and how does it affect the design of plumbing systems inside the building?
3. Why is high-temperature waste discharge prohibited from entering drainage piping?
4. Why do sewers back up into buildings and how can this be rectified?
5. Is a house trap necessary? Substantiate your answer.
6. When are fresh air inlets required?
7. Why should connections be made to the upper half of the house drain whenever possible?
8. What are the criteria for making branch connections at a stack offset?
9. What is the minimum pitch for pipes 3 in. and smaller? Why?
10. Why should a double sanitary tee never be used?
11. Why are cleanouts required? Where should they be installed?

Drainage System Sizing

The rate of flow in a fixture drain should be the same as the rate of flow discharged at the fixture outlet. The flow from the fixture outlet is comparable to flow from a water supply orifice discharging under flow pressure. For a fixture, the orifice can be taken as the minimum diameter of the waste outlet and the flow pressure is equivalent to the height of water above that minimum diameter. There is a decrease in the rate of flow from a fixture as the water level in the fixture decreases.

The average rate of flow from a fixture can be expressed fairly accurately by the formula

$$q = 13.17\, d^2\, h^{1/2} \tag{6-1}$$

where q = Rate of discharge, gpm
       d = Diameter of outlet orifice, in.
       h = Mean vertical height of water surface above the point of outlet orifice, ft

Rate of Flow in Branches

The rate of flow in a branch drain is obviously the sum of the flows from the fixtures connected to that branch. When the branch is of extended length (greater than 5 ft), to avoid the development of hydrostatic pressures at the required flow, the design of the branch is based upon the assumption that uniform flow conditions exist in the branch. Hydrostatic pressure manifests itself in the backup of water in the branch and surcharging. In relatively short branches, which is usually the case, the high velocities of flow from the fixtures produce surging flow conditions and higher capacities and velocities are produced than are produced under uniform flow conditions. Under no circumstances should the branch flow exceed the flow capacity of the stack or house drain to which it connects.

The discharge of a branch into a stack should not cause excessive interference with the sheet of water flowing down the stack, or back-pressure in the branch could result. The maximum permissible flow capacity must be related to the probable simultaneous rate of flow in the stack. The maximum probable flow in the stack is a function of the total flow into the stack and the number of branches connecting into it.

Fixture Drain Size

The minimum fixture drain size will be the minimum size of the fixture trap. Recommended minimum sizes of traps for various fixtures are given in Table 6-1. The plumbing engineer is warned, however, that he/she must comply with the code requirements in the locality of the job.

Sanitary Drainage Fixture Unit

The assigning of fixture unit values to fixtures to represent their load producing effect on the plumbing system was originally proposed in 1923. The fixture unit values were designed for application in conjunction with the probability of simultaneous use of fixtures so as to establish the maximum permissible drainage loads expressed in fixture units (FU) rather than in gallons per minute of drainage flow. Since the original proposal, various changes were proposed and made by Roy B. Hunter in the National Bureau of Standards Report, BMS 65 (1940). More recently, further changes have been recommended by researchers. Further study is presently being given to the assignment of fixture unit values to fixtures. Table 6-2 gives the original values. The plumbing engineer is again warned to conform to local codes.

Table 6-1   Minimum Size of Trap for Plumbing Fixtures

Fixture	Trap Size, In.
Bathtub (with or without overhead shower)	1½
Bidet	1½
Combination sink and tray	1½
Cup sink, laboratory	1½
Dental unit or cuspidor	1¼
Dishwasher, commercial	2
Dishwasher, domestic	1½
Drinking fountain	1
Funnel drain	1½
Kitchen sink, domestic	1½
Laboratory sink	2
Lavatory	1¼
Lavatory (barber, beauty parlor or surgeon)	1½
Lavatory, multiple type (wash fountain, wash sink)	1½
Laundry tray (1 or 2 compartments)	1½
Shower stall, domestic	2
Shower, gang	3
Sinks:	
Surgeon	1½
Flushing rim (with flush valve)	3
Service (trap standard)	3
Service (P trap)	2
Pot, scullery, etc.	1½
Sterilizers	2
Urinal, pedestal	3
Urinal, stall	2
Urinal, wall-hung	2

A fixture unit is a quantity in terms of which the load producing effects on the plumbing system of different kinds of plumbing fixtures are expressed on some arbitrarily chosen scale. The fixture unit flow rate is the total discharge flow in gallons per minute of a single fixture divided by 7.5 (gal/ft^3), which provides the flow rate of that particular plumbing fixture as a unit of flow. Fixtures are rated as multiples of this unit of flow. The fixture unit flow rate designates only the measured amount of water discharged through the waste plug of a fixture in gallons per minute and is not to be confused with the load producing effects as expressed by the fixture unit value, although they are related in the overall picture. Hunter conceived the idea of assigning a fixture unit value to represent the degree to which a fixture loads a system when used at the maximum assumed frequency. The sole purpose of the fixture unit concept is to make it possible to calculate the design load on the system directly when the system is composed of different kinds of fixtures, each having a loading

Table 6-2 Fixture Units Per Fixture or Group

Fixture Type	Fixture-unit value as load factors
1 bathroom group consisting of water closet, lavatory, and bathtub or shower stall:	
Tank water closet	6
Flush-valve water closet	8
Bathtub[1] (with or without overhead shower)	2
Bidet	3
Combination sink and tray	3
Combination sink and tray with food-disposal unit	4
Dental unit or cuspidor	1
Dental lavatory	1
Dishwasher, domestic	2
Drinking fountain	½
Floor drains[2]	1
Kitchen sink, domestic	2
Kitchen sink, domestic (with food-disposal unit)	3
Lavatory[3] small P.O.	1
Lavatory[3] large P.O.	2
Lavatory (barber, beauty parlor or surgeon)	2
Laundry tray (1 or 2 compartments)	2
Shower stall, domestic	2
Showers (group) per head	3
Sinks:	
Surgeon	3
Flushing rim (with flush valve)	8
Service (trap standard)	3
Service (P trap)	2
Pot, scullery, etc.	4
Urinal (pedestal, siphon jet, blowout)	8
Urinal, stall	4
Urinal, wall-hung	4
Urinal, trough (each 2-ft. section)	2
Wash sink (circular or multiple), each set of faucets	2
Water closet:	
Tank-operated	4
Valve-operated	6

[1] A shower head over a bathtub does not increase the fixture unit value.
[2] Size of floor drain shall be determined by the area of surface water to be drained.
[3] Lavatories with 1 1/4- or 1½-inch trap have the same load value: larger P.O plugs have greater flow rate.

characteristic different from the others.

Stack Sizing

The recommended maximum permissible flow in a stack is $7/24$ of the total cross-sectional area of the stack. The formula as previously stated (Equation 4-3) is:

$$q = 27.8\, r^{5/3}\, d^{8/3}$$

Referring to Table 4-1, where $r = 7/24$, the maximum permissible flow for the various sizes of pipe in gallons per minute can be determined. Table 6-3 shows the tabulation of maximum permissible fixture units to be conveyed by stacks of various sizes. The table was compiled by taking into account the probability of simultaneous use of fixtures. Reference to the table will verify that 500 FU is the maximum loading for a 4-in. stack. Thus, 145 gpm is equivalent to 500 FU. This is the total load from all branches. It is extremely important to note that there is a restriction as to the amount of flow permitted to enter the stack from any one branch when the stack is more than three stories in height.

If too large a flow is introduced into a stack from any one branch, the inflow will fill the stack at the level of branch discharge and the water will back up above the level of inflow causing violent pressure fluctuations in the stack with consequent loss of trap seals as well as sluggish flow in the branch. The problem then becomes how great a flow can be introduced into a stack at an intermediate level in a multistory building when there is a given flow coming down the stack from upper floors. Wyly and Eaton made a study of this problem at the National Bureau of Standards for the Housing and Home Finance Agency in 1950.

Figure 6-1 shows an illustration of a short section of a stack, including a horizontal branch connection using a sanitary tee fitting. Water is flowing down the stack from the upper floors and the branch is flowing at a great enough rate to completely fill the branch at its point of connection.

The discharge of the branch can enter the stack either by mixing with the vertical flow or by deflecting it. The deflection can only be accomplished by developing a hydrostatic head in the branch. This hydrostatic head is developed by the backing up of water in the branch until the head is great enough to change the momentum of the vertical stream sufficiently to permit the discharge of the branch to enter the stack. The limit for the maximum head allowable in the branch by the backing up of water was established at that height that would prevent the water from backing up into a stall shower or cause sluggish flow. This was set as the head measured at the axis of the pipe that will just cause the branch to flow full near its outlet.

When long-turn tee wyes are used for the connection of the branch to the stack, the discharge has a greater vertical velocity component as it enters the stack and back-pressures are appreciably less for the same stack and branch flows.

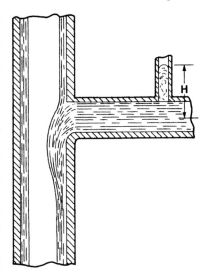

Figure 6-1 Diagrammatic Representation of Interference of Flows in Stack Fitting

The theoretical analysis of the problem is too involved and lengthy for coverage in this book, but the interested reader is referred to the original papers for study in depth.

Referring to Table 6-3 for a 4-in. stack of more than three stories in height, the maximum loading permitted is 500 FU. Although the maximum loading for a 4-in. branch is 160 FU as per Table 6-4, this load is limited by Table 6-3,

Table 6-3 Maximum Permissible F.U. Loads for Sanitary Stacks

Stack Diameter, inches	Stack three stories or less in height F.U.	Stacks more than three stories in height F.U.	Total discharge into one branch interval F.U.
2	10	24	6
2½	20	42	9
3	30*	60*	16**
4	240	500	90
5	540	1100	200
6	960	1900	350
8	2200	3600	600
10	3800	5600	1000
12	6000	8400	1500

* Not more than six water closets permitted.
** Not more than two water closets permitted.

Table 6-4 Maximum Permissible F.U. Loads for Sanitary Branches

Branch Diameter inches	Total Load F.U.
1½	3
2	6
2½	12
3	20*
4	160
5	360
6	620
8	1400

* Not more than two water closets permitted.

which permits only 90 FU to be introduced into a 4-in. stack in any one branch interval. *The stack would have to be increased in size to accommodate any branch load exceeding 90 FU!*

A branch interval is that part of a soil or waste stack, at least 8 feet in length between horizontal branch connections, within which one or more horizontal branches are connected. Figure 6-2 illustrates branch intervals.

Assume a 4-in. soil stack has a 4-in. branch connection, which is loaded to its maximum capacity. The total number of allowable fixture units on a 4-in. horizontal branch is 160 FU (Table 6-4). The total fixture units permitted to discharge into any one branch interval are 90 FU. Obviously the 90 FU limit is exceeded and it is necessary to increase the size of the stack to 5 in. or reduce the number of fixture units on the branch so as not to cause violent pressure fluctuations in the system.

Assume there are two horizontal branches connected to the 4-in. stack within one branch interval. One branch carries 80 FU and the other 40 FU for a total of 120 FU discharging into one branch interval. This exceeds the maximum permitted of 90 FU and the stack must be increased to 5 in. If one branch carries 30 FU and the other 60 FU, then the maximum is not exceeded and the 4-in. size of the stack is adequate.

Figure 6-2 Branch Interval

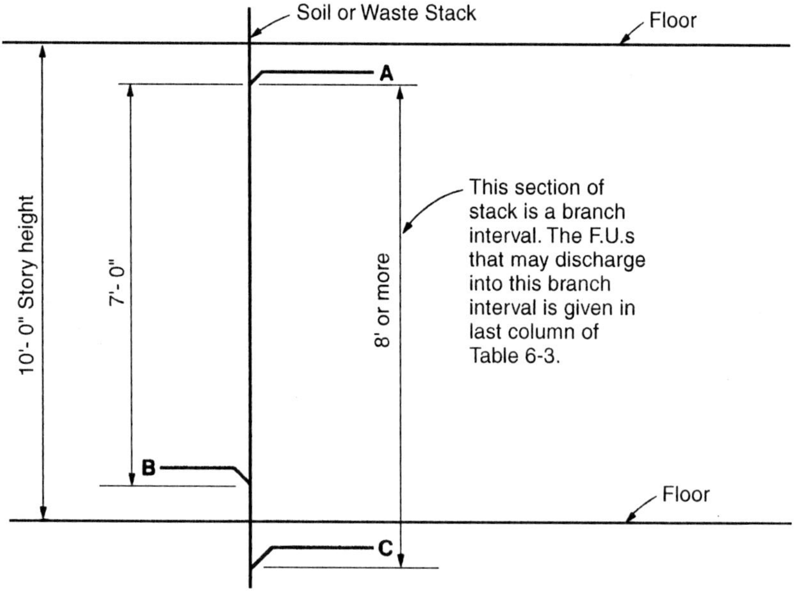

The distance of 7'- 0" between branches A & B is not considered a branch interval. The definition states "at least 8 feet in length."

Procedure for Sizing Stacks

The procedure for sizing a multistory stack is first to size the horizontal branches connected to the stack. This is done by totaling the fixture units connected to each branch and size in accordance with Table 6-4. Next, total all the fixture units connected to the stack and determine the size from Table 6-3, under the column "Stacks More than Three Stories in Height." (The minimum size of the stack must be at least equal to the largest connected branch.) Immediately check the next column, "Total Discharge into One Branch Interval," and determine that this maximum is not exceeded by any of the branches. If it is exceeded, the size of the stack as originally determined must be increased one size or the loading of the branches must be redesigned so that maximum conditions are satisfied.

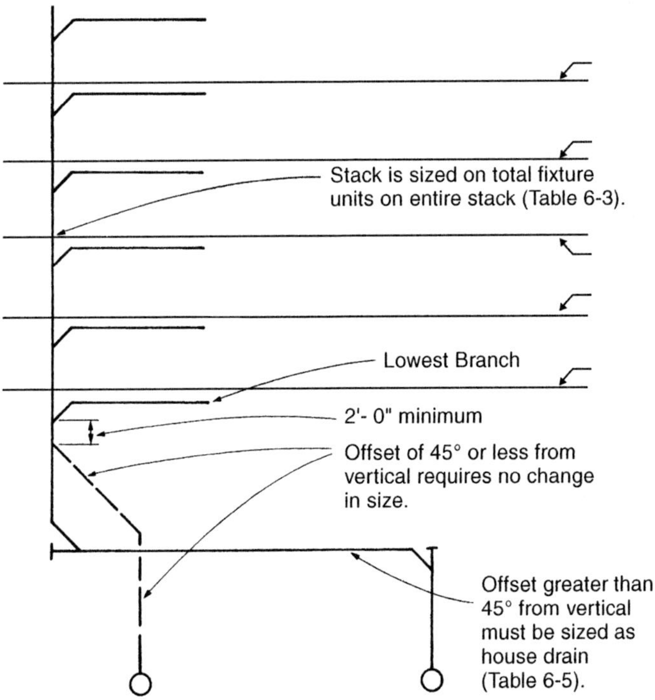

Figure 6-3 illustrates a typical stack and the correct procedure for sizing.

Sizing a stack that offsets more than 45° from the vertical can be done in a manner that will result in significant economics. The procedure (illustrated in Figure 6-4) is as follows:

Figure 6-3 Procedure for Sizing a Stack

- Stack is sized on total fixture units on entire stack (Table 6-3).
- Lowest Branch
- 2'- 0" minimum
- Offset of 45° or less from vertical requires no change in size.
- Offset greater than 45° from vertical must be sized as house drain (Table 6-5).

1. The portion of the stack above the offset is sized in the same manner as a stack that does not offset, based upon the total fixture units connected to the stack above the offset.
2. The horizontal offset portion of the stack is sized on the basis of sizing a house drain. (See Table 6-5.)
3. The portion of the stack below the offset is the same size as the offset or based upon the total connected fixture units of the entire stack (above and below the offset), whichever size is larger.

Example

Given a stack loaded as shown in Figure 6-5, the following procedure is followed to size all the drainage piping:

Engineered Plumbing Design

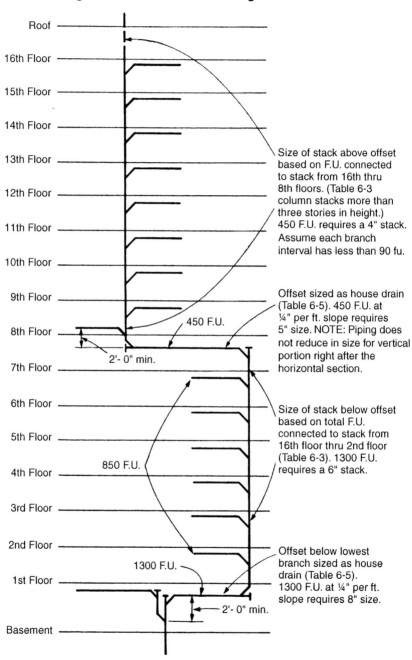

Figure 6-4 Procedure for Sizing an Offset Stack

Chapter 6—Drainage System Sizing

Step 1: Size all the branch piping (Table 6-4).

A	70 FU	=	4 in.	B	90 FU	=	4 in.	C	120 FU	=	4 in.
D	90 FU	=	4 in.	E	150 FU	=	4 in.	F	180 FU	=	5 in.
G	190 FU	=	5 in.	H	160 FU	=	4 in.	I	200 FU	=	5 in.

Step 2: Size the portion of stack above offset (Table 6-3).

J = Sum of all branches connected to the portion of the stack above offset.

A + B + C + D = 370 FU = 4 in.

Immediately check last column of Table 6-3 to ascertain that total discharge into any one branch interval of a 4-in. stack is not exceeded. Branch C is 120 FU, which exceeds the limits of 90 FU and the 4-in. stack must therefore be increased to 5 in. so as not to overload that

Figure 6-5 Sizing Example

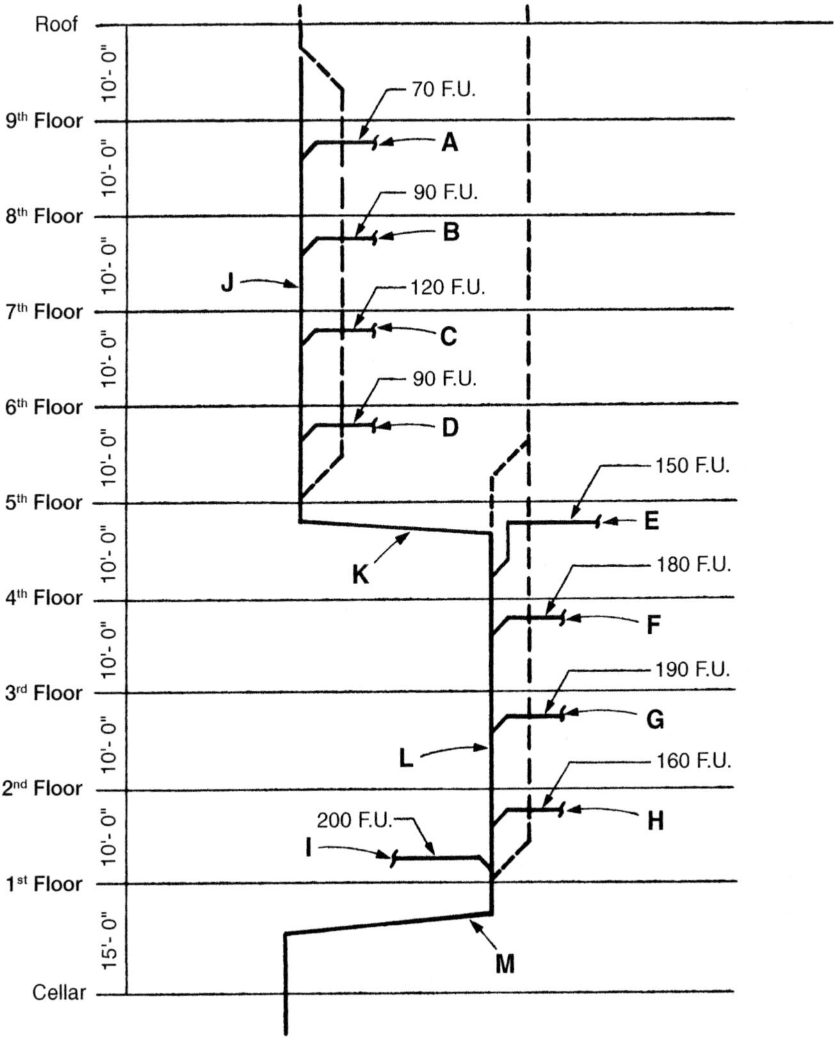

portion of the stack between branches C and D. Therefore stack J = 5 in.

Step 3: Size the offset (Table 6-5).
Offset K = 370 FU = 5 in. (whether the slope is ⅛ in. or ¼ in.)

Step 4: Size the portion of stack below offset (Table 6-3).
This equals the sum of all the branches, above and below the offset, connected to the stack.
A + B + C + D + E + F + G + H + I = 1250 FU = 6 in.
Immediately check the last column of Table 6-3. The maximum permitted into any one branch interval of a 6-in. stack is 350 FU. This value is not exceeded and so Stack L = 6 in.

Step 5: Size the runout to the house drain (Table 6-5).
M = 1250 FU = 8 in. (whether slope is ⅛ in. or ¼ in.).

It is interesting to note the sizing when certain variations are made to the stack shown in Figure 6-5. If the lowest branch does not connect to the stack but is run down to the cellar separately and connected directly to the house drain, then the load on the stack is 1250 - 200 = 1050 FU and the portion of the stack below the offset can be 5 in. instead of 6 in. If the last column of Table 6-3 is checked, it is found that 200 FU is the maximum for a 5-in. stack, and since no branch exceeds this discharge, stack L would remain 5 in. If the stack did not have an offset at the fourth floor ceiling, then for the load of 1250 FU the size of the stack would have to be 6 in. from the base through the roof.

House Drains

House drains are designed to flow at half-full capacity to a maximum of three-quarters full under uniform flow conditions so as to prevent violent pneumatic pressure fluctuations and the development of hydraulic pressures. A minimum slope of ¼ in./ft should be provided for 3-in. pipe and smaller, ⅛ in./ft for 4-in. through 6-in. pipe, and ¹⁄₁₆ in./ft for 8 in. and larger. These minimum slopes are required to maintain a velocity of flow greater than 2 fps for scouring action. Table 6-6 gives the approximate velocities for given slopes and diameters of horizontal drains.

A value of 2 FU can be assigned for each gallon per minute of flow for continuous or semicontinuous flow into the drainage system such as from sump pumps, ejectors, air-

Table 6-5 Maximum Permissible Fixture Unit Loads for Sanitary Building Drains and Runouts from Stacks

Pipe Diameter inches	Building Drain Slope (inches per foot)		
	⅛	¼	½
2	-	21	26
2½	-	24	31
3	20	27	36
4	180	216	250
5	390	480	575
6	700	840	1000
8	1600	1920	2300
10	2900	3500	4200
12	4600	5600	6700
15	8300	10000	12000

Chapter 6—Drainage System Sizing

conditioning equipment, water-cooled equipment or similar devices (e.g., a sump pump that discharges at the rate of 200 gpm is equivalent to $200 \times 2 = 400$ FU). Figure 6-6 shows a diagram of a sanitary system with an ejector and the correct sizes for all piping.

When sizing the house drain, always start at the point farthest from the point of exit from the building. The calculations are then simplified because additions are required only as progress in the direction of flow proceeds. The same procedure is followed in sizing each branch of the house drain. Figure 6-7 shows a simple drainage system with the loads noted with an example of how the system is sized.

Table 6-6   Approximate Flow Velocity of Sewage

Size, inches	Velocity, fps			
	1/16 in./ft	1/8 in./ft.	1/4 in./ft.	1/2 in./ft.
2	1.02	1.44	2.03	2.88
3	1.24	1.76	2.49	3.53
4	1.44	2.03	2.88	4.07
5	1.61	2.28	3.53	4.56
6	1.76	2.49	4.07	5.00
8	2.03	2.88	4.23	5.75
10	2.28	3.23	4.58	6.44

Figure 6-6   Sanitary System with an Ejector

Piping installed at 1/8" slope

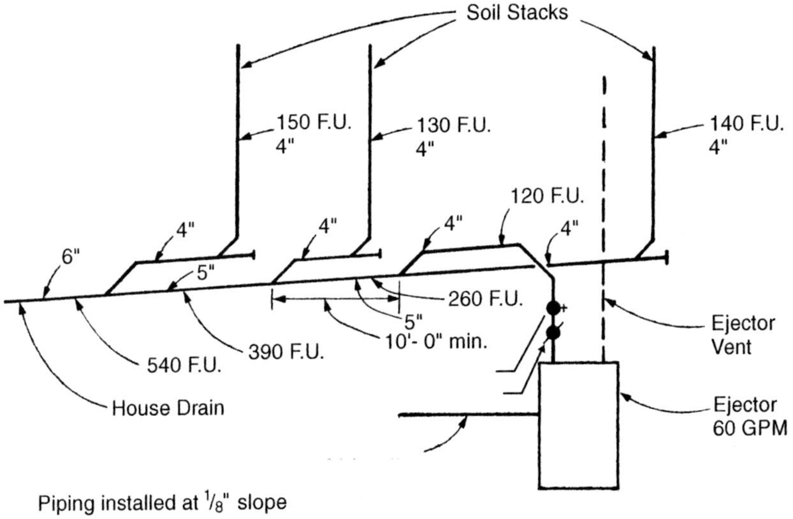

Example

Given a sanitary system with loads as shown in Figure 6-7, the sizing procedure is as follows: Starting at the point farthest from the point of exit from the building, total all fixture unit loads. Fixture unit values are obtained from Table 6-2, horizontal piping sizes from Table 6-5 and stack sizes from Table 6-3.

Engineered Plumbing Design

Figure 6-7   A Simple Drainage System

[Diagram: A simple drainage system showing stacks P-1 through P-10 with the following fixtures:
- P-1: 6 W.C., 2 Ur., 4 Lav.
- P-2: 6 Sinks
- P-3: 10 W.C., 2 Ur., 6 Lav.
- P-4: 18 W.C., 4 Ur., 18 Lav.
- P-5: 8 W.C., 8 Lav., 8 Sink, 8 B.T.
- P-6: 3 W.C., 3 Lav., 3 B.T., 3 Sh.
- P-7: 4 W.C., 4 Lav.
- P-8: 10 Sinks
- P-9: 12 Lav.
- P-10: 30 W.C., 6 Ur., 8 Sh.
Flow direction indicated at point S. Junction labels A through R shown along the main horizontal drain.]

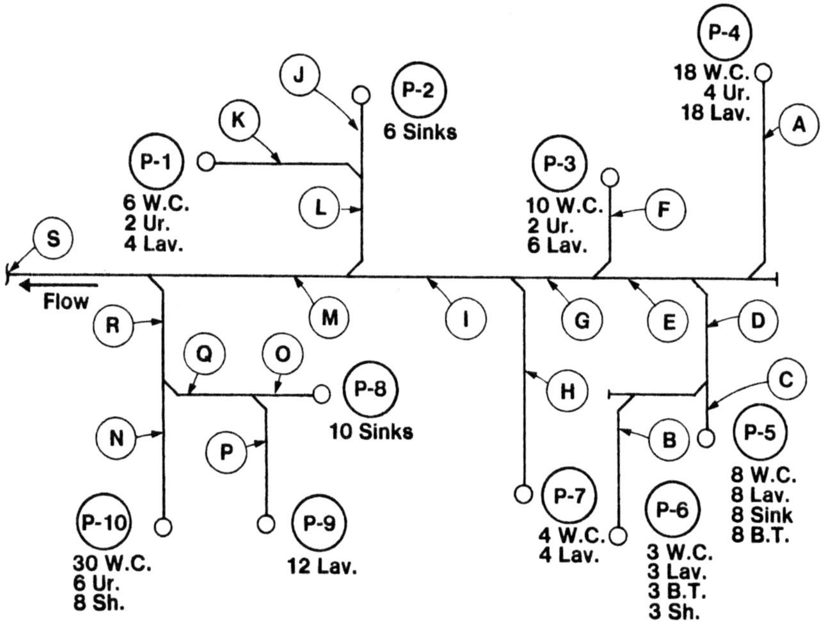

Note: Water closets are valve-operated (flushometers), urinals are wall hung, lavatories have small P.O., sinks are kitchen sinks and building is 7-story. All piping is installed at 1/8" per ft. slope.

Step 1. A Stack P-4:

    18 WC × 6 FU = 108 FU
    4 UR × 4   = 16
    18 LAV × 1   = 18
                   142 FU

    Stack P-4 = 142 FU = 4 in.
    A   = 142 FU = 4 in.

Step 2. B Stack P-6:

Note: A bathroom group consists of 1 WC, 1 Lav, and 1 bathtub or shower. In this case, there are three bathroom groups.

    3 Bathroom Groups × 8 = 24 FU
    3 SH × 2   = 6
                            30 FU

    Stack P-6 = 30 FU = 3 in.
    B   = 30 FU = 4 in.

Step 3. C Stack P-5:

Note: A bathroom group consists of 1 WC, 1 Lav, and 1 bathtub or shower. In this case, there are eight bathroom groups.

    8 Bathroom Groups × 8 = 64 FU
    8 BT × 2   = 16
                            80 FU

Chapter 6—Drainage System Sizing

          Stack P-5 = 80 FU = 4 in.
          C   = 80 FU = 4 in.

Step 4. D = B + C
          D = 30 + 80 = 110 FU = 4 in.

Step 5. E = A + D
          E = 142 + 110 = 252 FU = 5 in.

Step 6. F Stack P-3:
          10 WC × 6   = 60 FU
          2 UR × 4   =   8
          6 LAV × 1   =   6
                          74 FU

          Stack P-3 = 74 FU = 4 in.
          F   = 74 FU = 4 in.

Step 7. G = E + F
          G = 252 + 74 = 336 FU = 5 in.
          (If piping had been installed at a slope of ¼ in./ft, the size would be 5 in.)

Step 8. H Stack P-7:
          4 WC × 6   = 24 FU
          4 LAV × 1   =   4
                          30 FU

          Stack P-7 = 30 FU = 3 in.
          (Minimum size of stack serving water closets is 3 in.)
          H   = 30 FU = 4 in.

Step 9. I = G + H
          I = 336 + 30 = 366 FU = 5 in. (5 in. if ¼ in./ft slope)

Step 10. J Stack P-2:
          6 SK × 2   = 12 FU
          Stack P-2 = 12 FU = 2 in.
          J   = 12 FU = 3 in.

Step 11. K Stack P-1:
          6 WC × 6   = 36 FU
          2 UR × 4   =   8
          4 LAV × 1   =   4
                          48 FU

          Stack P-1 = 48 FU = 3 in. (Limit of 6 WC is not exceeded.)
          K   = 48 FU = 4 in.

Step 12. L = J + K
          L = 12 + 48 = 60 FU = 4 in.

Step 13. M = I + L
          M = 366 + 60 = 426 FU = 6 in.

Engineered Plumbing Design

Step 14. N Stack P-10:
    30 WC × 6   = 180 FU
    6 UR × 4   = 24
    8 SH × 2   = 16
                      220 FU
    Stack P-10 = 220 FU = 4 in.
    N   = 220 FU = 5 in.

Step 15. O Stack P-8:
    10 SK × 2 = 20 FU
    Stack P-8   = 20 FU = 2 in.
    O   = 20 FU = 3 in.

Step 16. P Stack P-9:
    12 LAV × 1 = 12 FU
    Stack P-9   = 12 FU = 2 in.
    P   = 12 FU = 3 in.

Step 17. Q = O + P
    Q = 20 + 12 = 32 FU = 4 in.

Step 18. R = N + Q
    R = 220 + 32 = 252 FU = 5 in.

Step 19. S = M + R
    S = 426 + 252 = 678 FU = 6 in.
    (If piping had been installed at a slope of ¼ in./ft, the size would be 6 in.)

Questions
1. What type of flow is assumed to exist in a branch drain line for purposes of sizing?
2. If back-pressure in the branch drain is caused by excessive interference with the stack flow at point of entry, what possible problems could be created?
3. What is the minimum fixture drain size for any fixture?
4. Define fixture unit.
5. What is the maximum permissible flow in a stack stated in terms of cross-sectional area of flow to stack area?
6. What will happen if too great a flow is introduced into a stack from any one branch?
7. Can more than one branch connect to a branch interval?
8. Why are limitations placed upon the maximum quantity of flow that may be introduced into any one branch interval of a specific size stack?
9. What is the procedure for sizing a stack?
10. What is the procedure for sizing a stack that offsets more than 45°?
11. Is a house drain sized on the basis of full flow? Why?
12. Why can't all sizes of the house drain be installed at ⅛ in./ft slope?
13. How do you convert continuous flows into fixture unit values?
14. What is the procedure for sizing a house drain?

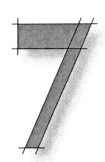

Storm Water Systems

Storm water is considered to be rainwater, surface runoff, groundwater, subsurface water, or similar clear liquid wastes, exclusive of sewage and industrial wastes.

Design of the storm drainage system is based upon the piping flowing full under uniform flow conditions. Leaders (also called conductors or downspouts) and horizontal piping can flow full as there is no necessity for maintaining pneumatic or hydraulic pressures within any fixed limits in the system, as is required in a sanitary drainage system.

Where there are separate public storm and sanitary sewers available, the storm drainage system shall not discharge into the sanitary sewer. This prohibition prevents additional loads being imposed upon the sewage disposal plants. Most municipalities are beginning to insist on separate systems. Where a public combined sewer system exists and separate sewers are planned for future construction, the storm drainage system should be isolated from the sanitary system inside the building and combined outside the building. If this procedure is followed, there will be no need to make alterations within the building when the time comes to connect to the future storm sewer system.

When only a combined sewer is available, the storm system can be connected to the sanitary system within the building at the most convenient locations. When a leader or any storm drain is connected to a sanitary system, the storm piping must be trapped. The purpose of the trap is to prevent foul odors present in the sanitary piping from escaping at the roof drains or areaways. If not for this nuisance, it would be advantageous not to install the traps and thus obtain the additional venting and air circulation provided by the open storm system. One trap may serve more than one leader. If the storm piping is run separately from the sanitary piping, one trap may serve the entire storm system before connection to the sanitary system or combined sewer. The size of leader traps shall be the same size as the horizontal runout from the leader to the house drain. Single–hand hole traps for the main traps are acceptable, but a two–hand hole trap is preferable. A fresh air inlet is not required on the storm system for the main trap.

Leaders shall never be used as soil, waste, or vent pipe or vice versa. It is good practice to connect leaders at least 10 ft downstream from any soil or waste connection on the combined house drain. If the connection is any closer it tends to impede the discharge of the soil or waste when the leader is discharging and could possibly cause backups. (See Figure 7-1.)

Due to the extremes of variable outside temperatures in relation to the fairly constant inside temperature, provisions should be made at the roof drains for

Engineered Plumbing Design

Figure 7-1 Connection of Leader of Combined Building Drain*
*Check Local Code

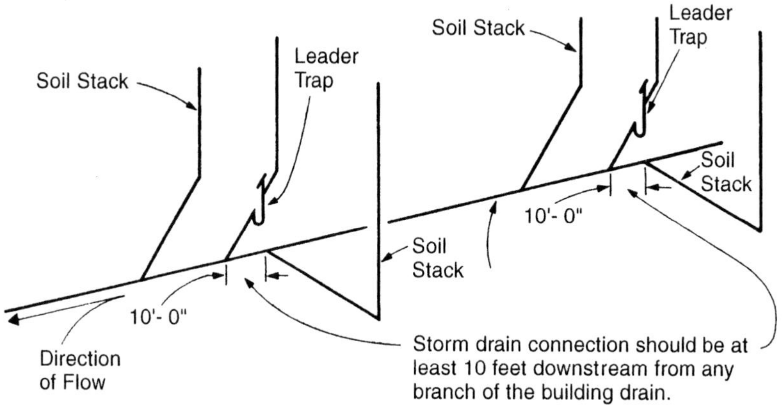

[Figure: Diagram showing Soil Stacks, Leader Traps, and direction of flow with 10'-0" spacing notations. Note: "Storm drain connection should be at least 10 feet downstream from any branch of the building drain."]

expansion and contraction. This can be accomplished by means of an offset as illustrated in Figure 7-2.

When an offset connection cannot be utilized due to architectural or structural limitations it is recommended that a roof drain with an integral expansion joint or a separate expansion joint be used. An expansion joint or offset should always be used at the connection to roof drains to prevent the pipe expansion from raising the roof drain and destroying the integrity of the waterproofing of the roof. Storm water piping is probably subjected to the most frequent movement of any plumbing system, although not necessarily to the maximum movement.

Low-temperature liquid flow in the storm water piping will cause condensation to form on the outside of the piping in the building. It is therefore advisable to insulate all storm water offsets and the bodies of all roof drains to prevent condensation from staining the ceiling. (See Figure 7-2.)

Collection Areas

Storm water should be conveyed from drainage areas at the same rate at which it collects on these areas. The required rate of discharge from any collection area, dependent upon the size of the area and the maximum rate of rainfall per hour, is to be employed as the design criterion. For many regions of the United States the maximum rainfall rate has been recorded at 4 in./hr. Using this rate of rainfall, it can be shown that 1 gpm will collect on 24 ft² of horizontal surface.

$$1 \text{ gpm} = 60 \text{ gph}$$
$$\text{converting to cubic feet: } 60 \div 7.5 = 8 \text{ ft}^3/\text{hr}$$
$$\text{for 4 in./hr rainfall: } 8 \div \tfrac{4}{12} = 24 \text{ ft}^2$$

Thus 24 ft² of horizontal area is equivalent to 1 gpm that must be conveyed from the surface. Sizing tables in most codes are formulated on this basis. If in any locality the maximum rate of rainfall is more or less than 4 in./hr., then the area that is equivalent to 1 gpm may be obtained by the simple conversion

Figure 7-2   A Simple Drainage System

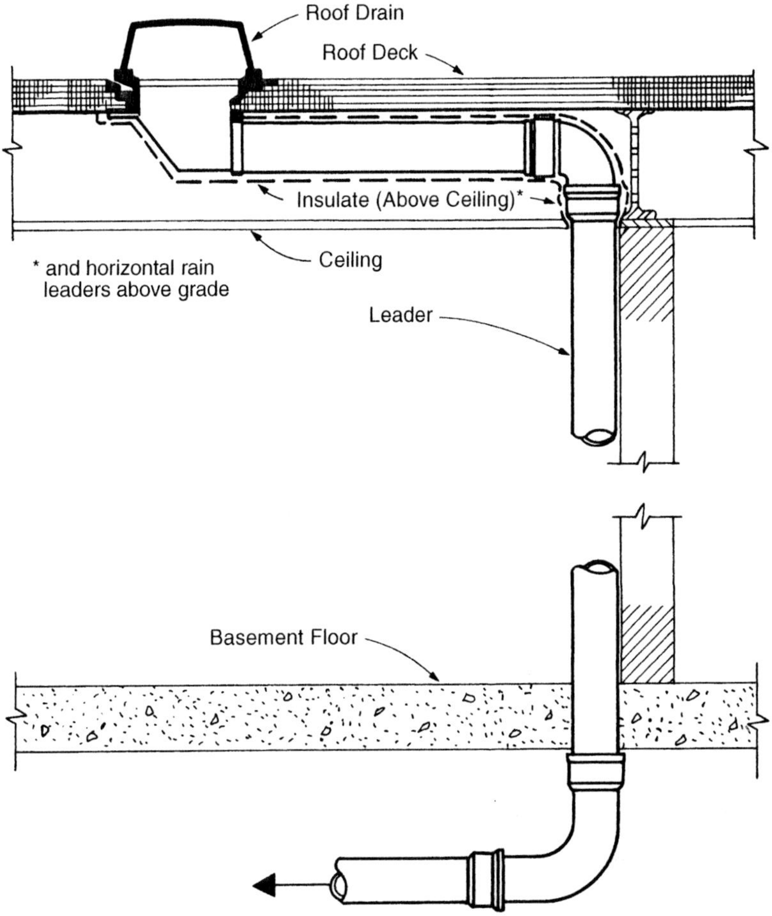

of multiplying 24 by 4 and dividing by the maximum rate of rainfall in inches per hour for that particular locality, e.g.,

$$4 \text{ in./hr rainfall: } 24 \text{ ft}^2 = 1 \text{ gpm}$$
$$5 \text{ in./hr rainfall: } 24 \times \tfrac{4}{5} = 19.2 \text{ ft}^2 = 1 \text{ gpm}$$
$$3 \text{ in./hr rainfall: } 24 \times \tfrac{4}{3} = 32 \text{ ft}^2 = 1 \text{ gpm}$$

Vertical Walls

Many authorities recommend that 50% of the vertical wall area adjacent to a drained area should be added to the horizontal drained area. It is extremely unusual for rain to fall in a perfectly vertical pattern. Depending upon wind conditions, the angle of rainfall could be as much as 60° to the vertical or possibly more. Under this condition, rain falls on the wall and is conducted down and added to the rain falling on the horizontal area to be drained. While this may be theoretically valid, it is the author's opinion that the theory does not hold up under actual conditions. If it were true, there would be such a depth of water on the sidewalks in front of buildings during a storm that people could not walk. Apparently the wind that drives the rain into the wall also whips the rain on the wall, away from the area. More research is required on this subject and the reader is advised to follow recommended practice until definitive data can be collected.

Sizing

Table 7-1 is a tabulation of pipe sizes required for leaders and horizontal storm piping in terms of drained areas based upon a maximum 4 in./hr. rainfall rate.

Roof Gutters

The size of semicircular gutters can be obtained from Table 7-2. Rectangular gutters can be selected on the basis of an equivalent cross-sectional area to the semicircular gutter. Rectangular leaders, because of the four walls and corners, offer greater frictional losses and this diminishes their carrying capacity. To compensate for this loss, a rectangular leader must be about 10% larger than a round leader to convey the same load. Table 7-3 gives the sizes of rectangular leaders that are equivalent to round leaders.

Roof Drains

There are three basic components that form the construction of a roof drain: the strainer, flashing ring (combined with a gravel stop where required), and drain body or sump. Mushroom, or domed, strainers should be used for roofs where leaves or other debris may accumulate. An open area for drainage is still maintained even though the leaves and debris may clog the lower portion of the strainer. The open area of the strainer should be one-and-a-half to two times the area of the pipe to which it connects. (Figure 7-3 illustrates a typical roof drain.)

Corner strainers are required when the drain is located at the corner of the roof and the parapet.

Table 7-1 Maximum Permissible Loads for Storm Drainage Piping

Pipe Diam., in.	Leaders Drained area, sq. ft.	Horizontal Piping Drained area, sq. ft., for various slopes		
		⅛ in.	¼ in.	½ in.
2	720	-	-	-
2½	1300	-	-	-
3	2200	822	1160	1644
4	4600	1880	2650	3760
5	8650	3340	4720	6680
6	13500	5350	7550	10700
8	29000	11500	16300	23000
10		20700	29200	41400
12		33300	47000	66600
15		59500	84000	119000

Table 7-2 Maximum Permissible Loads for Semicircular Gutters

Gutter Diam., in.	Drained Area, Area, sq. ft., for various slopes			
	¹⁄₁₆ in.	⅛ in.	¼ in.	½ in.
3	170	240	340	480
4	360	510	720	1020
5	825	880	1250	1770
6	960	1360	1920	2770
7	1380	1950	2760	3900
8	1990	2800	3980	5600
10	3600	5100	7200	10000

Table 7-3 Rectangular Leaders Equivalent to Round Leaders

Round Leader Diameter, inches	Rectangular Leader Dimension, inches
2	2 × 2
	1½ × 2½
3	2 × 4
	2½ × 3
4	3 × 4¼
	3½ × 4
5	4 × 5
	4½ × 4½
6	5 × 6

Chapter 7—Storm Water Systems

Figure 7-3 Typical Roof Drain

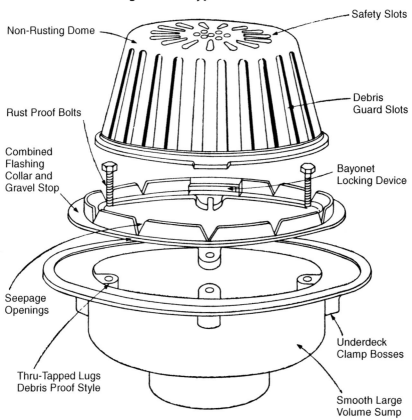

The connection between roofs and roof drains that pass through the roof and into the interior of the building should be made watertight by the use of proper flashing material. The flashing ring should clamp the flashing to the flashing collar without puncturing the flashing. Drain bodies may be secured to metal decks and plank roofs by means of underdeck clamps. Clamps are not necessary for poured-in-place concrete roofs.

Material

Inside leaders may be cast iron, galvanized steel, galvanized ferrous alloys, brass, copper, or plastic. Actually, any dependable piping material would be satisfactory so long as it complies with the requirements of the local code. Underground storm drainage piping may be cast iron soil pipe, ferrous alloy pipe, vitrified clay, concrete, bituminous fiber, or asbestos cement, depending upon the requirements of the local code. Roof drains may be cast iron, copper, lead, or other acceptable corrosion-resistant material.

Flow Velocity

While a minimum velocity of 2 fps is adequate for scouring action in sanitary piping, it has been found that a greater velocity is required for storm water piping. A minimum velocity of 3 fps is required to keep the grit, sand, and debris found in storm water in suspension.

Controlled-Flow Roof Drainage

The concept of conventional roof drainage design is to drain the water off the roof as rapidly as it collects on the roof surface. In 1960 an alternate concept was introduced which has proven to be extremely successful and has provided considerable benefits not obtainable by the conventional approach.

The new concept—*controlled-flow roof drainage*—instead of attempting to drain rainwater as fast as it comes down, drains the water at a much slower controlled rate. The excess water is permitted to accumulate on the roof under predetermined conditions and is drained off at the controlled rate after the storm has abated. This is the same principle that is applied in the use of dams for flood control, although on a much smaller scale. The result is a marked reduction in the maximum flow rate that leaders and sewers must accommodate. Due to the reduced flow rates, pipe sizes can be drastically reduced and the loads imposed on sewage treatment plants handling the effluent of combined sewers are also greatly reduced.

Controlled-flow roof drainage therefore not only provides more economically sized piping, but also affords a means of alleviating the chronic flooding conditions that are prevalent whenever a storm occurs. Many sewage treatment plants are forced to bypass huge quantities of untreated combined sewer effluent into our waterways during heavy storms. Controlled flow will help alleviate this problem and aid in the prevention of pollution of our streams, lakes, and oceans.

The application of required data for controlled-flow design is just as sophisticated as for flood control and all factors involved must be interrelated. Fortunately, various roof drain manufacturers have investigated the engineering parameters in great depth and have waded through the involved and sophisticated mathematical calculations required to produce simplified and straightforward design procedures. Each manufacturer's procedure is slightly different from the other's due to the configuration of the end product they manufacture and the method they utilize to control flow, but the engineering principles employed by all are valid and can be safely followed. The following criteria, however, are applicable to all controlled-flow drainage situations and should be followed to achieve a satisfactory and trouble-free system.

Roof Loading

A satisfactory design must incorporate adequate protection to prevent overloading of the roof structure. This protection can be accomplished by installing overflow scuppers at the correct height in the roof parapet walls. The recommended maximum storage depth of water on the roof is 3 in. for a dead-flat roof and a 3-in. average depth for a sloped roof. The maximum roof slope that could be tolerated, therefore, would be 6 in. from the low point to the high point of the roof construction.

The 3-in. design depth, as the maximum accumulated head, is recommended because it results in an applied load of 15.6 lb/ft^2 (psf) to the roof. In snow areas of the country, roof construction is generally designed for a 30 psf load, whereas in other areas a roof design load of 20 psf is permissible. By limiting the imposed water load to 15.6 psf (3-in. depth) an adequate factor of safety is provided for any roof.

Storm Intensity

Controlled-flow design is dependent upon the relationship of the rate of rainfall and the rate of controlled flow from the roof. Rainfall data are available for every area of the country from the U.S. Weather Service and are also included in the data of manufacturers' catalogs of controlled-flow roof drains. Rainfall intensities are given for 10, 25, 50, and 100-year storm frequencies. It is thus possible to predict the maximum rainfall intensity that will occur at least once every 10 years, every 25, 50, or 100 years.

Most public sewers in existence have been designed on the basis of a five-year storm and some have been designed for a 10-year storm. It is recommended that the 25-year storm be selected as the design criterion for roof drainage design. Thus, it is likely that only once in 50 or 100 years the system will be overloaded and excess water discharged through the overflow scuppers.

Drain-Down Time

Drain-down time is another factor that must always be incorporated in the design. Drain-down time is that factor, measured in hours, that indicates the time required for total drainage of the roof from the time the storm has reached its frequency peak. If the drain-down time is designed for 12 hr. based upon a storm frequency of 25 years, then after the 25-year frequency storm has attained its maximum intensity and duration and has ceased, it will require 12 hr. for total drainage of the accumulated water to be accomplished. The length of drain-down time can be varied by the designer in accordance with his/her preferences so long as the local code has not set any limitations. Some codes have established maximum drain-down times, which vary from 8 hr. in one city to 24 hr. in another city. It is recommended that 24 hr. be set as the maximum limit.

Design Configuration is the Key

The design configuration of the roof drain is the key to controlled flow. This is the "dam" that controls the rate of flow from the roof. The roof drain is constructed to incorporate a weir, which passes a predetermined rate of flow in proportion to the head of water on the roof. Manufacturers have employed various-shaped weirs to accomplish this purpose. One roof drain utilizes a parabolic weir, another an adjustable rectangular weir, and another a triangular weir. All adequately serve their purpose within the accuracies required for the storm water systems.

Suggested Code for Controlled-Flow Roof Drainage

Many municipalities have not yet adopted criteria for controlled-flow storm water systems. The following is a suggested format that may prove helpful.

Section No... Controlled-Flow Storm Water System. In lieu of sizing the storm drainage system in accordance with the methods previously described in this chapter, the roof drainage may be sized on controlled flow and storage of the storm water on the roof providing the following conditions are met:

1. The water of a 25-year frequency storm is not stored on the roof for more than 24 hr.
2. The water depth on flat roofs does not exceed 3 in. during the above storm and 3 in. average depth on sloped roofs.
3. The roof is equipped with 45° cants installed at any wall or parapet.
4. Roof design for controlled-flow roof drainage shall be based on a minimum of 20 psf loading to provide a safety factor above the 15.6 psf represented by the 3-in. design depth of water.
5. Flashing extends at least 6 in. above the roof level and scuppers are placed on the parapet walls at an invert elevation of ½ in. above the maximum designed head.
6. No fewer than two roof drains are installed in roof areas 10,000 ft^2 or less and at least four roof drains in areas over 10,000 ft^2.
7. Control of runoff from roofs shall be by weirs; no valves or mechanical devices shall be permitted.
8. Drainage from controlled flow, which is based upon gallons per minute of flow, shall be converted to equivalent square feet of roof or paved area, on the basis of each gallon per minute of flow being equivalent to 24 ft^2 of area. Drains not equipped with weirs—such as area or plaza drains—may be connected to the controlled-flow system provided the square feet of area, including the converted gallons per minute flow to square feet, are added together and the drain is sized to convey the sum of all loads.

Subsoil Drainage

When subsoil drains are placed under the cellar or basement floor or are placed around the perimeter of the building, they should be made of open joint, perforated, or porous pipe, not less than 4 in. in diameter. When the discharge of the subsoil drain is subject to backwater, the subsoil drain should be protected by a backwater valve that is accessibly located. The discharge of the subsoil drain should be into a sump. It is not required for this sump to be vented.

For further and more detailed information on subsoil (or subsurface) drainage, refer to ASPE *Data Book*, Volume 2 (2000), Chapter 4, "Storm-Drainage Systems."

Combined Storm and Sanitary System

When storm piping is connected with sanitary piping to form a combined system it is necessary to have a basis for converting the fixture unit load of the sanitary system into its equivalent square feet of drained area so that the combined system can be properly sized. The flow in the storm drainage piping during the maximum rainfall periods is almost always in excess of the flow in the sanitary piping at maximum design conditions. It is therefore desirable to base the sizing of the combined system on storm drainage design. A method of converting sanitary fixture unit loads to equivalent square feet of drained area has been recommended. The procedure is as follows:

1. When the total fixture unit load is 256 FU or less, use 1000 ft^2 as a minimum.

2. When the total fixture unit load is greater than 256 FU, multiply the fixture units by 3.9 ft² to convert to equivalent drained area.
3. For continuous or intermittent flow, multiply each gallon per minute of flow by 24 ft² to convert to equivalent drained area.

The above conversion rules are based upon a 4 in./hr. rainfall. Multiply by the correct factor for other rainfall rates.

By an examination of the water supply fixture unit loading tables it can be determined that 0.161 gpm of additional load is added for each fixture unit in excess of 256 FU. Assuming a load of 2000 FU, it is found that this load is equivalent to 321 gpm. One FU is then

$$321 \text{ gpm} \div 2000 \text{ FU} = 0.161 \text{ gpm}$$
$$\text{since } 24 \text{ ft}^2 = 1 \text{ gpm}$$
$$\text{then } 0.161 \times 24 = 3.9 \text{ ft}^2 \text{ per FU}$$

To size a combined drainage system, convert sanitary fixture units to equivalent drained area and add to the storm drainage area to obtain the total load. Use Table 7-1 to determine the appropriate size of the combined piping for the total combined load.

Example

To size the entire system as shown in Figure 7-4, assuming all piping is installed at ⅛ in./ft. slope, perform the following steps.

Step 1. P-2 = 256 FU = 4 in. (Table 6-3)
A = 256 FU = 5 in. (Table 6-5)

Step 2. L-2 = 3500 ft² = 4 in. (Table 7-1 Column "Leaders")
B = 3500 ft² = 6 in. (Table 7-1 Column "Horizontal Piping")

Step. 3. C = A + B
Convert sanitary FU to equivalent drained area:
256 = 1000 ft²
then
C = 1000 + 3500 ft² = 4500 = 6 in. (Table 7-1)

Step 4. P-1 = 180 FU = 4 in. (Table 6-3)
D = 180 FU = 4 in. (Table 6-5)

Step 5. L-1 = 4000 ft² = 4 in. (Table 7-1 Column "Leaders")
E = 4000 ft² = 6 in. (Table 7-1 "Horizontal Piping ⅛ in.")

Step 6. F = D + E
Convert sanitary fixture units to equivalent area. Any quantity less than 256 FU is assumed to be equivalent to 1000 ft²
Then 180 FU = 1000 ft² and
F = 1000 ft² + 4000 ft² = 5000 ft² = 6 in. (Table 7-1)

Step 7. G = C + F
G = 4500 ft² + (180 FU × 3.9 ft²/FU) + 4000 ft² = 9202 ft²
The value of 180 FU is not taken as 1000 ft² because stack P-2, 256 FU has already been converted to the 1000 ft² and any additional fixture units added are converted on the basis of 1 FU = 3.9 ft².

Engineered Plumbing Design

Figure 7-4 Plan of a Combined Storm and Sanitary System

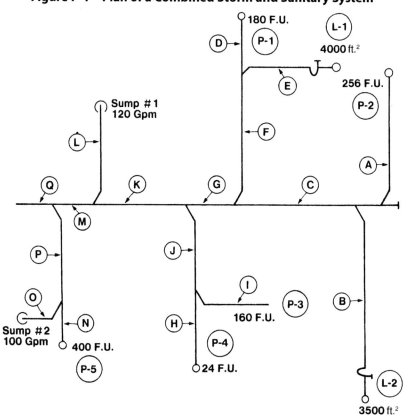

Step 8. P-4 = 24 FU = 3 in. (Table 6-3, assuming there are water closets on the stack).

Step 9. H = 24 FU = 4 in. (Table 6-5)

Step 10. P-3 = 160 FU = 4 in. (Table 6-3)

Step 11. I = 160 FU = 4 in. (Table 6-5)

Step 12. J = H + I
J = 24 FU + 160 FU = 184 FU = 5 in. (Table 6-5)

Step 13. K = G + J
K = 9202 ft² + (184 FU × 3.9 ft²/FU) = 9,19.6 ft
K = 8 in. (Table 7-1)

Step 14. L = 120 gpm × 24 ft²/gpm = 2880 ft²
L = 5 in. (Table 7-1)

Step 15. M = K + L
M = 9919.6 ft² + 2880 ft² = 12,799.6 ft²
M = 10 in. (Table 7-1)

Step 16. P-5 = 400 FU = 4 in. (Table 6-3)

Step 17. N = 400 FU = 6 in. (Table 6-5)

Step 18. O = 100 gpm × 2 = 200 FU
(The gpm discharge is converted to equivalent fixture units because it combines with a sanitary drain. In Step 14, the discharge was converted to the equivalent drained area because it was connected to a combined storm and sanitary drain.)

Step 19. P = N + O
P = 400 FU + 200 FU = 600 FU = 6 in. (Table 6-5)

Step 20. Q = M + P
Q = 12,800 ft² + (600 FU × 3.9 ft²/FU) = 15,140 ft²
Q = 10 in. (Table 7-1)

Questions

1. What is the difference between storm water and sanitary waste water?
2. Is the storm drainage system designed for full flow? Why?
3. Under what conditions can the storm drainage system connect with the sanitary system within the building?
4. What precautions must be taken when storm drainage is connected to sanitary drainage?
5. If a public storm sewer is available is it ever permissible to connect the storm water into the sanitary public sewer?
6. Can a leader be used as a waste or vent under some conditions?
7. Why should a leader be connected at least 10 ft downstream from any soil or waste connection to the combined house drain?
8. Should provisions be made for expansion and contraction of storm water lines?
9. When and where is it advisable to insulate storm drainage piping?
10. For a 4 in./hr. rainfall, how many gallons per minute will collect on 24 ft² of roof?
11. How do you convert the area on which 1 gpm will collect for a 4 in./hr. rainfall for other rates of rainfall?
12. What is the theory of controlled-flow roof drainage? Are there any advantages to this method?
13. How is continuous flow in gpm converted to equivalent drained area?

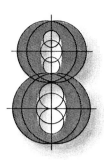

Vent Systems

Flow of air is the primary consideration in the design of a venting system for the ventilation of the piping and protection of the fixture trap seals of a sanitary drainage system. Since air is of such primary importance, it is essential that the plumbing engineer be familiar with certain physical characteristics that are pertinent to its behavior in a plumbing system.

Density of any substance is its mass per unit volume. The density of air is its weight in pounds per cubic foot of volume. The density of air is affected by temperature, moisture content, and pressure. The density of "standard air" is taken at atmospheric pressure and 68.4°F. It is equal to 0.075 lbm/ft^3. With a rise in temperature, the density of air decreases and with a lowering of temperature its density increases. The moisture content of air in the plumbing system has a negligible effect on its density and can be disregarded in all calculations. Pressure has an appreciable effect; the higher the pressure the greater the density, and the lower the pressure the less the density.

Specific Weight of a fluid is not an absolute property, but depends upon the local gravitational field (gravitational acceleration on earth is g=32.2 ft/sec^2) and the properties of the fluid itself. Commonly called "density" when concerning gravitational force, the numerical value of specific weight (lbf/ft^3) is equal to density (lbm/ft^3).

Elasticity is the ability of a substance to assume its original characteristics after the removal of a force that has been applied. Air is a perfectly elastic substance. From the scientific definition of elasticity it becomes clear that a rubber band is really a very inelastic material. If a weight is suspended from a rubber band and left for a few hours, then the weight is removed, the rubber band will spring back, but definitely not to its original length. If a force is applied to air, the force can be applied for days or years, and when it is removed, the air will return exactly to its original condition.

Air is compressible. There is an increase in pressure when air is compressed. In the plumbing system, only an extremely small change in pressure can be tolerated. For a pressure of 1 in. of water column (0.036 psi), the volume of air will be compressed by $\frac{1}{400}$ of its original volume. Assuming an original volume of 400 ft^3 of air at atmospheric pressure and the application of a pressure of 1 in. of water column, the air will be compressed by $400 \times \frac{1}{400} = 1$ ft^3. It is obvious that a comparatively small change in volume can very easily cause the accepted design limitation of ±1 in. of pressure to be exceeded with the consequent danger of destroying the trap seals. The vent piping must be designed to permit the air to flow freely without compression or expansion except for the small amount necessary to overcome friction.

Static Head

Static head is the pressure exerted at any point by the weight of the substance above that point. The pressure can be stated in feet of the substance, i.e., when the substance is water the static head is 100 ft of water, or if the substance is air, 100 ft of air. To convert from feet of head to pounds-force per square inch:

(8-1)
$$p = \frac{\gamma h}{144} \text{ and } h = \frac{144p}{\gamma}$$

where p = Pressure, lbf./in²
  γ = Specific weight of substance, lbf/ft³
  h = Static head, ft

Pneumatic Effects in Sanitary Systems

As water flows in contact with air in vertical or horizontal piping, there is friction between the air and water. The frictional effect causes the air to be dragged along with the water and at practically the same velocity as the water. When the cross-sectional area of the water occupying the pipe is suddenly increased, such as at the hydraulic jump or where a branch discharges into the stack, the air passage is constricted. This constriction acts exactly the same as a stoppage or a blockage to the flow of air. This causes a buildup of pressure, the highest pressure occurring at the constriction and diminishing upstream. It is for this reason that excessive pressure usually develops at the lower floors of a building and at offsets of the stack. It is important to always be aware that protection from the entry of sewer gases is afforded by the 2-in. trap seal, and the design of plumbing systems must be such as to maintain pressure variations within ±1 in. column of water.

Rate of Flow from Outlets

The velocity at which air flows out of an outlet to the atmosphere (at the roof terminal of a stack) is due to the total pressure available in the vent pipe at the outlet. This pressure is the flow pressure, which is equal to the static pressure less the pressure lost in friction. The maximum rate of discharge in practice is expressed as:

(8-2)
$$q_D = c_D q_I$$

where q_D = Actual quantity of discharge, gpm
  q_I = Ideal quantity of discharge, gpm
  c_D = Coefficient of discharge

Utilizing the formula $q = AV$ and substituting,

(8-3)
$$q_D = c_D (2.448 \, d_o^2 \, V_I)$$

where d_o = Outlet diameter, in.
  V_I = Ideal velocity, fps

Velocity is equal to $\sqrt{2gh}$, where g = acceleration due to gravity and h = height (or head) of air column.

$$q_D = c_D (2.448 \, d_o^2 \sqrt{2gh}) \tag{8-4}$$

$$= c_D (19.65 \, d_o^2 \sqrt{h}) \tag{8-5}$$

Using 0.67 as an acceptable coefficient of discharge, per Equation 6-1,

$$q_D = 13.17 \, d_o^2 \, h^{1/2}$$

Static Pressure of Air

The design criterion of maintaining pneumatic pressure fluctuations within ± 1 in. of water column is constantly stressed throughout this book. It should prove interesting to state this pressure in terms of an equivalent column of air. The formula for any substance is, per Equation 8-1:

$$P = \frac{\gamma h}{144}$$

then

$$P = \frac{\gamma_W h_W}{144} = \frac{\gamma_A h_A}{144} \tag{8-6}$$

where γ_W = Specific weight of water, lbf/ft³
 h_W = Static head of water, ft
 γ_A = Specific weight of air, lbf/ft³
 h_A = Static head of air, ft

Transposing and using 1 in. of water column,

$$h_A = \frac{\gamma_W h_W}{\gamma_A} = \frac{(62.408)(1/12)}{0.07512 \text{ (at 70° F.)}}$$

$$h_A = 69.23 \text{ ft of air column}$$

A column of air 69.23 ft will exert the same pressure as a column of water 1 in. high. Stated another way, a static head of 1 in. of water will support a column of air 69.23 ft high.

The rate of discharge from a vent outlet can now be determined when the pressure at the outlet is 1 in. of water or 69.23 ft of air.

$$q_D = 13.17 \, d_o^2 \, h^{1/2}$$
$$= 13.17 \, d_o^2 \, (69.23)^{1/2}$$
$$= (13.17)(8.32) \, d_o^2$$
$$= 109.57 \, d_o^2$$

The gallons per minute (cubic feet per minute) discharge rate for various diameters of vent pipe at a flow pressure of 1 in. of water column is given in Table 8-1.

Table 8-1 Discharge Rates of Air (1 Inch Water Pressure)

Outlet Diam, d_o inches	Air Discharge, q_D gpm (cfm)
2	438.3 (58.6)
2½	684.8 (91.5)
3	986.1 (131.8)
4	1753.0 (234.3)
5	2739.0 (366.1)

Friction Head Loss

When air flows in a pipe there is a pressure loss which occurs due to the friction between the air and pipe wall. This loss of pressure can be expressed by the Darcy formula:

Chapter 8—Vent Systems

(8-7)
$$h = \frac{fLV^2}{D2g}$$

where h = friction head loss, ft. of air column
f = coefficient of friction
L = length of pipe, ft.
D = diameter of pipe, ft.
V = velocity of air, ft/sec
g = gravitational acceleration 32.2 ft./sec^2

Air Flow in Stacks

The complete venting of a sanitary drainage system is very complicated as evidenced by the variety of vents employed. There are so many variables that produce positive and negative pneumatic pressure fluctuations that it is not feasible to prepare tables of vent sizing for each particular design. Recognizing this, authorities base the formulation of venting tables for vent stacks and horizontal branches on the worst conditions that may reasonably be expected. To determine the maximum lengths and minimum diameters for vent stacks it would be valuable to review the conditions of flow in the drainage stack.

At maximum design flow, the water flows down the stack as a sheet of water occupying 7/24 of the cross-sectional area of the stack. The remaining 17/24 is occupied by a core of air. As the water falls down the stack, it exerts a frictional drag on the core of air and as this air is dragged down it must be replaced by an equivalent quantity of air so as not to develop negative pressures in excess of –1 in. of water. This is accomplished by extending the soil stack through the roof so that air may enter the stack to replenish the air being pulled down the stack. This is why stacks must be extended full size through the roof and also why soil stacks may not be reduced in size even though the load is less on the upper portions of the stack than it is at the lower portions. Any restriction in the size before terminating at the atmosphere would cause violent pressure fluctuations.

As the water flows down the stack and enters the horizontal drain there is a severe restriction to the flow of air as the hydraulic jump occurs. The air is compressed and pressure buildup may become very high. A vent stack is provided in this area of high pressure to relieve the pressure by providing an avenue for the flow of air. Obviously, the vent stack must be large enough to permit the maximum quantity of air dragged down the drainage stack to discharge through it and to the atmosphere without exceeding ±1 in. of water fluctuation The rate of air discharge that must be accommodated for various sizes of drainage stacks flowing at design capacity is tabulated in Table 8-2.

Table 8-2 Air Required by Attendant Vent Stacks (Drainage Stack Flowing 7/24 Full)

Diameter of Drainage Stack, inches	Water Flow, gpm	Air Flow, gpm (cfm)
2	23.5	57.1 (7.6)
3	70	170.1 (22.7)
4	145	352.4 (47.1)
5	270	656.1 (87.7)
6	435	1057.1 (141.3)
8	920	2235.6 (298.9)
10	1650	4009.5 (536)
12	2650	6439.5 (860.8)

Air Flow in Horizontal Drains

It is assumed that the drainage branch flows half full at design conditions and the air in the upper half of the pipe is flowing at the same velocity and capacity. Table 8-3 tabulates these values for various slopes of drain.

Table 8-3   Rate of Air In Horizontal Drains

Diameter of Drain, inches	Slope, inches per foot	Rate of Flow, gpm (cfm)
1½	¼	6.0 (.80)
2	¼	8.8 (1.2)
2½	¼	15.5 (2.1)
3	¼	25.5 (3.4)
4	⅛	38.0 (5.1)
5	⅛	69.0 (9.2)
6	⅛	112.0 (15)
8	⅛	240.0 (32.1)

Permissible Length of Vent Pipe

The maximum length of vent piping, for any particular size with a pressure drop of 1 in. of water, is established by computing the pressure loss for various rates of flow in vents of various diameters. Combining Darcy's pipe friction formula (Equation 8-7) and the flow formula, and converting the terms of the equations to units generally used in plumbing:

$$h = \frac{f L V^2}{D \, 2g} \tag{8-8}$$

$$q = 2.448 \, d^2 V \tag{8-9}$$

$$V = \frac{q}{2.448 d^2}$$

Substituting V in the Darcy equation,

$$h = \frac{f L q^2}{(d/12)(64.4)(2.448)^2 (d)^4}$$

Solving for L,

$$L = \frac{h \, d^5}{0.013109 \, f q^2} = \frac{2226 \, d^5}{f q^2} \tag{8-10}$$

where L = Length of pipe, ft
 d = Diameter of pipe, in.
 f = Coefficient of friction
 q = Quantity rate of flow, gpm

Gravity Circulation

The principle of gravity circulation of air is utilized to keep the entire sanitary system free of foul odors and the growth of slime and fungi. The circulation is induced by the difference in head (pressure) between outdoor air and the air in the vent piping. This difference of head is due to the difference in temperature, and thus the difference in density, of each and the height of the air column in the vent piping. The cool air, being more dense, tends to displace the less dense air of the system and circulation of the air is induced. The formula is

(8-11)
$$H = 0.1925 (\gamma_o - \gamma_1) H_s$$

where H = Natural draft pressure, in. of water
γ_o = Specific weight of outside air, lbf/ft³
γ_1 = Specific weight of air in pipe, lbf/ft³
H_s = Height of air column or stack, ft

Under conditions of natural draft, the rate of flow will be just great enough to overcome losses due to friction.

Vent Stacks

Every drainage stack should be extended full size through the roof. The pipe from the topmost drainage branch connection through the roof to atmosphere is called the "vent extension." The vent extension provides the air that is dragged down the stack and also provides means for the gravity circulation of air throughout the system. Vent extensions may be connected with the vent stack before extending through the roof or may be connected together with other vent extensions or vent stacks in a vent header and the header extended through the roof as a single pipe.

Every drainage stack should have an attendant vent stack. The purpose of installing a vent stack is to prevent the development of excessive pressures in the lower regions of the drainage stack by relieving the air as rapidly as it is carried down by the discharge of the drainage stack. The most effective location for the vent stack is below all drainage branch connections and preferably at the top of the horizontal drain immediately adjacent to the stack base fitting. It is at this location that pressure is at its maximum and the danger of closure due to fouling is at its minimum. Figure 8-1 illustrates acceptable methods of vent stack connections.

The vent stack should extend undiminished in size through the roof or connect with the vent extension of the drainage stack at least 6 in. above the overflow of the highest fixture or connect to a vent header.

Vent Terminals

Vent terminals should not be located within 10 ft of any door, window, or ventilation intake unless they are extended at least 2 ft above such openings. Terminals should be at least 6 in. above roof level and at least 5 ft above when the roof is used for other purposes. When it is impractical to extend the vent through the roof, it is permissible to terminate through a wall, but the terminal must turn down and be covered with a wire screen. The terminal should never be located beneath a building overhang.

Fixture Trap Vents

The water seal of all fixture traps should be protected against siphonage or blowout by the proper installation of a venting system. When drainage stacks are provided with an adequate supply of air at the terminal and an adequate vent stack is provided to relieve excess pressures at the base of the drainage stack, the only additional vent protection required to prevent water seal loss

Figure 8-1 Various Vent Stack Connections

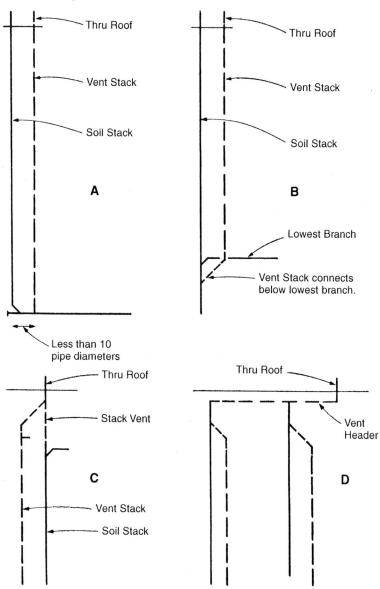

in fixture traps is that necessary to prevent self-siphonage when the fixture discharges and to relieve excessive pneumatic effects in the branch drains when other fixtures discharge into the branch. Some municipalities require that every fixture trap be individually vented, but most localities permit alternate methods such as

1. Wet venting
2. Stack venting
3. Circuit and loop venting

4. Combination waste and vent venting

Distance of Vent from Trap

The most comprehensive investigations of conditions under which fixture traps will be safe from self-siphonage have been conducted by the National Bureau of Standards in the United States and by the Building Research Station in England. The recommended maximum distances of a vent from the weir of the trap to the vent connection are tabulated in Table 8-4.

As illustrated in Figure 8-2, the vent pipe opening, except for water closets and similar fixtures, must never be below the weir of the fixture trap. A fixture drain that slopes more than one pipe diameter between vent opening and trap weir has a greater tendency to self-siphon the trap seal than a fixture drain installed at a slope of not more than one pipe diameter.

Table 8-4 Distance of Vent from Fixture Traps

Size of Fixture Drain inches	Maximum Distance of Vent to Trap inches
1¼	30
1½	42
2	60
3	72
4	120

Figure 8-2 Vent Pipe Opening

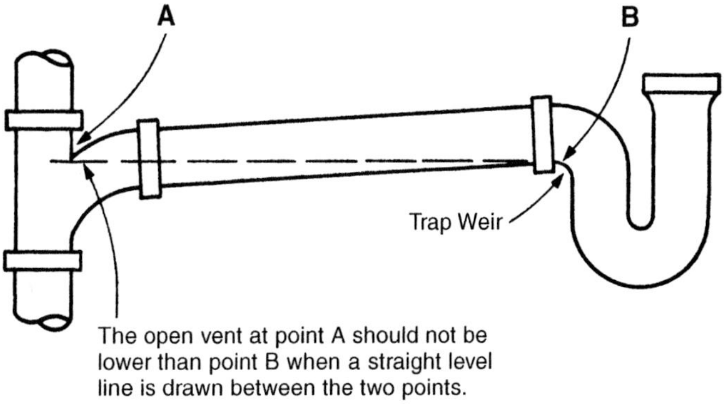

The open vent at point A should not be lower than point B when a straight level line is drawn between the two points.

Self-Siphonage of Fixture Traps, National Bureau of Standards Building Materials and Structures Report BMS 126 (1951), prepared by John L. French and Herbert N. Eaton, is a very thorough study of self-siphonage Some of the conclusions drawn by French and Eaton as a result of their investigations are very illuminating and are quoted herewith:

1. Increasing the diameter of the outlet orifice of a lavatory from 1⅛ in. to 1¼ in. increases the trap seal loss greatly, frequently more than 100%, owing to the increased discharge rate.

2. Flat-bottomed fixtures cause smaller trap seal losses than do round fixtures, owing to the greater trail discharge from the former.

3. With a 1½-in. fixture trap and drain, an 18-in. by 20-in. lavatory gave greater trap-seal losses than did a 20-in. by 24-in. lavatory, presumably owing to the greater trail discharge of the latter. When a 1¼ in. trap and drain were

used, no particular difference was noted in the trap seal losses caused by the two lavatories.

4. The elimination of the overflow in lavatories will increase the trap seal losses substantially.
5. The effect on trap seal losses of varying the vertical distance from the fixture to the trap from 6 in. to 12 in. appears to be negligible.
6. For a given rate of discharge from a lavatory, decreasing the diameter of the drain will increase trap seal losses.
7. An increase in slope or a decrease in diameter of the fixture drain will tend to cause increased losses due to self-siphonage, and these two dimensions are fully as important as the length of fixture drain in causing self-siphonage.
8. Trap seal losses are usually much greater when a long-turn stack fitting is used than when a short-turn or straight-tee fitting is used. No significant difference between the behavior of short-turn and straight-tee fittings was observed. Thus, since it is known that a long-turn fitting is more effective in introducing water from a horizontal branch into the stack than is either the short-turn or straight-tee fitting, the characteristics of these fittings are contradictory in these respects. The fitting that is most advantageous from the standpoint of introducing the water into the stack is the least advantageous from the standpoint of self-siphonage.
9. Trap seal losses are increased if the internal diameter of a P-trap is less than that of the fixture drain. Thus, if we are to prevent excessive trap seal losses for a P-trap due to self-siphonage, we should use a trap having a fairly large internal diameter. Furthermore, siphonage of the trap due to pressure reductions caused by the discharge of other fixtures on the system can be rendered less harmful by using a trap with a large depth of seal. While increasing the depth of seal may lead to greater trap seal losses, it also results in a greater remaining trap seal than if a trap with a shallow seal were used.
10. The test results on the self-siphonage of water closets have indicated that the unvented length of drain for these fixtures need not be limited because of self-siphonage.
11. Standardization of the dimensions of fixture traps and especially of lavatory traps, with regard to internal diameter and depth of trap seal is highly desirable. Minor restrictions on these dimensions can lead to substantially increased lengths of fixture drains.
12. Standardization of the hydraulic characteristics of fixtures is desirable, at least for lavatories, sinks, and combination fixtures. Substantially increased permissible unvented lengths of fixture drains can be obtained for a moderate decrease in the discharge rates of the fixtures.
13. Increase in depth of trap seal above the 2-in. minimum commonly permitted by codes will make it possible to increase appreciably the maximum permissible unvented lengths of fixture drains.

These conclusions clearly illustrate various approaches in the effort to make plumbing systems less costly without affecting efficiency. The proper design of fixtures and fixture drain lines and limiting the maximum discharge rates of faucet-controlled fixtures could result in longer unvented lengths of drains.

Various Methods of Fixture Trap Venting

Figure 8-3 illustrates various fixture trap vents and their proper nomenclature. When venting one trap the vent is called an "individual" or "back vent." If fixtures are back to back or side by side and one vent is used for the two traps, the vent is a "common vent." Any connection from the vent stack is a "branch vent."

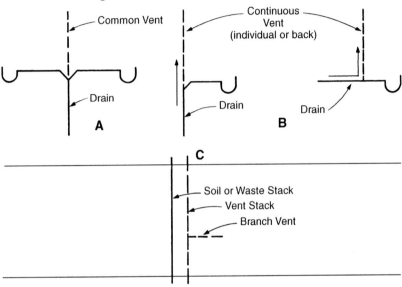

Figure 8-3 Various Fixture Trap Vents

All vent piping should be graded to drain back to the drainage piping by gravity. The vent should be taken off above the centerline of the drainpipe and rise vertically or at an angle of not more than 45° from the vertical. The horizontal run of the vent should be at least 6 in. above the flood rim of the fixture. (See Figure 8-4.)

Relief Vents

Pressures in the drainage and vent stacks of a multistory building are constantly fluctuating. The vent stack connection at the base of the drainage stack and the branch vent connections to the branch drains cannot always eliminate these fluctuations. It then becomes extremely important to balance pressures throughout the drainage stack by means of "relief vents" located at various intervals. The fluctuations in pressure may be caused by the simultaneous discharge of branches on various separated floors. Drainage stacks in buildings having more than ten branch intervals should be provided with a relief vent at each tenth interval, counting from the topmost branch downward. The lower end of the relief vent should connect to the drainage stack below the drainage branch connection and the upper end should connect to the vent stack at least 3 ft above the floor level. (See Figure 8-5.)

Relief vents are required where a drainage stack offsets at an angle of more than 45° to the vertical. Such offsets are subject to high pneumatic pressure

Engineered Plumbing Design

Figure 8-4   Horizontal Run of Vent

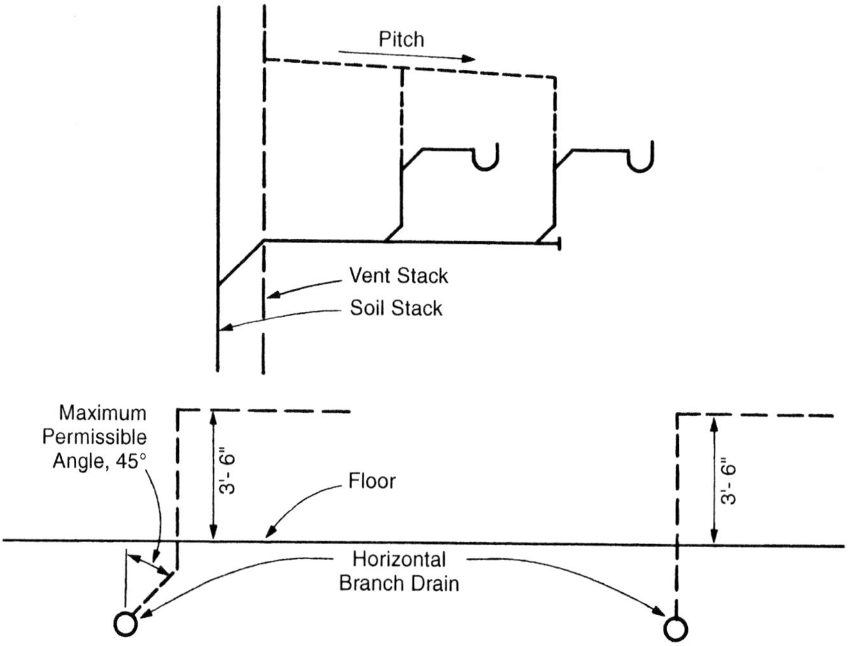

increases and extreme surging flow conditions. The methods of installing relief vents are illustrated in Figure 8-6.

Continuous Venting

A system of individual or common vents for every trap is called "continuous venting." Every fixture trap is provided with a vent. It is the most expensive system but provides positive protection of all trap seals.

Wet Venting

A "wet vent" is a vent that vents a particular fixture and at the same time serves as a waste to receive the discharge from other fixtures. The objective of using wet vents is to minimize the vent piping required by employing one pipe to serve two functions. There are three fundamental rules to follow when utilizing a wet vent:

At top floor:
1. No more than 1 FU is discharged into a 1½-in. wet vent nor more than 4 FU into a 2-in. wet vent.
2. Length of drain does not exceed maximum permissible distance between trap and vent.
3. Branch connects to the stack at the water closet connection level or below. (See Figure 8-7.)

At lower floors:

The rules are the same except that the water closets must be vented and the wet vent must be 2 in. minimum. Water closets below the top story need not be individually vented if a 2-in. wet vented waste pipe connects

Figure 8-5 Venting for Stacks Having More Than 10 Branch Intervals

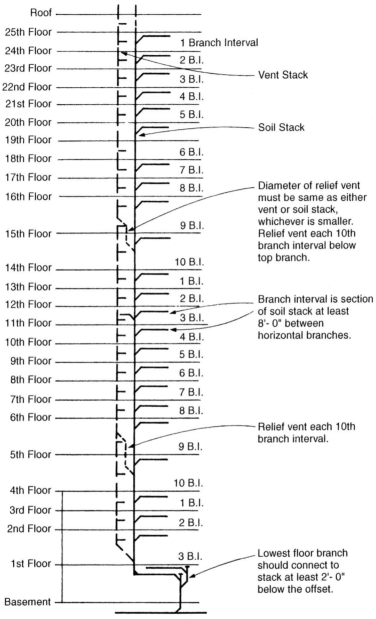

directly to the upper half of the horizontal water closet drain at an angle no greater than 45° from the angle of flow. (See Figure 8-8.)

"Stack venting" finds its general application in one-family homes and the top floor of multistory buildings. (See Figures 8-7 and 8-9.)

Engineered Plumbing Design

Figure 8-6 Venting at Stack Offsets

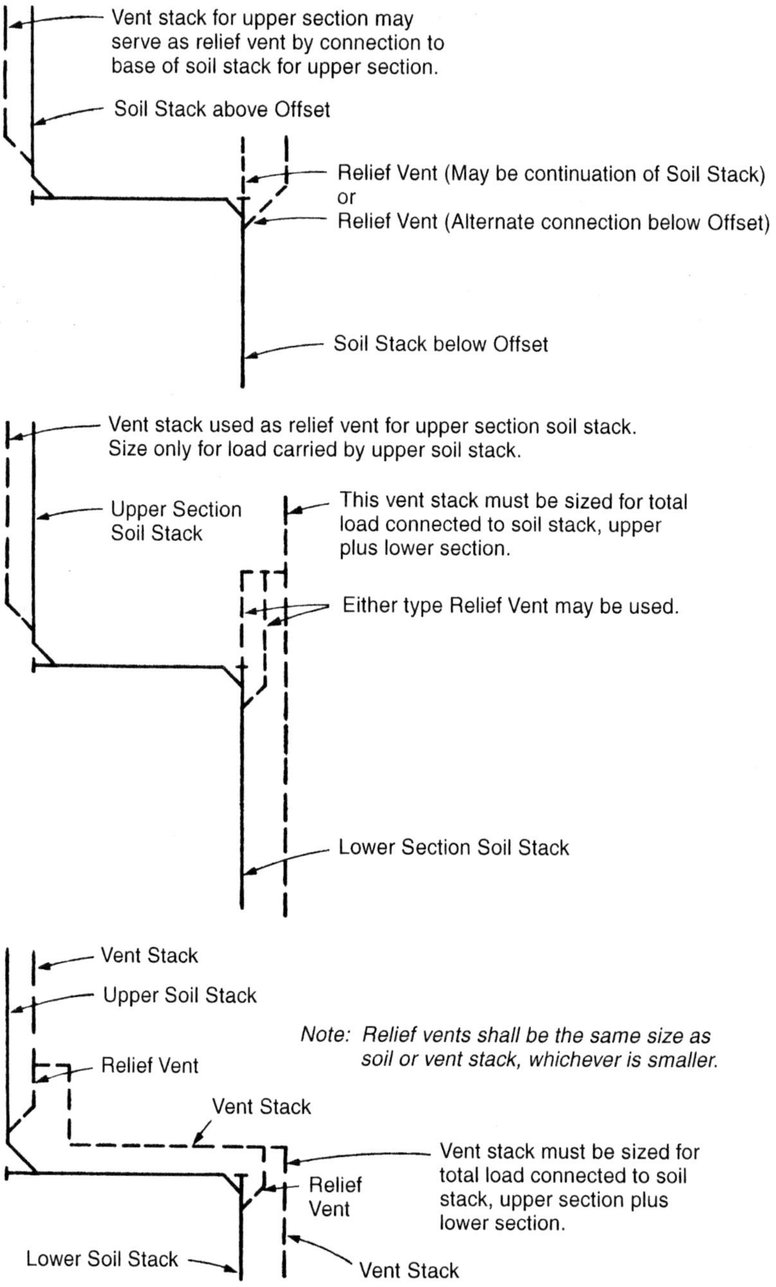

- Vent stack for upper section may serve as relief vent by connection to base of soil stack for upper section.
- Soil Stack above Offset
- Relief Vent (May be continuation of Soil Stack) or
- Relief Vent (Alternate connection below Offset)
- Soil Stack below Offset

- Vent stack used as relief vent for upper section soil stack. Size only for load carried by upper soil stack.
- Upper Section Soil Stack
- This vent stack must be sized for total load connected to soil stack, upper plus lower section.
- Either type Relief Vent may be used.
- Lower Section Soil Stack

- Vent Stack
- Upper Soil Stack
- Relief Vent
- Vent Stack
- Note: Relief vents shall be the same size as soil or vent stack, whichever is smaller.
- Vent stack must be sized for total load connected to soil stack, upper section plus lower section.
- Relief Vent
- Lower Soil Stack
- Vent Stack

Combination Waste and Vent Venting

"Combination waste and vent venting" is used primarily for venting floor drains and laboratory and work tables. The drainage piping is oversized at least two sizes larger than required for draining purposes only and the drainage

Chapter 8—Vent Systems

Figure 8-7 Wet Venting at Top Floor

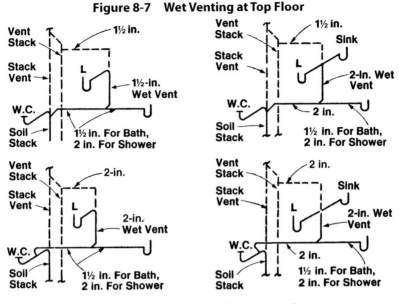

Figure 8-8 Wet Venting Below Top Floor

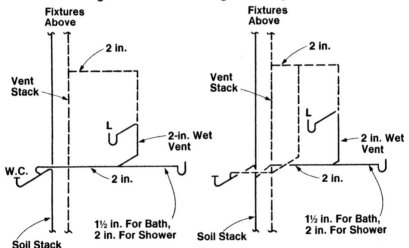

branch and stack should be provided with vent piping. This type of venting is employed when it is impractical to employ the other methods.

Circuit and Loop Venting

There has developed a tendency to call all "circuit venting" by the name applicable to a special installation of circuit venting. A circuit vent is a branch vent that serves two or more floor outlet fixtures, except blowout water closets, and extends from in front of the last fixture connection on the horizontal drain to the vent stack. A "loop vent" is the same, except that it is employed on the topmost floor serving fixtures and is connected to the vent extension of the drainage stack instead of to the vent stack. (See Figure 8-10.) When wall outlet fixtures are connected to the branch drain serving the floor outlet fixtures, the former must be provided with individual vents that can connect to the circuit vent or loop vent.

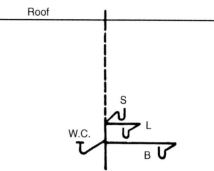

Figure 8-9   Stack Vented Unit

Common Vents

Where two fixtures are connected to a vertical branch at the same level, a "common vent" may be employed. When one of the fixtures connects at a different level than the other, observe the following procedure. If fixture drains are the same size, increase the vertical drain one size. If fixture drains are of different sizes, connect the smaller above the larger connection and maintain the vertical size up to the top connection.

Suds Pressure

The prevalent use of high-sudsing detergents in washing machines, dishwashers, laundry trays, and kitchen sinks has created serious problems in all residential buildings and especially in high-rise buildings. Until manufacturers are forced to market only detergents without sudsing characteristics, the plumbing engineer must understand and cope with the dangers created in the sanitary system by the presence of suds. (An interesting sidelight is that suds, in and of themselves, do not enhance the cleaning ability of soaps or detergents in any way.)

When the flow of wastes from upper floors contains detergents, the suds-ingredients are vigorously mixed with the water and air in the stack as the waste flows down the stack and further mixing action occurs as other branch waste discharges meet this flow. These suds flow down the stack and settle in the lower sections drainage system and at any offsets greater than 45 degrees in the stack. Investigation has shown that when sudsing wastes are present, the sanitary and vent stacks are laden with suds and this condition was found to exist for extended periods of time.

Liquid wastes are heavier than suds and easily flow through the suds-loaded drainage piping without carrying the suds along with the flow. Everyone is aware of the difficulty of flushing the suds out of a sink. The water simply flows through the suds and out the drain, leaving the major portion of the suds behind. The same action occurs in the lower sections of the drainage system except for one important difference—air, as well as water, is now flowing in the piping. This air, which is carried down with the waste discharge, compresses the suds and forces them to move through any available path of relief. The relief path may be the building drain, any branches connected to the building drain, the vent stack, branch vents, individual vents or combinations of the foregoing. A path of relief may not always be available or could be cut off or restricted by the hydraulic jump, or a path may just be inadequate because of location or size. If one or more of these conditions exist, excessively high

Chapter 8—Vent Systems

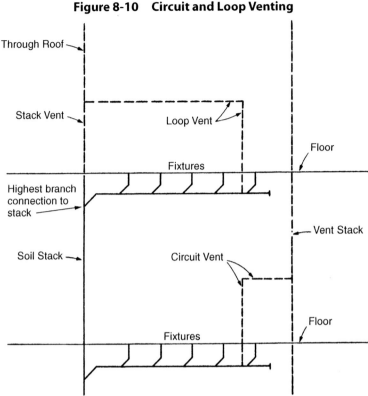

Figure 8-10   Circuit and Loop Venting

suds pressure can develop and blow the seals of traps with the accompanying appearance of suds in fixtures.

High suds pressure zones occur at every change in direction, vertically or horizontally, that is greater than 45°. Where vent stack base connections, relief vents, branch vents, or individual vents serve as the relief path for the high suds pressure, they are usually found to be inadequate in size with resultant suds conditions appearing at the fixtures. The vent pipe sizing tables in practically every code are calculated on the basis of air flow capacity and do not in any way provide for the more demanding flow of suds. Sizes that are based on these code tables are inadequate to accommodate suds flow and thus are incapable of providing adequate suds pressure relief.

Suds are much heavier than air and consequently do not flow with the same ease. They produce a much greater friction head loss for the same rate of flow. The density of old or regenerated suds varies from 2 lb/ft^3 to a high of 19 lb/ft^3, depending upon the detergent used. For equal rates of flow and pressure loss, the vent pipe diameter for suds relief flow must be from 20 to 80% greater than for air flow.

Whenever a soil or waste stack receives washing machines, dishwashers, laundry trays, kitchen sinks, or other fixtures where sudsing detergents are used, the drainage and vent piping for the lower-floor fixtures or for the fixtures above offsets must be arranged to avoid connection to any zone where suds pressure exists.

Suds pressure zones exist in the following areas:
1. At a soil or waste stack offset greater than 45°: 40 stack diameters upward and 10 stack diameters horizontally from the base fitting for the upper stack section. A pressure zone also exists 40 stack diameters upstream from the top fitting of the lower stack section.
2. At the base of a soil or waste stack: The suds pressure zone extends 40 stack diameters upward from the base fitting.
3. In the horizontal drain from the base of a stack: The suds pressure zone extends 10 stack diameters from the base fitting, and where an offset greater than 45° in the horizontal occurs, the pressure zones extend 40 stack diameters upstream and 10 diameters downstream from the offset fitting.
4. In a vent stack connected to a suds pressure zone: The suds pressure zone exists from the vent stack base connection upward to the level of the suds pressure zone in the soil or waste stack.

Figure 8-11 illustrates all the above zones.

Vapor Vents (Local Vents)

Years ago water closets and urinals were equipped with connections for venting the fixture to the outdoors to eliminate foul odors. Fixture design has been improved so that these vents are no longer required. The use of "vapor vents" is now applied to sterilizing equipment and bedpan washers. This application is also rapidly disappearing as new methods of condensing the foul vapors are being built into the equipment. When a vapor vent is used, it must be isolated from the sanitary venting system. The base of a vapor vent stack should terminate in a trap, to prevent the escape of vapors, and spill to a trapped, vented, and water-supplied receptacle. The stack should extend through the roof.

An individual vapor vent drip can be connected through an air gap to the inlet of the trap serving the fixture. Vapor vents for bedpan washers and bedpan sterilizers must not connect with the vapor vents of other fixtures.

Sizing of the vapor vent stack may be by empirical methods or the rational approach may be used. The minimum size of the stack should be 1¼ in.

Ejector and Sump Vents

Ejectors, other than the pneumatic type, operate at atmospheric pressure and receive drainage discharge under gravity flow conditions. An ejector is installed when the level of fixture discharge is below the level of the public sewer. It is convenient to view an ejector system as being exactly similar to the gravity sanitary system and all of the requirements for the proper design of the sanitary system are applicable. Thus, the air required to be conveyed by the vent piping is the same as the maximum rate at which sewage enters or is pumped out of the receiver.

Chapter 8—Vent Systems

Figure 8-11   Suds Pressure Zones

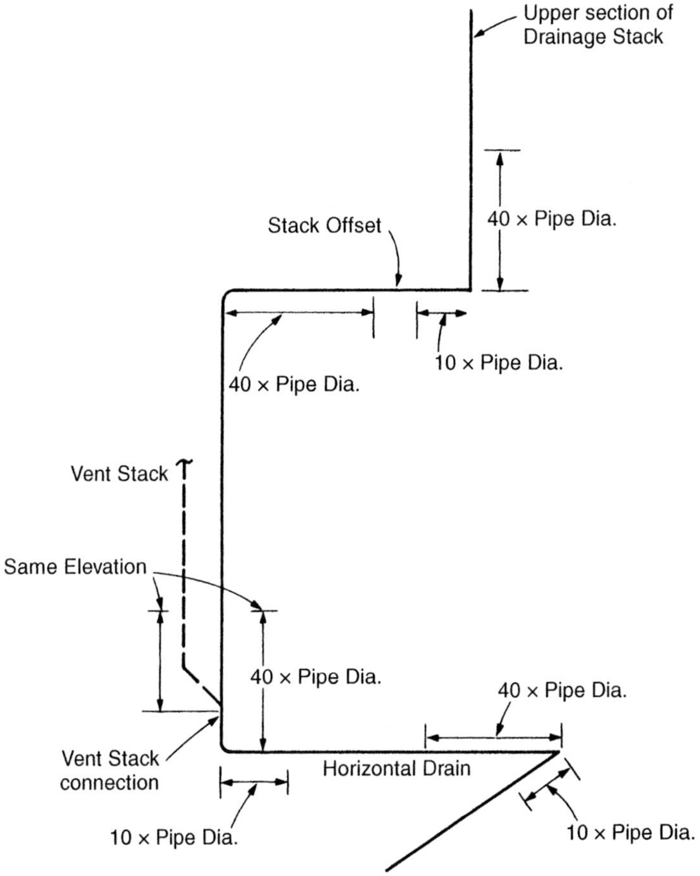

(Figure showing drainage stack with stack offset, vent stack connections, horizontal drain, and labeled distances: 40 × Pipe Dia. and 10 × Pipe Dia. at various points; upper section of drainage stack indicated.)

The ejector vent can be determined by reference to Equation 8-10:

$$L = 2226 \left(\frac{d^2}{fq^2}\right)$$

and using Table 8-1, which gives air discharge in gpm for various pipe diameters. It has been found in practice that 3 in. is adequate except for extremely large installations.

Frost Closure

Where the danger of frost closure of vent terminals is present, the minimum size of the vent stack or vent extension through the roof should be 3 in. When a vent stack must be increased in size going through the roof, the increase should be made inside the building at least 1 ft below the roof.

The National Bureau of Standards has investigated the problem of frost closure both theoretically and experimentally. It was demonstrated that a 3-in. vent terminal froze up solidly at −30°F only over an extended period of time. Closure occurs at the rate of 1½ in. for every 24 hr. that the temperature remains at −30°F.

It can be seen that frost closure presents a real problem only in the far northern regions. The problem is serious in Canada, and they have devised various methods of overcoming it:

1. Vent terminal to extend only 1 in. or 2 in. above the roof. The more pipe exposed to the atmosphere, the greater the problem. Snow covering the vent terminal has proven to cause no trouble. The snow is porous enough for the passage of air and melts rather rapidly at the outlet.
2. Enlargement of the stack below the roof. The increased diameter decreases the chance of complete closure and the stream of air tends to flow through the enlarged portion without touching the walls of the enlarged pipe.
3. Install cap flashing at the terminal and counterflashing to leave an air space from the heated building.

Frost closure depends upon the: (1) outside temperature, (2) temperature and humidity of inside air, (3) wind velocity, (4) length of exposed pipe, (5) diameter of exposed pipe, and (6) velocity of air flow. There is very little danger of frost closure unless the outside temperature falls below $-10°F$ and remains there for several days. It has been found that if frost closure does occur, siphonage of traps is reduced or prevented by connecting the drainage and vent stacks together before extending through the roof. An analysis of air flow under these conditions will convince the plumbing engineer of its validity, as it can be seen that air forced into the vent stack at its base will be introduced into the soil stack at the top connection.

Tests of Plumbing Systems

The complete storm and sanitary system should be subjected to a water test and proven watertight upon completion of the rough piping installation and prior to covering or concealment. The test pressure should be a minimum of a 10-ft column of water except for the topmost 10 ft of pipe. The test pressure should never exceed a maximum of a 100-ft column of water. Any greater pressure will cause the test plugs used to seal temporarily open piping in the system to blow. If the system is higher than 100 ft, test tees may be installed at appropriate heights so as to test the building in sections. Very rarely in practice are more than seven stories tested at one time.

If it is not possible to perform a water test, an air test is acceptable. The air test shall be made by attaching an air compressor testing apparatus to any suitable opening, and, after closing all other inlets and outlets to the system, forcing air into the system until there is a uniform gage pressure of 5 psi (34.5 kPa) or a pressure sufficient to balance a column of mercury 10 in. (254 mm) in height. The pressure shall be held without introduction of additional air for a period of at least 15 min.

Upon completion of the sanitary system and after all fixtures are installed with traps filled with water, the system should be subjected to an additional test and proved gastight.

An alternate test is the smoke test. The smoke test is performed by introducing pungent, thick smoke produced by smoke bombs or smoke machines. When smoke appears at the roof terminals, each terminal is sealed and a smoke pressure of 1-in. column of water is maintained to prove the system gastight. This test is not practical and is seldom used.

Another alternate test is the peppermint vapor test. At least 2 oz. of oil of peppermint are introduced into each roof terminal and vaporized by immediately pouring 10 qt of boiling water down the stack. The terminals are promptly sealed. Oil of peppermint and any person coming in contact or handling the oil must be excluded from the interior of the building for the duration of the test. Leakages will be detected by the peppermint odor at the source. However, it is very difficult to pinpoint the leak by this method. This test is not practical and is seldom used.

Questions

1. How does the density of air vary with temperature and pressure?
2. Is air elastic or inelastic? Explain.
3. How much will 1 ft^3 of air be compressed when a pressure of 1 in. water column is applied? Assume temperature to remain constant.
4. What happens to the air when the hydraulic jump occurs in a drain line? Assume no venting provisions have been made.
5. How high a column of air will a static pressure of l in. water column support?
6. Describe the flow of air in a soil stack when branches are discharging into the stack.
7. How is the sanitary system kept free of foul odors and the growth of slime and fungi?
8. What is the sole purpose of a vent stack?
9. What are the restrictions as to location of vent terminals?
10. Name five different methods of venting fixture traps and explain each.
11. Illustrate the maximum distance of the vent from the weir of the fixture for any size fixture drain.
12. Where should relief vents be provided in a 22-story building?
13. Where should relief vents be provided for a stack offset?
14. Where do suds pressure zones exist? How far upstream and downstream does excess pressure exist in each of the zones?
15. Can ejector vents be connected to the sanitary venting system?
16. Explain various methods of diminishing the danger of frost closure of vent terminals.
17. What tests are required for the sanitary and storm systems?

Vent Sizing

For any given airflow to be conveyed by a vent stack, the total allowable length is computed using Equation 8-10:

$$L = \frac{2226\,d^5}{fq^2}$$

This length is the equivalent length of run, not the developed length. Developed length is the distance measured along the centerline of the pipe and fittings. Fittings create a greater friction head loss than does pipe of equal diameter. Converting a fitting to an equivalent length of pipe that will produce the same pressure loss permits the utilization of applicable formulas. The equivalent length of the fittings is added to the developed length of run and the result is the *equivalent length of run* (ELR). In the average venting system, the equivalent length of fittings has been found to be approximately 50% of the developed length. The permissible length of run is then two-thirds of the equivalent length of run as determined by the formula.

The developed length of a vent stack is measured from the point of connection to the drainage system at its base to the point of discharge at the outlet above the roof. (See Figure 9-1.) Table 9-1 gives sizes and lengths for vent stacks in terms of fixture unit loading of the drainage stacks they service. The lengths given in this table and in all codes are the developed lengths.

The minimum size of a vent stack is one-half the diameter of the drainage stack it serves. See Figure 9-2 for an example of sizing.

Sizing Vent Extensions and Terminals

Vent extensions and stack terminals must be of adequate size to permit air to enter the top of the stacks without causing any air pressure reduction in the upper portions of the system. Terminals should be no smaller in size than the stack served. The minimum size to allow for frost closure is 3 in.. Each code establishes various minimum sizes.

Sizing Vent Headers

Stack vents and vent stacks may be connected into a common vent header at the top of the stacks then extended through the roof at one point. This header is sized in accordance with Table 9-1, the fixture unit loading being the sum of the fixture unit loading of all the stacks connected to it. The developed length is the longest vent length from the intersection at the base of the most distant stack to the vent terminal as a direct extension of one stack.

See Figure 9-3 for an illustrative example of the procedure described above.

Figure 9-1 Developed Length of a Vent Stack

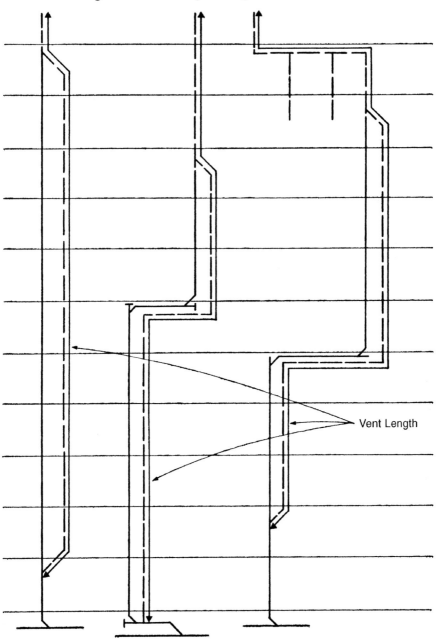

Table 9-1 Size and Length of Vent

Size of Soil or Waste Stack (in.)	Drainage Fixture Units Connected	Maximum Length of Vent (ft) — Diameter of Vent Required (in.)								
		1¼	1½	2	2½	3	4	5	6	8
1¼	2	30								
1½	8	50	150							
2	10	30	100							
2	12	30	75	200						
2	20	26	50	150						
2½	42		30	100	300					
3	10		30	100	200	600				
3	30			60	200	500				
3	60			50	80	400				
4	100			35	100	260	1,000			
4	200			30	90	250	900			
4	500			20	70	180	700			
5	200				35	80	350	1,000		
5	500				30	70	300	900		
5	1,100				20	50	200	700		
6	350				25	50	200	400	1,300	
6	620				15	30	125	300	1,100	
6	960					24	100	250	1,100	
6	1,900					20	70	200	700	
8	600						50	150	500	1,300
8	1,400						40	100	400	1,200
8	2,200						30	80	350	1,100
8	3,600						25	60	250	800
10	1,000							75	125	1,000
10	2,500							50	100	500
10	3,800							30	80	350
10	5,600							25	60	250

Note: Twenty percent of the total shown may be installed in a horizontal position.

Sizing Individual Vents and Branch Vents

The main purpose of the individual vent is to provide sufficient airflow to disrupt any siphonic action of the fixture trap. The minimum size vent is 1¼ in. and no less than one-half the size of the drainage pipe, except for water closets where the vent may be no less than 1½ in. However, some codes allow no less than 2-in. vents for water closets.

Apparently it has not been generally recognized that the requirements for venting horizontal drain branches do not approach those necessary for venting stacks. The velocity of flow of waste water in branches is considerably less than that in stacks; consequently, the flow of required air is considerably less. Currently there is only one code in the United States that recognizes this engineering fact and that has incorporated a sizing table for this application.

The National Bureau of Standards has performed extensive research on the subject of branch venting. The results of its studies are presented in Table 9-2. The designer is cautioned to use this table only where an applicable code does not exist. The size and length of horizontal vent branches are based upon the

Chapter 9—Vent Sizing

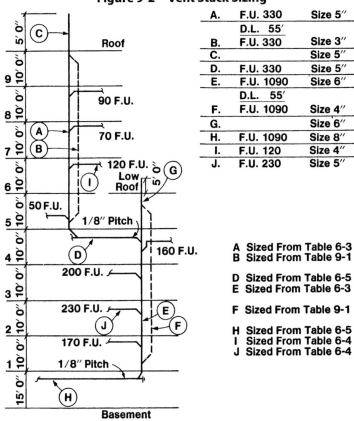

Figure 9-2 Vent Stack Sizing

A.	F.U. 330	Size 5"
	D.L. 55'	
B.	F.U. 330	Size 3"
C.		Size 5"
D.	F.U. 330	Size 5"
E.	F.U. 1090	Size 6"
	D.L. 55'	
F.	F.U. 1090	Size 4"
G.		Size 6"
H.	F.U. 1090	Size 8"
I.	F.U. 120	Size 4"
J.	F.U. 230	Size 5"

A Sized From Table 6-3
B Sized From Table 9-1

D Sized From Table 6-5
E Sized From Table 6-3

F Sized From Table 9-1

H Sized From Table 6-5
I Sized From Table 6-4
J Sized From Table 6-4

Note: Size of "A" must be increased from 4" to 5" due to exceeding maximum F.U. permitted in any one branch interval. ("I" exceeds 90 F.U.) For the same reason "E" must be increased from 5" to 6" ("J" exceeds 200 F.U.).

rate of waste water flow in the horizontal drain computed from Manning's formula for half-full flow. The rate of airflow is then assumed to be the same. It is this assumption that provides a more than adequate factor of safety.

Sizing Relief Vents

The relief vent should be the same size as the vent stack or drainage stack, whichever is smaller.

Sizing Circuit Vents

The major portion of the developed length of a circuit vent is horizontal; therefore, the allowable developed length is considerably less than that for vertical vents since natural circulation does not occur. The minimum size is half the drainage pipe size. Table 9-3 gives sizes for various fixture unit loadings and developed lengths.

Caution: All data, tables, illustrations, and formulas given in this text are based on firm engineering principles and their use will result in an economical

Table 9-2 Maximum Permissible Lengths of Vents for Horizontal Branches

Diameter of Horizontal Drainage Branch (in.)	Slope of drainage branch (in./ft.)	Maximum Length of Vent (ft) — Diameter of Vent (in.)							
		1¼	1½	2	2½	3	4	5	6
1¼	¼	a							
1½	¼	a	a						
2	⅛	a	a	a					
2	¼	a	a	a					
2½	⅛	a	a	a	a				
2½	¼	805	a	a	a				
3	⅛	660	a	a	a	a			
3	¼	335	710	a	a	a			
4	⅛	132	364	a	a	a	a		
4	¼	69	147	600	a	a	a		
5	⅛	—	94	355	845	a	a	a	
5	¼	—	47	160	412	a	a	a	
6	⅛	—	35	124	320	a	a	a	a
6	¼	—		60	155	a	a	a	a

a More than 1,000 ft.

Table 9-3 Horizontal Circuit and Loop Vent Sizing Table

Soil or Waste Pipe Diam. (in.)	Fixture Units (max no.)	Max. Horizontal Length (ft) — Diameter of Circuit or Loop Vent (in.)					
		1½	2	2½	3	4	5
1½	10	20					
2	12	20	40				
2	20	10	30				
3	10	—	20	40	100		
3	30	—	—	40	100		
3	60	—	—	16	80		
4	100	—	7	20	52	200	
4	200	—	6	18	50	180	
4	500	—	—	14	36	140	
5	200	—	—	—	16	70	200
5	1,100	—	—	—	10	40	140

and properly functioning system. The designer is strongly warned, however, to disregard this information and design his/her project in accordance with the code requirements in force at the location of the particular project. If there are no codes, the designer may use all information in this text with the utmost confidence.

Chapter 9—Vent Sizing

Figure 9-3 Sizing Vent Headers

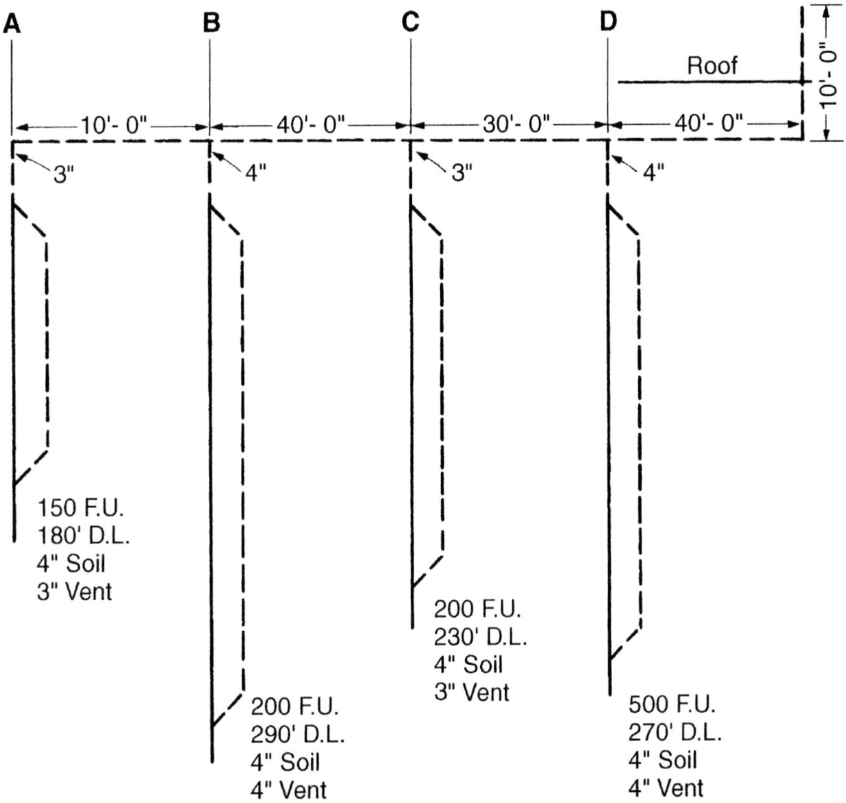

```
A          B              C            D
|          |              |            |          Roof        ↑
|          |              |            |─────────────────     10'-0"
|←─10'-0"─→|←──40'-0"────→|←──30'-0"──→|←────40'-0"─────→     ↓
\3"        \4"            \3"          \4"
 \          \              \            \
  \          \              \            \
   \          \              \            \
    \          \              \            \
     \          \              \            \
      |          \              \            \
      |           \              \            \
150 F.U.           |              \            \
180' D.L.          |               \            \
4" Soil            |                |            \
3" Vent            |                |            |
                   |                |            |
                200 F.U.         200 F.U.     500 F.U.
                290' D.L.        230' D.L.    270' D.L.
                4" Soil          4" Soil      4" Soil
                4" Vent          3" Vent      4" Vent
```

Fixture Units		Developed Length (D.L.)	Soil Stack	Vent Stack	Header
Stack A	150	180'	4"	3"	-
A-B	150	180 + 10 = 190'	-	-	4"
Stack B	200	290'	4"	4"	-
B-C	150 + 200 = 350	290 + 40 = 330'	-	-	4"
Stack C	200	230'	4"	3"	-
C-D	350 + 200 = 550	330 + 30 = 360'	-	-	5"
Stack D	500	270'	4"	4"	-
D-Outlet	550 + 500 = 1050	360 + 40 + 10 = 410'	-	-	5"

Soil stack sized per Table 6-3.
Vent stack and header sized per Table 9-1.
Developed length of vent stack is measured from connection to soil stack at base to point of connection to header.

Sumps and Ejectors

A subdrain is that portion of a drainage system that cannot drain by gravity into the building sewer. The wastes from a subdrain flow by gravity into a receiving tank, the contents of which are automatically lifted and discharged into the gravity drainage system. If the subdrain receives wastes containing oil, gasoline, or other flammable liquids, the discharge must first go through an oil interceptor before entering the receiving basin. A subdrain should not be confused with a subsoil drain (see Chapter 7). Each serves a different purpose.

Terminology

When the receiving basin collects the discharge of sanitary wastes, it is called an *ejector basin* and the automatic lifting and discharge equipment is called an *ejector pump*. When the wastes received are storm water and other clear water discharges, it is called a *sump basin* and the lifting and discharge equipment a *sump pump*. Confusion has been created in many codes and technical articles by loose terminology and not making a clear distinction between ejectors and sumps. *An ejector handles sewage and a sump handles clear water.*

Sump pumps are sometimes called *bilge pumps*. Ejectors and/or sumps for large projects are called *lift stations*. Thus, an ejector will be called a *sewage lift station* and a sump pump will be called a *storm water lift station*.

Ejector Basin

An ejector basin must be of airtight construction and must be vented. It is airtight to prevent the escape of foul odors generated by the wastes from the subdrainage system. Being airtight, a vent is required to relieve the air in the basin as wastes discharge into it and also to supply air to the basin while the contents are being discharged to the sanitary gravity drainage system. A sump basin need not be airtight and therefore does not require a vent. But if it is made airtight (as required by some codes), then a vent is required.

Basin Materials

The receiver, whether ejector or sump, may be constructed of cast iron, steel, fiberglass, or concrete. When a concrete basin is selected, it should be constructed by the general contractor, but the plumbing contractor should provide the frame and cover. The interior of the concrete basin must be completely waterproofed, and watertight sleeves must be provided for all pipe entries into the basin.

Cast iron or fiberglass basins are recommended for direct burial in the ground. Steel basins should never be buried unless the outside of the basin is thoroughly painted with a protective coating of bitumastic paint to prevent corrosion. Steel basins are sometimes buried in a concrete envelope or set in a concrete pit.

Lifting Devices

There are two types of lifting devices employed for the automatic discharge of the contents of ejector or sump basins:
1. Centrifugal pumps
2. Pneumatic ejectors.

When centrifugal pumps are utilized, several installation choices are available:
1. Submersible motor and pump
2. Top-mounted motor with submersible pump
3. Top-mounted motor and pump
4. Dry pit.

Figures 10-1 through 10-4 illustrate each type of centrifugal pump installation.

The selection of the pneumatic-type ejector is indicated as follows:
- When the discharge rate is 100 gpm or less; it may be advantageous up to 300 gpm
- Where large solids must be handled
- Where ground conditions are such that depth below floor is restricted.

Centrifugal pumps are very inefficient at low flows, often below 50%.

From the standpoint of safety, sump and ejector basins are often considered to be hazardous wet wells. As such, they are confined spaces into which personnel cannot safely enter unless proper safety precautions are taken. Submersible pumps and motors should be installed with a removal system. Such systems include guide rails, disconnecting pump discharge fittings, brackets, and guide plates. When included as part of the basin installation, removal systems allow the pumps to be removed from the basin without personnel entering the basin.

Operation of a Pneumatic Ejector

A pneumatic ejector consists of a hermetically sealed receiver with check valves on inlet and discharge. Two rigid stainless steel electrodes of unequal length suspended in the receiver automatically control the operating cycle (other methods of control can be utilized). The opening to the discharge line is only a few inches from the bottom of the receiver and the receiver is flushed in every cycle. There is no waste retention and no sludge accumulation. The ejector is designed to complete a cycle in 1 min. The liquid wastes are never exposed to atmosphere and no odors or gases are released from the hermetically sealed receiver.

The following diagrams and accompanying text explain the operation. In Figure 10-5(a), the inlet check valve is held open in the filling position by the weight of the incoming liquid, while the weight of the liquid in the lift line holds the discharge check valve closed. Notice that the exhaust valve is open, and air in the receiver is vented to the low-level manhole or other venting point.

Rising sewage in Figure 10-5(b) closes a circuit between the long and short electrode, energizing a relay, which starts the air compressor. As the compressor

Figure 10-1 Submersible Motor and Pump

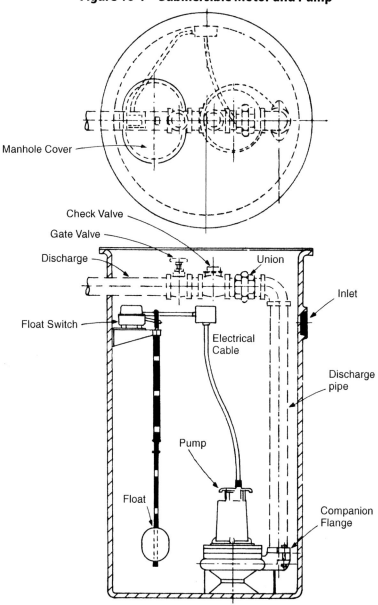

starts, the exhaust valve is closed by air pressure from the compressor cylinder. The inlet valve is also closed, while the discharge valve is open.

Air pressure on the surface of the sewage in Figure 10-5(c) forces the receiver contents out through the discharge valve and up through discharge piping to the high-level gravity sewer or other point of disposal. When the sewage falls below the end of the long electrode the compressor stops, the check valve near the compressor closes, and the diaphragm exhaust valve opens to exhaust the compressed air. The inlet and discharge valve then assume the filling position for

Figure 10-2 Top-Mounted Motor with Submersible Pump

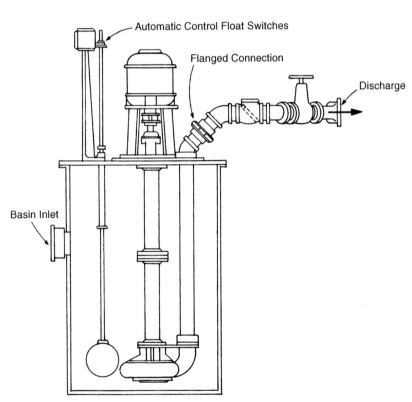

a repetition of the cycle. The compressed air can be supplied from an individual or duplex compressor selected for the exclusive operation of the pneumatic ejector, or the supply can be taken from the building central compressed air system (if available).

Ejector Pump Sizing (Centrifugal)

Use the water supply fixture unit method to determine the flow into the receiver. The following example illustrates the sizing procedure. Assume 10

Figure 10-3 Top-Mounted Motor and Pump

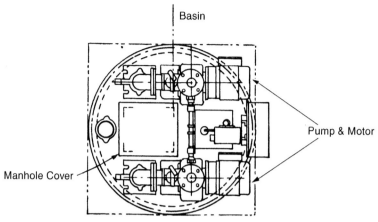

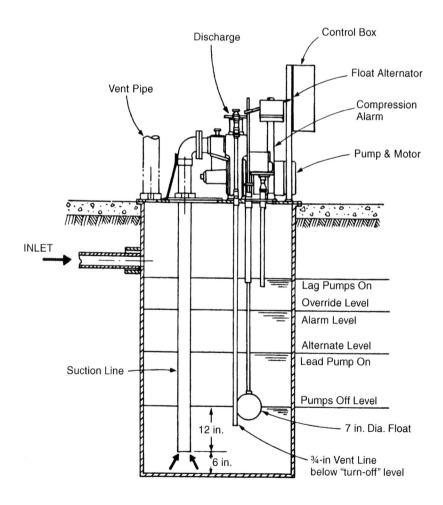

Figure 10-4 Dry Pit

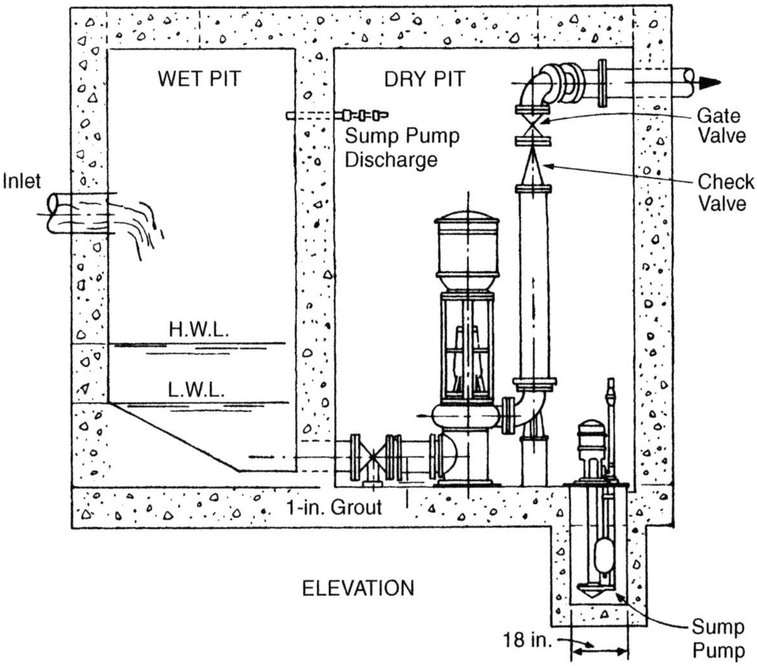

[Figure showing elevation view of wet pit and dry pit with sump pump discharge, gate valve, check valve, 1-in. grout, H.W.L., L.W.L., inlet, and 18 in. sump pump dimension]

flush valve-operated water closets (WC), 4 urinals (UR), 10 lavatories (Lav), 2 service sinks (SS), and 6 kitchen sinks (SK) are connected to the subdrain. As previously stated, a value of 1 gpm can be assigned for each 2 fixture units. Then:

$$
\begin{aligned}
10 \text{ WC} \times 10 \text{ FU} &= 100 \text{ FU} \\
4 \text{ UR} \times 5 &= 20 \\
10 \text{ Lavs.} \times 2 &= 20 \\
2 \text{ SS} \times 3 &= 6 \\
6 \text{ SK} \times 4 &= \underline{24} \\
170 \text{ FU} &= 85 \text{ gpm (peak load)}
\end{aligned}
$$

The pump capacity should be at least equal to the peak inflow. If it is critical that fixtures be kept in operation at all times, then a duplex pump should be specified in lieu of a simplex so that in the case of failure of one pump the other pump can carry the load until repairs are made. Each pump in a duplex set is sized to satisfy the peak inflow. The difference in cost between a simplex and a duplex is usually not great enough to warrant the specifying of a simplex system. The added safety provided by a duplex system is generally worth the nominal additional expenditure.

To determine the head of the pump, add the static head and the friction head loss in the discharge piping. The static head is measured from the low water line in the receiver up to the highest point pumped. The friction head loss can be calculated by using Equation 8-7:

$$h = \frac{fLV^2}{2gD} \tag{10-1}$$

where h = head loss due to friction, ft
 f = coefficient of friction
 L = length of piping, ft
 D = diameter of pipe, ft
 V = velocity of flow, ft/sec
 g = gravitation acceleration, 32.2 ft/sec^2

For short runs of piping (which is usually the case), a friction head loss of 20 ft can be assumed without introducing significant error. For runs of piping exceeding 50 ft, it is advisable to calculate the friction head loss.

The length of run is measured only from the pump outlet to the highest point. From the high point, the pipe is installed at ⅛ in. or ¼ in. pitch for gravity flow and thus does not impose a head on the pump. The piping from the pump to the high point should be the same size as the pump discharge outlet. From the high point to the connection with the gravity drain, the piping is sized as required for the gravity house drain. (See Table 6-5.)

Ejector Basin Sizing

The ejector basin must be of adequate capacity to prevent excessive cycling of pumps and the resultant wear and tear on the pump and motor. A minimum of 5 min of pump capacity between high and low water levels will provide satisfactory pump cycling. Assuming the same 85 gpm flow into the receiver utilized in the previous example, then 85 × 5 min = 425 gallons minimum capacity. Utilizing a standard diameter for a duplex pump system of 4 ft, the depth of the tank is calculated as follows:

$$.7854D^2 \times Depth \times 7.5 \text{ gal/ft}^3 = \text{gal}$$

$$Depth = \frac{425}{.7854 \times 4^2 \times 7.5} = 4.5 \text{ ft between high and low level}$$

The high water level should be maintained at least 3 in. below the invert of the entering pipe so as not to flood the subdrain piping and to provide air circulation throughout the system. The low level should always be maintained at least 6 in. above the bottom of the basin so that the suction intake of the pump is always covered by water. Verify this dimension with the manufacturer's recommendations. Following the above parameters, the depth of the basin below the invert of the entering pipe can be determined:

```
        3 in.  below the influent invert
  4 ft  6 in.  minimum storage
        6 in.  to cover pump suction
  4 ft 15 in. = 5 ft 3 in.
```

To this 5 ft 3 in. must be added to the distance from the subdrain invert to the finished floor to obtain the total basin depth. If it is assumed that the invert elevation of the subdrain is 2 ft below the finished floor in our example, the selected basin would then be 4 ft 0 in. diameter × 7 ft 3 in. depth (5 ft 3 in. + 2 ft 0 in.).

**Figure 10-5 Operation of a Pheumatic Ejector
(a) Filling Position**

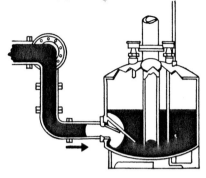

(b) Rising Sewage

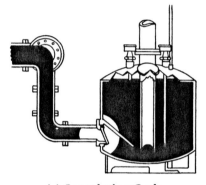

(c) Completing Cycle

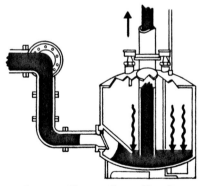

Courtesy: Yeoman Pump-Clow Corp.

It is good practice always to maintain a minimum depth of 3 ft below the invert of the subdrain pipe. Ejector basins should be sized so as not to retain the waste discharge for a period exceeding 12 hr.

Controls

For a simplex ejector pump system, level sensors start the pump at the high level and shut the pump off at the low level. A high-level alarm should always be installed to alert personnel that there is a malfunction. For a duplex system, the controls are the same except that provision should be made for automatic alternation of pumps to maintain equal wear on each pump. In addition, the controls should be set so that if the capacity of one pump is exceeded due to some unforeseen circumstances, the second pump will be cut in to aid in carrying the load.

Level sensors can be floats, sealed pressure actuators or electrodes, or mercury switch floats. If floats are used and the basin is more than 5 ft in depth, the float rod should be enclosed to prevent binding due to bending.

Installation

The criteria for design of the subdrain system up to the point of discharge into the receiver are exactly the same as those for the regular gravity drainage system, including all venting requirements. The ejector vent of the basin can be connected to the venting system for the rest of the building or it can be run independently to the atmosphere. Criteria for design and installation are the same as those for the gravity venting system used for the rest of the building. When the ejector vent is connected to the building venting system, it is necessary to convert the gpm discharge rate to FU to determine the additional load imposed on the building system. Assign 2 FU for each gpm of pump discharge. A rule of thumb that has proved satisfactory for sizing the ejector vent is to assign a size that is one-half the size of the subdrain discharge pipe into the basin.

A check valve and shut-off valve should always be installed in each pump discharge line. A swing check should always be used and it is recommended the weight-loaded type be specified. A weight-loaded check valve will eliminate water hammer when the pump shuts off and reverse flow occurs due to the water in the discharge piping. The shut-off valve should always be located on the discharge side of the check valve so that it can be closed and the check valve serviced.

Sump Basin Sizing

The sump basin receives storm water and other clear water wastes. Subsoil drains, because of their low elevations, are generally piped to a sump basin. A settling basin should always be provided so that sand, grit, and other deleterious material that could cause damage to the sump pump is settled out before entering the sump basin. The rate of discharge from a subsoil drainage system can be calculated by assigning a value of 2 gpm for each 100 ft^2 of area where the soil is sandy and 1 gpm per 100 ft^2 for clay soils. (See Chapter 7 for further discussion of subsoil drainage systems.)

Where the runoff from paved and impervious areas is collected in a sump, a value of 1 gpm/24 ft^2 can be assigned for locations where the annual rate of rainfall is 4 in./hr. When the total gpm discharge into the sump basin has been determined, calculation for the size of the basin is exactly the same as that for the ejector basin.

Sump Pump Sizing

The head on a sump pump is calculated exactly the same as it is for the ejector pump. It is good practice to select a capacity of 1¼ times the maximum inflow rate to provide an adequate factor of safety for possible periods when calculated maximum flows are exceeded by unusual storm conditions. Due to the ever-present danger of flooding caused by unusual rainfall conditions or the failure of one pump, it is recommended that a duplex sump pump system always be provided. Each pump should be sized for a capacity 1¼ times the maximum inflow.

The occurrence of a power failure is more likely during a storm than at other times—just when the sump pump is most needed. It is therefore advisable to provide a source of emergency power for sump pumps.

Controls

Controls, alarms, alternation, etc., are exactly the same as those for the ejector system.

A General Rule for a Subdrainage System

It should be a cardinal rule that any drainage that can possibly flow by gravity to a sewer should never be connected to the subdrainage system. The subdrainage system is dependent upon mechanical equipment (pumps and controls) for its operation and is therefore subject to failure, whereas the gravity drainage system will always remain in service during most types of emergencies.

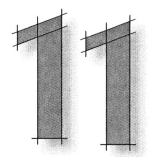

Flow in Water Piping

Hydraulics can be defined as the study of the principles and laws that govern the behavior of liquids at rest or in motion. *Hydrostatics* is the study of liquids at rest and *hydrokinetics* is the study of liquids in motion.

Although this text deals exclusively with water, all the data developed can be applied to any liquid.

Physical Properties of Water

The weight of water, or its *density*, varies with its temperature and purity. Water has its greatest specific weight (weight per cubic foot) at a temperature of 39.2°F. If this phenomenon did not occur, lakes would start freezing from the bottom up instead of from the top down. Table 11-1 tabulates densities of pure water at various temperatures. For the normal range of temperatures met in plumbing systems, the density of water is very close to 62.4 lbm/ft³ and this value can be used for all calculations without any significant error.

Viscosity can be defined as the internal friction, or internal resistance, to the relative motion of fluid particles. It can also be defined as the property by which fluids offer a resistance to a change of shape under the action of an external force. Viscosity varies greatly from one liquid to another. It approaches the conditions of a solid for highly viscous liquids and approaches a gas for the slightly viscous liquids. Viscosity decreases with rising temperatures. For example, #6 oil is a solid at low temperatures and begins to flow as it is heated.

Water is perfectly *elastic*, compressing when pressure is imposed and returning to its original condition when the pressure is removed. The compressibility of water may be expressed as $1/K$, where K is the coefficient of compressibility and is equal to 43,200,000 lb/ft². It can be seen that if a pressure of 100 lb/ft² were applied, the volumetric change would be $100/43,200,000$. The change is of such negligible significance that water is always treated as incompressible for all calculations in plumbing design.

The temperature at which water boils varies with the pressure to which it is subjected. At sea level—14.7 psi—water boils at 212°F. At an elevation above sea level, where the atmospheric pressure is less than 14.7 psi, water will boil at a lower temperature. In a closed system, such as that found in the domestic hot

Table 11-1 Density of Pure Water at Various Temperatures

Temperature °F	Density lbm/ft³	Temperature °F	Density lbm/ft³
32	62.416	100	61.988
35	62.421	120	61.719
39.2	62.424	140	61.386
40	62.423	160	61.006
50	62.408	180	60.586
60	62.366	200	60.135
70	62.300	212	59.843
80	62.217		

water system where the pressure is generally around 50 psi above atmospheric pressure, the water will not boil until a temperature of 300°F is reached.

Types of Flow

When water is moving in a pipe, two types of flow can exist. One type is known by the various names of *streamline, laminar,* or *viscous.* The second is called *turbulent* flow. At various viscosities (various temperatures), there is a certain critical velocity for every pipe size above which turbulent flow occurs and below which laminar flow occurs. This critical velocity occurs within a range of Reynolds numbers from approximately 2100 to 4000. Reynolds formula is:

$$R_e = \frac{DV\rho}{\mu g_c} \quad (11\text{-}1)$$

where R_e = Reynolds number, dimensionless
D = Pipe diameter, ft
V = Velocity of flow, ft/sec
ρ = Density, lbm/ft³
μ = Absolute viscosity, lbf · sec/ft²
g_c = Gravitational constant, 32.2 lbm·ft/lbf·sec²

Within the limits of accuracy required for plumbing design, it can be assumed that the critical velocity occurs at a Reynolds number of 2100. In laminar flow, the roughness of the pipe wall has a negligible effect on the flow but the viscosity has a very significant effect. In turbulent flow, the viscosity has an insignificant effect but the roughness of the pipe wall has a very marked effect on the flow.

Very rarely is a velocity of less than 4 ft/sec employed in plumbing design. The Reynolds number for a 3 in. pipe and a velocity of flow of 4 ft/sec would be

$$R_e = \frac{(0.250\text{ft})(4\text{ft/sec})(62.4\text{lbm/ft}^3)}{(2.35\times 10^{-5}\text{ lbf·sec/ft}^2)(32.2\text{ lbm·ft/lbf·sec}^2} = 82,500$$

(which is well above the critical number of 2,100)

It can be seen that all plumbing design is with turbulent flow and only when very viscous liquids or extremely low velocities are encountered does the plumbing engineer deal with laminar flow. Critical velocities of ½, 1, and 2-in. pipe at 60°F are 0.61, 0.31, and 0.15 ft/sec, respectively, and at 140°F they are 0.25, 0.13, and 0.06 ft/sec, respectively.

Velocity of Flow

When the velocity of flow is measured across the section of pipe from the center to the wall, it is found that there is a variation in the velocity, with the greatest velocity at the center and a minimum velocity at the walls. The average velocity for the entire cross-section is approximately 84% of the velocity as measured at the center. The plumbing engineer is concerned only with the average velocity, and all formulas are expressed in average velocity. Whenever and wherever the term *velocity* is used, it is the average velocity of flow that is meant.

Since water is incompressible within the range of pressures met in plumbing design, a definite relationship can be expressed between the quantity flowing past a given point in a given time and the velocity of flow. This can be expressed as (Equation 3-2):

Chapter 11—Flow in Water Piping

$$Q = AV$$

where Q = quantity of flow (volumetric flow rate), ft³/sec
  A = cross-sectional area of flow, ft²
  V = velocity of flow, ft/sec

The units employed in this flow formula are inconvenient for use in plumbing design. The plumbing engineer deals in gallons per minute and inches for pipe sizes. Converting to these terms, the flow rate becomes (Equation 8-8):

$$q = 2.448 \, d^2 V$$

where q = quantity of flow (flow rate), gpm
  d = diameter of pipe, in.
  V = velocity of flow, ft/sec

Potential Energy

One of the most fundamental laws of thermodynamics is that energy can be neither created nor destroyed; it can only be converted from one form to another. The energy of a body due to its elevation above a given level is called its *potential energy* in relation to that datum. Work had to be performed to raise the body to that elevation and this work is equal to the product of the weight of the body and the height it was raised. This can be expressed as:

(11-2)
$$E_p = wh = \frac{mgh}{g_c}$$

where E_p = potential energy, ft lbf
  w = weight of the body, lbf
  h = height raised, ft
  g = gravitational acceleration, 32.2 ft/s²
  g_c = gravitational constant, 32.2 lbm·ft/lbf·s²

When the weight is equal to 1 lb the formula becomes

(11-3)
$$E_p = \frac{hg}{g_c}$$

where E_p = potential energy per pound weight

Kinetic Energy

The energy of a body due to its motion is called *kinetic energy* and is equal to one-half its mass and the square of its velocity. Mass is equal to the weight of the body divided by its acceleration imposed by gravity.

(11-4)
$$m = \frac{wg_c}{g_c}$$

(11-5)
$$E_K = \tfrac{1}{2} \times \frac{wg_c}{g_c} \times \frac{V^2}{g_c} = \frac{w}{2g} \times V^2$$

where E_K = kinetic energy, ft lbf
  w = weight of body, lbf
  m = mass of the body, lbm
  g = gravitational acceleration, 32.2 ft/sec²
  g_c = gravitational constant, 32.2 lbm·ft/lbf·sec²
  V = velocity, ft/sec

Engineered Plumbing Design

When the body weighs 1 lb the formula becomes

$$E_K = \frac{V^2}{2g_c} \tag{11-6}$$

where E_K = kinetic energy per pound weight

Static Head

At any point below the surface of water that is exposed to atmospheric pressure, the pressure (head) is produced by the weight of the water above that point. The pressure is equal and effective in all directions at this point and is proportional to the depth below the surface. This pressure is variously called *static head, static pressure, hydrostatic head,* or *hydrostatic pressure*. It is the measure of the potential energy. Because pressure is a function of the weight of the water, it is possible to convert the static head expressed as feet of head into pounds per square inch. (See Table 11-2.)

The pressure developed by the weight of a column of water 1 in.² in cross-sectional area and h ft high may be expressed as

Table 11-2 Heads of Water in Feet Corresponding to Pressure in Pounds per Square Inch

PSI	0	1	2	3	4	5	6	7	8	9
0		2.3	4.6	6.9	9.2	11.6	13.9	16.2	18.5	20.8
10	23.1	25.4	27.7	30.0	32.3	34.7	37.0	39.3	41.6	43.9
20	46.2	48.5	50.8	53.1	55.4	57.8	60.1	62.4	64.7	67.0
30	69.3	71.6	73.9	76.2	78.5	80.9	83.2	85.5	87.8	90.1
40	92.4	94.7	97.0	99.3	101.6	104.0	106.3	108.6	110.9	113.2
50	115.5	117.8	120.1	122.4	124.7	127.1	129.4	131.7	134.0	136.3
60	138.6	140.9	143.2	145.5	147.8	150.2	152.5	154.8	157.1	159.4
70	161.7	164.0	166.3	168.6	170.9	173.3	175.6	177.9	180.2	182.5
80	184.8	187.1	189.4	191.7	194.0	196.4	198.7	201.0	203.3	205.6
90	207.9	210.2	212.5	214.8	217.1	219.5	221.8	224.1	226.4	228.7
100	231.0	233.3	235.6	237.9	240.2	242.6	244.9	247.2	249.5	251.8
110	254.1	256.4	258.7	261.0	263.3	265.7	268.0	270.3	272.6	274.9
120	277.2	279.5	281.8	284.1	286.4	288.8	291.1	293.4	295.7	298.0
130	300.3	302.6	304.9	307.2	309.5	311.9	314.2	316.5	318.8	321.1
140	323.4	325.7	328.0	330.3	332.6	335.0	337.3	339.6	341.9	344.2
150	346.5	348.8	351.1	353.4	355.7	358.1	360.4	362.7	365.0	367.3
160	369.6	371.9	374.2	376.5	378.8	381.2	383.5	385.8	388.1	390.4
170	392.7	395.0	397.3	399.6	401.9	404.3	406.6	408.9	411.2	413.5
180	415.8	418.1	420.4	422.7	425.0	427.4	429.7	432.0	434.3	436.6
190	438.9	441.2	443.5	445.8	448.1	450.5	452.8	455.1	457.4	459.7
200	462.0	464.3	466.6	468.9	471.2	473.6	475.9	478.2	480.5	482.8
210	485.1	487.4	489.7	492.0	494.3	496.7	499.0	501.3	503.6	505.9
220	508.2	510.5	512.8	515.1	517.4	519.8	522.1	524.4	526.7	529.0
230	531.3	533.6	535.9	538.2	540.5	542.9	545.2	547.5	549.8	552.1
240	554.4	556.7	559.0	561.3	563.6	566.0	568.3	570.6	572.9	575.2
250	577.5	579.8	582.1	584.4	586.7	589.1	591.4	593.7	596.0	598.3

Notes: 1. To use the chart, find the point corresponding to the specific pressure (psi) by adding incremental values in the top line to the base values in the extreme left column. For example, to find head in ft. corresponding to 25 psi, follow the line of figures to the right of 20 psi and read 57.8 ft under 5 psi. 2. Head values in the body of the chart were calculated by multiplying psi by 2.31. To convert ft of head to psi, multiply by 0.433, or use the chart in reverse.

$$p = \frac{\gamma}{144} \times h \quad (11\text{-}7)$$

where p = pressure, lbf/in²
γ = specific weight of water, lbf/ft³
h = static head, ft

At 50°F, the pressure expressed in pounds per square inch for a 1-ft column of water is then:

$$p = \frac{62.408}{144} \times 1 = 0.433 \text{ lbf/in}^2$$

Conversely, the height of a column of water that will impose a pressure of 1 lb/in.² is

$$h = p \times \frac{144}{\gamma}$$

$$h = 1 \times \frac{144}{62.408} = 2.31 \text{ ft}$$

To convert from feet of head to pounds per square inch, multiply the height by 0.433. To convert pounds per square inch to feet of head, multiply the pounds per square inch by 2.31.

Velocity Head

In a piping system with the water at rest, the water has potential energy. When the water is flowing it has kinetic energy as well as potential energy. To cause the water to flow some of the available potential energy must be converted to kinetic energy. The decrease in the potential energy, or static head, is called the *velocity head*.

In a freely falling body, the body is accelerated by the action of gravity at a rate of 32.2 ft/sec². The height of the fall and the velocity at any moment may be expressed as:

$$h = \frac{gt^2}{2} \quad (11\text{-}8)$$

$$V = gt \quad (11\text{-}9)$$

$$\text{or } t = \frac{V}{g}$$

where h = velocity head, ft
t = time, sec
g = gravitational acceleration, 32.2 ft/sec²
V = velocity, ft/sec

Substituting t = V/g in the first equation,

$$h = \frac{g}{2} \times \frac{V^2}{g^2}$$

$$h = \frac{V^2}{2g} \quad (11\text{-}10)$$

The foregoing illustrates the conversion of the potential energy of a body (static head) due to its height into kinetic energy (velocity head). The velocity head, $V^2/2g$, is a measure of the decrease in static head expressed in feet of column of water.

Bernoulli's Theorem

As previously stated, energy can be neither created nor destroyed. Bernoulli developed an equation to express this conservation of energy as it is applied to a flowing liquid. The liquid is assumed to be frictionless and incompressible.

(11-11)
$$\frac{Zg}{g_c} + \frac{Pg_c}{\rho g} + \frac{V^2}{2g_c} = E_T$$

where E_T = total energy ft·lbf/lbm
Z = height of point above datum, ft
P = pressure, lbf/ft^2
ρ = density, lbm/ft^3
V = velocity, ft/sec
g = gravitational acceleration, 32.2 ft./sec^2
g_c = gravitational constant, 32.2 lbm·ft/lbf·sec^2

The term $Pg_c/\rho g$ is equal to the static head or height of the liquid column. Substituting in the equation it becomes

Figure 11-1 Bernoulli's Theorem (Disregarding Friction)

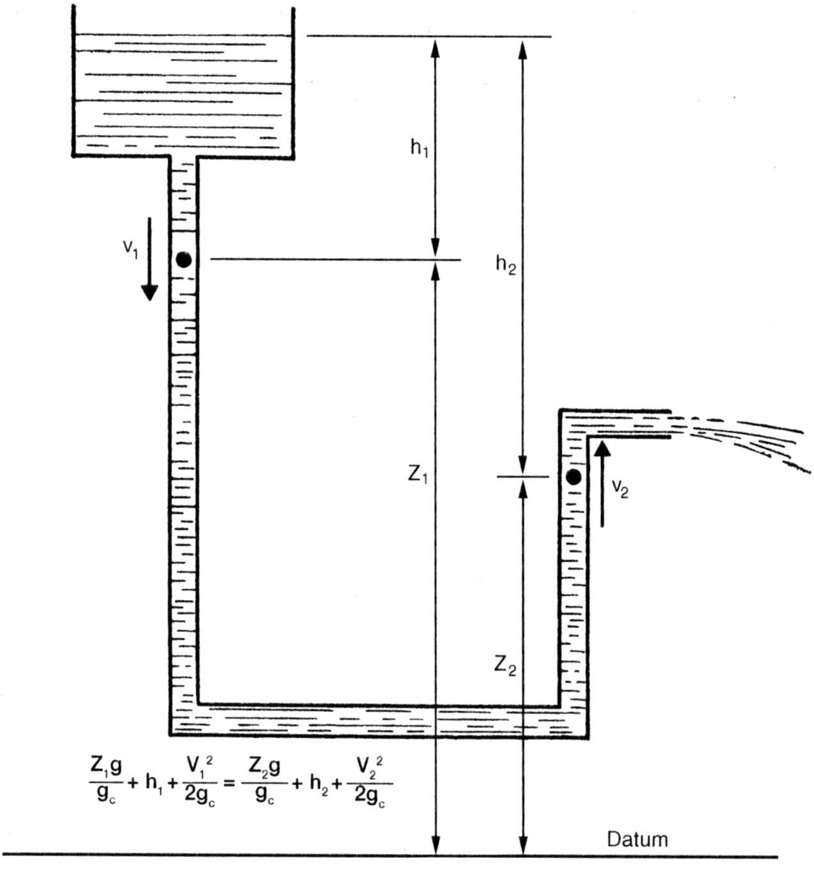

$$\frac{Z_1 g}{g_c} + h_1 + \frac{V_1^2}{2g_c} = \frac{Z_2 g}{g_c} + h_2 + \frac{V_2^2}{2g_c}$$

According to Bernoulli's theorem of conversion of energy, the energy of a mass particle at one point is equal to its energy at any other point in a fluid system. In the absence of friction losses.

$$\frac{Zg}{g_c} + h + \frac{V^2}{2g_c} = E_T \tag{11-12}$$

For any two points in a system, we may then write:

$$\frac{Z_1 g}{g_c} + h_1 + \frac{V_1^2}{2g_c} = \frac{Z_2 g}{g_c} + h_2 + \frac{V_2^2}{2g_c} \tag{11-13}$$

Figure 11-1 illustrates the application of this equation.

Friction

When water flows in a pipe, friction is produced by the rubbing of water particles against each other and against the walls of the pipe. This friction generates heat, which is dissipated in the form of a rise in the temperature of the water and the piping. This temperature rise in plumbing systems is insignificant and can safely be ignored in plumbing design. It requires a potential energy of 778 ft-lbf to raise 1 lb of water 1°F. The friction produced by flowing water also causes a pressure loss along the line of flow, which is called *friction head*. By utilizing Bernoulli's equation this friction head loss can be expressed as:

$$h_F = \left[\frac{Z_1 g}{g_c} + h_1 + \frac{V_1^2}{2g_c}\right] - \left[\frac{Z_2 g}{g_c} + h_2 + \frac{V_2^2}{2g_c}\right]$$

Flow from Outlets

Experiments to determine the velocity of flow from an outlet in the side of an open tank were performed by Toricelli in the 17th century. The result of these experiments was expounded in the theorem: "Except for minor frictional effects, the velocity is the same as if the fluid had fallen freely from the surface through a vertical distance to the outlet." This can be expressed as:

$$V = \sqrt{2gh} \tag{11-15}$$

It is graphically shown in Figure 11-2.

If friction, size, and shape of the opening and entrance losses are disregarded, the ideal velocity is the same as the maximum velocity and is equal to the velocity attained by free fall. The actual velocity, however, is always less than the ideal. All the factors, previously ignored, when taken into consideration can be expressed as the coefficient of discharge, C_D. The actual velocity can then be written:

$$V = C_D \sqrt{2gh} \tag{11-16}$$

For most outlets encountered in a plumbing system an average coefficient of discharge of 0.67 can be safely applied.

Flow in Piping

The velocity of flow at any point in a system is due to the total energy at that point. This is the sum of the potential and kinetic energy, less the friction head loss. The static head is the potential energy, but some of it was converted to kinetic energy to cause flow and some of it was used to overcome friction. It is for these reasons that the *pressure during flow is always less than the static*

Figure 11-2 Toricelli's Theorem

In the absence of friction, the discharge velocity of a fluid sample moving downward inside tank would equal velocity of a similar sample dropping the same distance outside container. This relationship is based on Toricelli's theorem.

pressure. The pressure measured at any point while water is flowing is called the flow *pressure*. This is the pressure that is read on a pressure gauge installed in the piping.

The kinetic energy of water flowing in a plumbing system is extremely small. Very rarely is the design velocity for water flow in plumbing systems greater than 8 ft/sec. The kinetic energy (velocity head) at this velocity is $V^2/2g$ or $8^2/64.4$. This is equal to 1 ft or 0.433 psi, which is less than 0.5 psi. It can be seen that such an insignificant pressure can be safely ignored in all calculations. The maximum rate of discharge from an outlet can now be determined from the flow pressure and the diameter of the outlet (using Equations 8-2 to 8-5):

$$q_D = C_D q_1$$
$$q_D = C_D \times 2.448 \, d^2 V_1$$
$$q_D = C_D \times 2.448 d^2 \sqrt{2gh}$$
$$q_D = C_D \times 19.65 d^2 \sqrt{h} \text{ or}$$
$$q_D = C_D \times 29.87 d^2 \sqrt{p}$$
(11-17)

where q_D = actual quantity of discharge, gpm
q_1 = ideal quantity of discharge, gpm
C_D = coefficient of discharge, dimensionless
d = diameter of outlet, in.
V_1 = ideal velocity, ft/sec
h = flow pressure, ft
p = flow pressure, psi

If 0.67 is used for the coefficient of discharge, then, per Equation 6-1,

$$q_D = 13.17 d^2 \sqrt{h}$$
and
$$q_D = 20 d^2 \sqrt{p}$$
(11-18)

Friction in Piping

As stated previously, whenever flow occurs, there is a continuous loss of pressure along the piping in the direction of flow. The amount of this head loss because of friction is affected by

1. Density and temperature of the fluid
2. Roughness of the pipe
3. Length of run
4. Velocity of the fluid

Experiments have demonstrated that the friction head loss is inversely proportional to the diameter of the pipe, proportional to the roughness and length of the pipe, and varies approximately with the square of the velocity. Darcy expressed this relationship as:

$$h = \frac{fLV^2}{D \times 2g} \text{ or}$$
$$p = \frac{\rho fLV^2}{144 D \times 2g_c}$$
(11-19)

where h = friction head loss, ft
p = friction head loss, lbf/in^2
ρ = density of fluid, lbm/ft^3
f = coefficient of friction, dimensionless
L = length of pipe, ft
D = diameter of pipe, ft
V = velocity of flow, ft/sec
g_c = gravitational constant, 32.2 lbm ft/lbf sec^2

Values for the coefficient of friction are given in Table 11-3.

It can be seen from Table 11-3 that steel pipe is much rougher than brass, lead or copper. It follows that there will be a greater head loss in steel pipe than in the other material.

For ease of application for the plumbing engineer, the formula for friction head loss can be reduced to a simpler form. Assuming an average value for the coefficient of friction of 0.02 for brass and copper and 0.04 for steel, the formula becomes:

For brass and copper

$$h = 0.000623 q^2 \times \frac{L}{d^5} \quad (11\text{-}20)$$

$$p = 0.00027 q^2 \times \frac{L}{d^5} \quad (11\text{-}21)$$

and for steel

Table 11-3 Average Values for Coefficient of Friction, f

Nominal Pipe Size, in.	Brass, Copper, or Lead	Galvanized Iron or Steel
½	0.022	0.044
¾	0.021	0.040
1	0.020	0.038
1¼	0.020	0.036
1½	0.019	0.035
2	0.018	0.033
2½	0.017	0.031
3	0.017	0.031
4	0.016	0.030

$$h = 0.00124 q^2 \times \frac{L}{d^5} \quad (11\text{-}22)$$

$$p = 0.00539 q^2 \times \frac{L}{d^5} \quad (11\text{-}23)$$

These formulas can be rearranged in another useful form:
For brass and copper

$$q = 40.1\, d^{2\frac{1}{2}} \left(\frac{h}{L}\right)^{\frac{1}{2}} \quad (11\text{-}24)$$

$$q = 60.8\, d^{2\frac{1}{2}} \left(\frac{p}{L}\right)^{\frac{1}{2}} \quad (11\text{-}25)$$

and for steel

$$q = 28.3\, d^{2\frac{1}{2}} \left(\frac{h}{L}\right)^{\frac{1}{2}} \quad (11\text{-}26)$$

$$q = 43.0\, d^{2\frac{1}{2}} \left(\frac{p}{L}\right)^{\frac{1}{2}} \quad (11\text{-}27)$$

where q = quantity of flow, gpm
 d = diameter of pipe, in.
 h = pressure, ft
 p = pressure, psi
 L = length of pipe, ft

The terms h/L and p/L represent the loss of head due to friction for 1 ft of pipe length and is called the *uniform friction loss*. Values of $d^{2\frac{1}{2}}$ for various diameters of pipe and various materials are given in Table 11-4.

In all water flow formulas, the term L (length of run in feet) is always the *equivalent length of run (ELR)*. Every fitting and valve imposes more frictional resistance than the pipe itself. To take this additional friction head loss into account, the fitting or valve is converted to an equivalent length of pipe of the same size that will impose an equal friction loss, e.g., a 4-in. elbow is equivalent to 10 ft of 4-in. pipe. Thus, if the measured length of run of 4-in. piping with one elbow is 15 ft, then the equivalent length of run is 15 + 10 = 25 ft. The

Table 11-4 Values of $d^{2\frac{1}{2}}$

Nominal Size, In.	Brass or Copper Pipe	Copper Type K	Copper Type L	Galvanized Iron or Steel
½	0.31	0.20	0.22	0.31
¾	0.61	0.48	0.55	0.62
1	1.16	0.99	1.06	1.13
1¼	2.19	1.73	1.80	2.24
1½	3.24	2.67	2.78	3.29
2	6.17	5.37	5.55	6.14
2½	9.88	9.25	9.54	9.58
3	16.41	14.41	14.87	16.48
4	32.00	29.23	30.13	32.53

length of pipe measured along the centerline of pipe and fittings is the *developed length*. Table 11-5 shows equivalent lengths of pipe for valves and fittings of various sizes. Note that the larger the pipe size, the more significant the equivalent length of run becomes. In the design phase of piping systems, the size of the piping is not known and the equivalent lengths cannot be accurately determined. A rule of thumb that has worked exceptionally well is to assume 50% of the developed length as an allowance for fittings and valves. Once the sizes are determined, the accuracy of the assumption can be checked.

All equipment imposes a friction head loss and must be carefully considered in the design and operation of a system. The pressure drop through any piece of equipment can be obtained from the manufacturer. The knowledgeable engineer is careful to specify the maximum pressure drop he/she will permit through a piece of equipment.

Table 11-5 Equivalent Pipe Length for Valves and Fittings

Nominal Pipe Size inches	Gate Valve Full Open	Angle Valve Full Open	Globe Valve Full Open	Swing Check Full Open	45° Elbow	Long Sweep Elbow or Run of Tee	Std. Elbow	Std. Tee Thru Side Outlet
½	0.35	9.3	18.6	4.3	0.78	1.11	1.7	3.3
¾	0.44	11.5	23.1	5.3	0.97	1.4	2.1	4.2
1	0.56	14.7	29.4	6.8	1.23	1.8	2.6	5.3
1¼	0.74	19.3	38.6	8.9	1.6	2.3	3.5	7.0
1½	0.86	22.6	45.2	10.4	1.9	2.7	4.1	8.1
2	1.10	29.0	58.0	13.4	2.4	3.5	5.2	10.4
2½	1.32	35.0	69.0	15.9	2.9	4.2	6.2	12.4
3	1.60	43.0	86.0	19.8	3.6	5.2	7.7	15.5
4	2.10	57.0	113.0	26.0	4.7	6.8	10.2	20.3
5	2.70	71.0	142.0	33.0	5.9	8.5	12.7	25.4
6	3.20	85.0	170.0	39.0	7.1	10.2	15.3	31.0
8	4.30	112.0	224.0	52.0	9.4	13.4	20.2	40.0

Velocity Effects in Piping

Hydraulic shock is commonly and erroneously referred to as "water hammer." The two terms are not synonymous. Water hammer is just one manifestation of the harmful effects created by hydraulic shock and one symptom of a very dangerous condition. Hydraulic shock occurs when fluid flowing through a pipe is subjected to a sudden and rapid change in velocity. The kinetic energy of the fluid is converted into a dynamic pressure wave that travels at the rate of 3,000 miles per hour (mph). This tremendous velocity produces a terrific impact, rebounding back and forth in the piping until the energy is dissipated. When the piping is not adequately secured or supported, or when the pipe runs are exceptionally long, these rebounding waves cause the piping to vibrate or hit against the building structure. This creates the noise that is commonly called "water hammer."

Noise, of course, is a nuisance but is not inherently dangerous. Of much greater importance than the noise of water hammer is the hydraulic shock. The latter can, and does, expand and burst pipe; cause weakening of the joints, which eventually leads to leaks; vibrate piping, causing pipe hangers to tear loose; wear out valves and faucets; rupture tanks and heaters; damage meters, gauges, pressure and temperature regulators; and generally accelerate the deterioration of the entire piping system. The result is costly repair, maintenance, and replacement. (See Figure 12-1.)

Because most runs of piping within a building are relatively short and well supported, hydraulic shock generally occurs without any noticeable or alarming noise. Under these conditions, it can virtually destroy a system before the danger is recognized.

The most common causes of hydraulic shock are the starting and stopping of pumps, improper check valves and rapid closure of a valve. The speed of valve closure time, particularly in the last 15% of movement, is directly related to the intensity of the surge pressure. Quick valve closure can be defined as a closure equal to or less than 2L/a seconds, where "L" is the length of pipe, in ft, from point of closure to point of relief (point of relief is usually a larger pipe riser, or main, or water tank) and "a" is the velocity of propagation of elastic vibration in the pipe, in ft/sec.

The expression "2L/a" is the time interval required for the pressure wave to travel from the point of closure to the relief point and back to the point of closure. The magnitude of the pressure wave can be expressed as

Figure 12-1   Illustrations of a Shock Wave

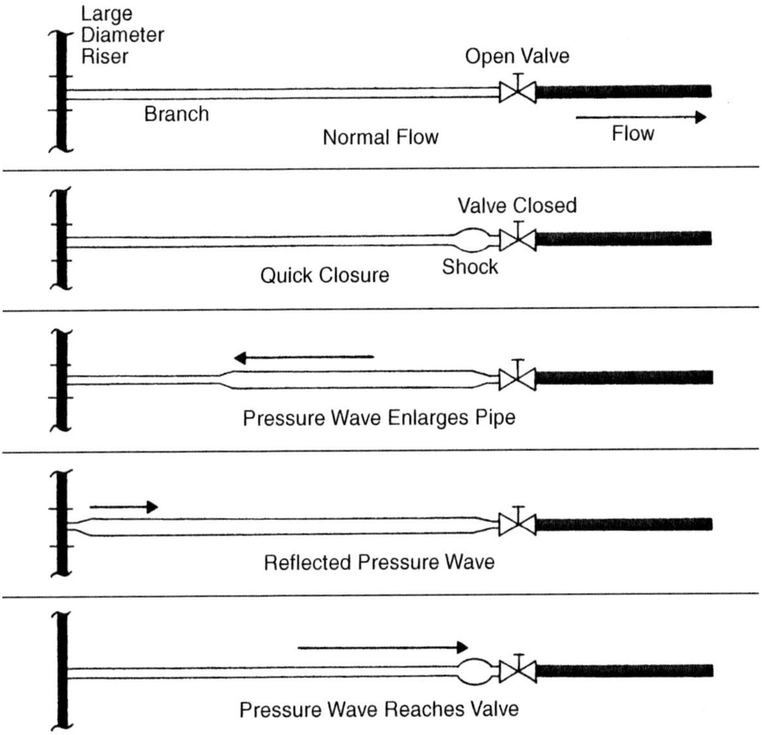

Illustration adapted from Plumbing & Drainage Institute, Standard PDI-WH201.

$$P = \frac{\gamma a V}{144 g} \tag{12-1}$$

where P = pressure in excess of flow pressure, ft³
γ = specific weight of liquid, lbf/ft³
a = velocity of propagation of elastic vibration in the pipe, ft/sec
V = change in velocity of flow, ft/sec
g = gravitational acceleration, 32.2 ft/sec²

The value of "a" can be determined by

$$a = \frac{4660}{(1 + KB)^{1/2}} \tag{12-2}$$

where 4660 = the velocity of sound in water, ft/sec
K = ratio of modulus of elasticity of fluid to the modulus of elasticity of the pipe
B = ratio of pipe diameter to wall thickness

Values of K for water and various materials are as follows:

Material	K
Cast Iron	0.020
Copper	0.017
Steel	0.010
Brass	0.017

Malleable cast iron   0.012

As stated previously, quick closure is the time interval, T, required for the pressure wave to travel back and forth in the pipe:

$$T = \frac{2L}{a} \tag{12-3}$$

When the valve closing time, Tv, is shorter than T, the returning pressure wave runs against the closed valve with the maximum intensity P. When Tv is longer than T, the returning pressure wave runs against a partially open valve and thus minimizes the effect of hydraulic shock.

A simplified equation, which may be used to determine the magnitude of hydraulic shock, is

$$P = \frac{0.027LV}{t} \tag{12-4}$$

where $P = lbf/in^2$
          L = pipe run, ft
          V = velocity, ft/sec
          t = time of valve closure, sec

A rule of thumb that has given satisfactory approximations is to multiply the velocity of flow by 60. This does not apply to exceptionally long runs of piping. It can be seen that if the velocity of flow is 10 ft/sec, then the hydraulic shock would be in the range of 600 lbf/in²!

For many years, air chambers have been utilized as a means of controlling hydraulic shock. The unit consists of a capped piece of pipe the same diameter as the line it serves and between 12 and 24 in. long. They have proven to be less than satisfactory and in many cases absolutely worthless. Unless they are of the correct size and contain an adequate volume of air, they are not suitable for even temporary control of shock. Although a correctly sized air chamber will temporarily control shock to within safe limits of pressure, adequate performance is effective only during the period in which the air chamber retains its initial charge of air. In practice, however, this initial charge of air is rapidly depleted and the chamber becomes waterlogged, completely losing its ability to control shock. Recognizing the inability of air chambers to perform their function, engineers have turned to the engineered or manufactured shock absorber.

Engineered or manufactured devices utilize a cushion of inert gas or air to absorb and control hydraulic shock, but the gas or air is permanently sealed in the unit and never dissipated. This construction provides many years of effective operation. (See Figure 12-2.)

Swing check valves should never be used in the discharge line of pumps. When the pump stops, there is a reversal of flow and the check slams closed, causing a sudden change in velocity. Spring-loaded check valves should always be installed in lieu of swing checks. The spring-loaded check is designed so that it will close at the exact moment the water flow comes to rest. There is

Figure 12-2 Shock Absorber

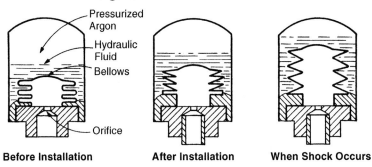

| Before Installation | After Installation | When Shock Occurs |

Before installation, engineered shock absorber bellows are held in fully compressed position by pressurized gas in upper part of chamber. After installation, line pressure extends and flexes bellows until pressures inside and outside are equalized. When hydrostatic shock occurs, increased line pressure extends bellows, absorbing the shock. After shock, bellows returns to normal installed position.

Illustration courtesy of Josam Mfg. Co.

no change in the velocity of flow when the check closes and thus no hydraulic shock produced.

Figure 12-3 shows the results of tests conducted by the United States Testing Company of Hoboken, New Jersey. Curve 1 is for the commonly used air chamber. It is 24 in. in height and one pipe size larger than the line served. Initially it controlled the surge at approximately 240 lbf/in² and then its control worsened. Curve 2 represents the average performance of a calculated air chamber (adequate size and air volume), which initially controlled the surge at approximately 145 lbf/in² but rapidly failed. Curve 3 is for a manufactured shock absorber, which initially controlled the surge below 150 lbf/in² and maintained the control for 10,000 cycles of valve closures.

Figure 12-3 Pressure Surge Control Curves

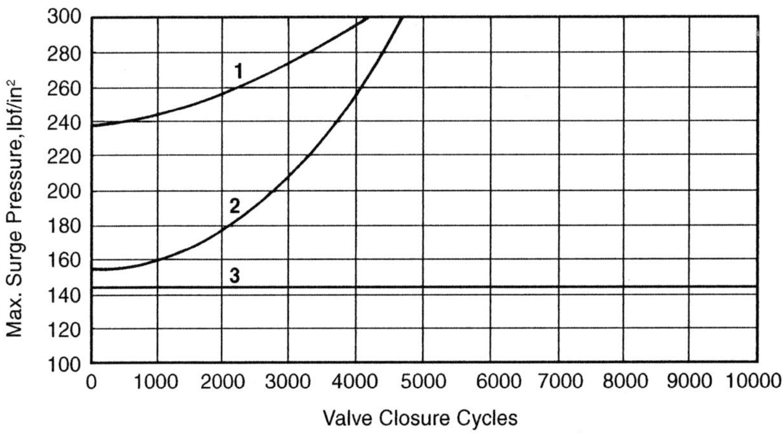

Erosion, Noise, and Cavitation

The pressure loss of flowing fluid due to friction varies approximately with the square of the velocity. This loss is also directly related to the roughness of the pipe wall. As the velocity of flow is increased, the abrasive effect upon the pipe wall increases and erosion of the pipe occurs. The extent of erosion caused by velocity is dependent upon the physical characteristics of the pipe material and the buildup of any deposits on the pipe walls.

When the flow velocity is high, line noises may be produced in the form of a whistling sound. When the fluid strikes protruding high spots in the pipe wall, energy is transferred into the pipe, which can cause it to move or vibrate. Generally, this vibration is dampened or absorbed by the piping. When the piping arrangement is such that resonance develops, however, the vibration can gain sufficient amplitude to cause noise.

When the direction of flow is sharply changed and the velocity of flow is high, the phenomenon of *cavitation* can occur. Cavitation is always accompanied by noise that sounds like gravel bouncing in the pipe or the popping of balloons. Fluids flowing around a short radius bend at a high velocity are subject to cavitation. The centrifugal force developed causes an increase of pressure at the outer bend with a resultant lowering of the pressure at the throat. This low-pressure zone can drop below atmospheric to a pressure that corresponds to the boiling point of the flowing fluid. Under this condition, the "cavity" that forms at the inside of the bend permits the fluid to flash into vapor or steam bubbles. Once these bubbles flow past the low-pressure zone into the normal pressure area downstream, the bubbles collapse. The rapid volumetric changes in bubble formation and collapse cause intense noise effects as well as stresses in the piping. Cavitation is a very serious problem in pump operation as well as in line flow. It can literally tear a pump apart.

Most of the noise problems (water hammer, whistling, and cavitation) can be greatly alleviated, if not completely eliminated, by maintaining flow velocities in any part of the pipe system below a velocity of 10 ft/sec.

The Copper Development Association recommends that water velocity in copper tubing not exceed 5–8 ft/sec. Under most operating conditions, it is recommended that water velocity in thermoplastic piping not exceed 5 ft/sec.

Water System Design

The objective in designing the water supply systems for any project is to ensure an adequate water supply at adequate pressure to all fixtures and equipment at all times and to achieve the most economical sizing of the piping.

There are at least six important reasons that proper design of water distribution systems is absolutely essential:

1. **Health.** This is of irrefutable and paramount importance. Inadequate or improper sizing can cause decreases in pressure in portions of the piping system, which in turn can cause contamination of the potable water supply by backflow or siphonage. There are too many well-documented deaths attributable to this cause.
2. **Pressure.** It is essential to maintain the required flow pressures at fixtures and equipment or improper operation will result.
3. **Flow.** Proper and adequate quantities of flow must be maintained at fixtures and equipment for obvious reasons.
4. **Water Supply.** Improper sizing can cause failure of the water supply due to corrosion or scale buildup.
5. **Pipe Failure.** Pipe failure can occur due to the relation of the rate of corrosion with excessive velocities.
6. **Noise.** Velocities in excess of 10 ft/sec will cause noise and increase the danger of hydraulic shock.

Of all the complaints resulting from improperly designed water systems, the two that occur most frequently are (1) lack of adequate pressure and (2) noise.

Noise may not be detrimental to the operation of a water distribution system but it is very definitely a major nuisance. The lack of adequate pressure, however, can have very serious repercussions in the operation of any water system.

Flow Pressure

It is essential that the term *flow pressure* be thoroughly understood and not confused with *static pressure*. Flow pressure is that pressure that exists at any point in the system when water is flowing at that point. It is always less than the static pressure. To have flow, some of the potential energy is converted to kinetic energy and additional energy is used in overcoming friction, which results in a flow pressure that is less than the static pressure.

When a manufacturer lists the minimum pressure required for the proper operation of a flush valve as 25 psi, it is the flow pressure requirement that is being indicated. The flush valve will not function at peak efficiency (if at all) if the engineer has erroneously designed the system so that a static pressure of 25 psi exists at the inlet to the flush valve.

Flow at an Outlet

There are many times when the engineer must determine how many gallons per minute are being delivered at an outlet. This can easily be determined by installing a pressure gauge in the line adjacent to the outlet and reading the gauge while flow is occurring. With the flow pressure known, the following formula can be used:

$$q = 20d^2 p^{1/2} \qquad (13\text{-}1)$$

where q = rate of flow at the outlet, gpm
d = actual inside diameter (ID) of outlet, in.
p = flow pressure, psi

Assume a faucet with a ⅜-in. supply and the flow pressure is 16 psi. Then:

$$\begin{aligned} q &= 20 \times (3/8)^2 \times (16)^{1/2} \\ &= 20 \times 9/64 \times 4 \\ &= 11.25 \text{ gpm} \end{aligned}$$

The flow for a ¼-in. and ⅛-in. supply at the same pressure would be 5 gpm and 1.25 gpm, respectively.

Constant Flow

Pressures in the various parts of the piping system are constantly fluctuating depending upon the quantity of flow at any moment. Under these conditions the rate of flow from any one outlet will vary with the change of pressure. In industrial and laboratory projects there is some equipment that must be supplied with a fixed and steady quantity of flow regardless of line pressure fluctuations. This feature is also desirable in any type of installation.

This criterion can easily be achieved by the utilization of an automatic flow-control orifice. A flow control is a simple, self-cleaning device designed to deliver a constant volume of water over a wide range of inlet pressures. (See

Figure 13-1   Flow Control

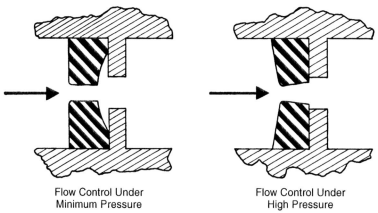

Flow Control Under
Minimum Pressure

Flow Control Under
High Pressure

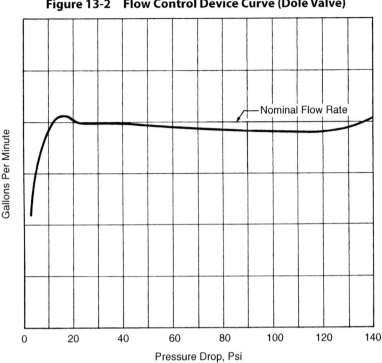

Figure 13-2 Flow Control Device Curve (Dole Valve)

Figures 13-1 and 13-2.) The automatic controlling mechanism consists of a flexible orifice that varies its cross-sectional area inversely with the pressure so that a constant flow rate is maintained under all conditions. Until the inlet pressure reaches the threshold pressure (12–15 psi), the flexible insert acts as a fixed orifice. When the threshold pressure is exceeded, the cross-sectional area of the orifice is decreased by the flexure of the insert. This causes a pressure drop that is equal to whatever pressure is necessary to absorb the energy not required to overcome system friction and to sustain the rated flow. The curve shown in Figure 13-2 is typical of most flow controls regardless of the rated flow, which is why no figures are shown for the gallons per minute axis. It is possible to approximate the flow of a specific flow control by using the line marked "Nominal Flow Rate" as the desired rate.

Assume a piece of equipment requires the fixed flow of 40 gpm and there is considerable line pressure fluctuation. A flow control would be specified to deliver 40 gpm. By use of the curve in Figure 13-2 the deviation from 40 gpm at various pressures can be read by assigning a value of 40 to the nominal flow rate line on the vertical scale and zero to the baseline. Standard flow controls are available in sizes from ¼ in. to 2½ in. and flow rates from ¼ to 90 gpm. They are ideal for use in limiting the maximum rate of flow to any fixture.

It is not unusual in a water distribution system to experience fluctuating discharges at fixtures and equipment due to other fixtures and equipment starting up or shutting down. Flow controls will minimize these problems because they automatically compensate for changes in the line pressure to hold the rate of water delivery from all outlets to a preselected number of gallons per minute. One very important word of caution—a flow control is not designed to perform the function of pressure regulation and should never be used where a pressure-regulating valve is required.

Material Selection

Before the type of material for the piping of a water distribution system can be selected, certain factors must be evaluated:
1. The characteristics of the water supply must be known. What is the degree of alkalinity or acidity? A pH above 7 is alkaline and below 7 is acidic. A pH of 7 represents neutral water. What is the air, carbon dioxide, and mineral content? The municipal water supply department can usually furnish all this information. If it is not available, a water analysis should be made by a qualified laboratory.
2. What are the relative costs of the various suitable materials?
3. Ease of replacement—can the material be obtained in a reasonable time or must it be shipped from localities that might delay arrival for months?
4. Actual inside dimensions of the same nominal size of various materials differ. This variation in ID can have a significant effect on sizing because of the variation in quantity rates of flow for the same design velocity. Table 13-1 shows the actual ID for various materials.
5. The roughness or smoothness (coefficient of friction) of the pipe will

Table 13-1 Actual Inside Diameter of Piping, in Inches

Nominal Pipe Size, In.	Iron or Steel Pipe, Sch. 40	Brass or Copper Pipe	Copper Water Tube, Type K	Copper Water Tube, Type L
½	0.622	0.625	0.527	0.545
¾	0.824	0.822	0.745	0.785
1	1.049	1.062	0.995	1.025
1¼	1.380	1.368	1.245	1.265
1½	1.610	1.600	1.481	1.505
2	2.067	2.062	1.959	1.985
2½	2.469	2.500	2.435	2.465
3	3.068	3.062	2.907	2.945
4	4.026	4.000	3.857	3.905
5	5.047	5.062	4.805	4.875
6	6.065	6.125	5.741	5.845
8	7.981	8.001	7.583	7.725
10	10.020	10.020	9.449	9.625

have a marked effect on pipe sizes.

Parallel Circuits

There are many parallel pipe circuits in the water distribution system of any job. An arrangement of parallel pipe circuits is one in which flow from a single branch divides and flows in two or more branches which again join in a single pipe. Figure 13-3 illustrates a simple two-circuit system. The total flow entering point A is the same leaving point A with a portion flowing through branch 1 and the rest through branch 2. Flows q_1 and q_2 must equal q and the total

Chapter 13—Water System Design

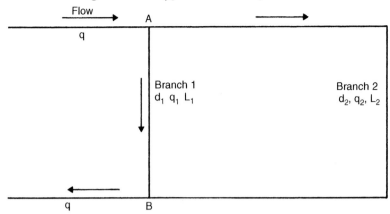

Figure 13-3 Typical Parallel Pipe Circuit

pressure drop from A to B is the same whichever branch is traversed. The rate of flow through each branch becomes such as to produce this equal pressure drop. The division of flow in each branch can then be expressed as:

(13-2)

$$\frac{q_1}{q_2} = \sqrt{\frac{L_2}{L_1}\left(\frac{d_1}{d_2}\right)^5}$$

Assume there is a flow in a 3-in. pipe of 160 gpm entering point A and leaving point B as shown in Figure 13-4. The length of branch 1 is 20 ft and branch 2 is 100 ft. The size of branch 1 is 2 in. and branch 2 is 3 in. To determine the quantity of flow in each branch, the basic formula is applied, and:

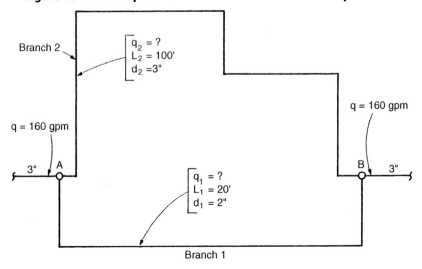

Figure 13-4 Example of Division of Flow in a Parallel Pipe Circuit

$$\frac{q_1}{q_2} = \sqrt{\frac{100}{20}\left(\frac{2}{3}\right)^5}$$
$$= \sqrt{5(.66)^5}$$
$$= \sqrt{5 \times 0.125}$$
$$= 0.79$$
$$q_1 = 0.79 q_2$$

since $q_1 + q_2 = 160$
then $0.79 q_2 + q_2 = 160$
$$1.79 q_2 = 160$$
$$q_2 = 89.4 \text{ gpm}$$

and $q_1 = 160 - 89.4 = 70.6$ gpm
or $q_1 = 0.79 \times 89.4 = 70.6$ gpm

Inadequate Pressure

As previously noted, lack of adequate pressure is one of the most frequent complaints and could be the cause of serious troubles. The pressure available for water distribution within a building can come from various sources. Municipalities usually maintain water pressure in their distribution mains within the range of 35–45 psi. There are localities where the pressure maintained is much less or greater. The local utility will furnish the information as to their minimum and maximum operating pressures. When utilizing only the public water main pressure for the water distribution system within a building, it is very important to determine the pressure available in the mains during the summer months. Huge quantities of water are used during this period for sprinkling of lawns and for air-conditioning cooling tower makeup water, which usually cause excessive pressure loss in the mains. Future growth of the area must also be analyzed. If large housing, commercial, or industrial development is anticipated, the pressure available will certainly decrease as these loads are added to the public mains. It is good practice to assume a pressure available for design purposes as 10 psi less than the utility quotes.

If the pressure from the public mains is inadequate for building operation, other means must be provided for increasing the pressure to an adequate level. There are three basic methods available:
1. Gravity tank system
2. Hydropneumatic tank system
3. Booster pump system

Each system has its own distinct and special advantages and disadvantages. All three should be evaluated in terms of capital expenditure, operating costs, maintenance costs, and space requirements. Depending upon which criteria are the most important, this will dictate which system is selected.

Flow Definitions

Maximum flow or *maximum possible flow* is the flow that will occur if the outlets on all fixtures are opened simultaneously. *Average flow* is that flow likely to occur in the piping under normal conditions. *Maximum probable flow* is the maximum flow that will occur in the piping under peak conditions. It is also called *peak demand* or *peak flow*.

Demand Types

Some outlets impose what is called a *continuous demand* on the system. They are differentiated from outlets that impose an *intermittent demand*. Outlets such as hose bibbs, lawn irrigation, air-conditioning makeup, water cooling, and similar flow requirements are considered to be continuous demands. They occur over an extended period of time. Plumbing fixtures draw water for a relatively short period of time and are considered as imposing an intermittent demand.

Each fixture has its own singular loading effect on the system, which is determined by the rate of water supply required, the duration of each use, and the frequency of use. The water demand is related to the number of fixtures, type of fixtures, and probable simultaneous use.

Estimating Demand

The basic requirements for estimating demand call for a method that
1. Produces estimates that are greater than the average demand for all fixtures or inadequate supply will result during periods of peak demand.
2. Produces an accurate estimate of the peak demand to avoid oversizing.
3. Produces estimates for demand of groups of the same type of fixtures as well as for mixed fixture types.

Design Loads

Arriving at a reasonably accurate estimate of the maximum probable demand is complicated due to the intermittent operation and irregular frequency of use of fixtures. Different kinds of fixtures are not in uniform use. Bathroom fixtures are most frequently used on arising or retiring and, not surprisingly, during television commercials. Kitchen sinks find heavy usage before and after meals. Laundry trays and washing machines are most likely to be used in the late morning. During the period from midnight to 6 P.M. there is very little fixture use. Luckily, fixtures are used intermittently and the total time in operation is relatively small so it is not necessary to design for the maximum potential load. Maximum flow is therefore of no real interest to the designer. Average flow is also of no concern, for if a system were designed to meet this criterion it would not satisfy the conditions under peak flow. It is therefore necessary to consider only the maximum probable demand (peak demand) imposed by the fixtures on a system.

Table 13-2 Demand at Individual Fixtures and Required Pressure

Fixture	Flow Pressure, psi	Flow Rate, gpm
Ordinary lavatory faucet	8	3.0
Self-closing lavatory faucet	12	2.5
Sink faucet, ⅜ in.	10	4.5
Sink faucet, ½ in.	5	4.5
Bathtub faucet	5	6.0
Laundry tub faucet, ½ in.	5	5.0
Shower head	12	5.0
Water closet flush tank	15	3.0
Water closet flush valve, 1 in.	10–25	15–45
Urinal flush valve, ¾ in.	15	15.0
Hose bibb or sill cock, ¾ in.	30	5.0

Two methods have evolved in the United States that, when used where applicable, have proven to give satisfactory results. They are the *empirical method* and *method of probability*. The empirical method is based upon arbitrary decisions arrived at from experience and judgment. It is useful only for small groups of fixtures. The method of probability is based upon the theory of probabilities and is most accurate for large groups of fixtures.

In the past, certain demand rates became generally accepted as standard. These rates are tabulated in Table 13-2 for the common types of fixtures and the average pressure necessary to deliver this rate of flow. The actual pressure for a specific fixture will vary with each manufacturer's design, some requiring a greater or lesser pressure than others.

Although the flow rates shown in Table 13-2 have been used by engineers, they are hopelessly outdated. Water conservation measures being mandated by federal regulations and model codes make the flow rates shown in Table 13-2 unreasonable for use in the design of systems. The federal Energy Policy Act (EPACT92) established the following criteria for water use by fixture:

Water closets: 1.6 gal/flush
Urinals: 1.5 gal/flush
Showers: 2.5 gpm
Lavatories: 2.5 gpm
Sinks: 2.5 gpm

Manufacturers offer fixtures meeting these and more stringent requirements. Lavatories with 0.5 gpm flow rates and urinals with 1.0 gal/flush have been installed in thousands of buildings with satisfactory results. However, there is a need for research to determine the actual minimum flow required, for each type of fixture, to satisfy psychological requirements of the user and provide the necessary sanitary requirements.

Water Supply Fixture Units

A standard method for estimating the water demand for a building has evolved through the years and has been accepted almost unanimously by plumbing designers. It is a system based on weighting fixtures in accordance with their water supply load-producing effects on the water distribution system. The National Bureau of Standards has published report BMS 65, *Methods of Estimating Loads in Plumbing Systems,* by the late Dr. Roy B. Hunter, which gives tables of load-producing characteristics (fixture unit weights) of commonly used fixtures, along with probability curves that make it possible to apply the method easily to actual design problems.

The method of probability should not be used for a small number of fixtures. Although the design load, as computed by this method, has a certain probability of not being exceeded, it may nevertheless be exceeded on rare occasions. When a system contains only a few fixtures, the additional load imposed by one fixture more than has been calculated by the theory of probability can easily overload the system. When a system contains a large number of fixtures, one or several additional fixture loadings will have an insignificant effect on the system.

In developing the application of the theory of probability to determine design loads on a domestic water distribution system, Hunter assumed that the operation of the fixtures in a plumbing system could be viewed as purely random events. He then determined the maximum frequencies of use of the fixtures. He obtained the values of the frequencies from records collected in hotels and apartment houses during the periods of heaviest usage. He also determined characteristic values of the average rates of flow for different fixtures and the time span of a single operation of each.

If only one type of fixture were used in a building, the application of the theory of probability would be very simple and straightforward. When dealing with systems composed of various types of fixtures that must be combined, the process becomes extremely involved and too complicated to be of any practical use. Faced with this dilemma, Hunter devised an ingenious method to circumvent the problem by a simple process which yields results within ½% accuracy of the more involved and laborious calculations required. He conceived the idea of assigning "fixture loading factors" or "fixture unit weights" to the different kinds of fixtures to represent the degree to which they loaded a system when used at their maximum assumed frequency. A fixture unit weight of 10 was arbitrarily assigned by Hunter to a flush valve, and all other fixtures were assigned values based on their load-producing effect in relation to the flush valve. All fixtures have been converted, in essence, to one fixture type and the application of the theory of probability is greatly simplified.

Hunter assigned water supply fixture unit (FU) values for different kinds of fixtures, which are given in Table 13-3. Conversion of fixture unit values to equivalent gallons per minute, based on the theory of probability of usage developed by Hunter, is given in Table 13-4. A graphic representation of this table is shown by Figures 13-5 and 13-6 (Hunter's Curve). Figure 13-7 gives

Table 13-3 Demand Weight of Fixtures, in Fixture Units

Fixture or Group	Occupancy	Type of Supply Control	Fixture Units		
			Hot	Cold	Total
Water closet	Public	Flush valve	—	10	10
Water closet	Public	Flush tank	—	5	5
Pedestal urinal	Public	Flush valve	—	10	10
Stall or wall urinal	Public	Flush valve	—	5	5
Stall or wall urinal	Public	Flush tank	—	3	3
Lavatory	Public	Faucet	1.5	1.5	2
Bathtub	Public	Faucet	3	3	4
Shower head	Public	Mixing valve	3	3	4
Service sink	Office, etc.	Faucet	3	3	4
Kitchen sink	Hotel or restaurant	Faucet	3	3	4
Water closet	Private	Flush valve	—	6	6
Water closet	Private	Flush tank	—	3	3
Lavatory	Private	Faucet	.75	.75	1
Bathtub	Private	Faucet	1.5	1.5	2
Shower head	Private	Mixing valve	1.5	1.5	2
Bathroom group	Private	Flush valve W.C.	2.25	6	8
Bathroom group	Private	Flush tank W.C.	2.25	4.5	6
Separate shower	Private	Mixing valve	1.5	1.5	2
Kitchen sink	Private	Faucet	1.5	1.5	2
Laundry tray	Private	Faucet	2	2	3
Combination fixture	Private	Faucet	2	2	3

Table 13-4 Conversion of Fixture Units to Equivalent gpm

Demand (Load) Fixture Units	Demand (Load), gpm System with Flush Tanks	Demand (Load), gpm System with Flush Valves
1	0	—
2	1	—
3	3	—
4	4	—
5	6	—
6	5	—
8	6.5	—
10	8	27
12	9	29
14	11	30
16	12	32
18	13	33
20	14	35
25	17	38
30	20	41
35	23	44
40	25	47
45	27	49
50	29	52
60	32	55
70	35	59
80	38	62
90	41	65
100	44	68
120	48	73
140	53	78
160	57	83
180	61	87
200	65	92
225	70	97
250	75	101
275	80	106
300	85	110
400	105	126
500	125	142
750	170	178
1,000	208	208
1,250	240	240
1,500	267	267
1,750	294	294
2,000	321	321
2,250	348	348
2,500	375	375
2,750	402	402
3,000	432	432
4,000	525	525
5,000	593	593
6,000	643	643
7,000	685	685
8,000	718	718
9,000	745	745
10,000	769	769

a graphic representation of the conversion from fixture units to gallons per minute for a mixed system. An examination of the curves and tables reveals that demand for a system utilizing flush valves is much greater than that for flush tanks for small quantities. The difference in demand for each system decreases as the fixture unit load increases until 1,000 FUs are reached. At this loading and beyond, the demand for both types of systems is the same.

For hot water piping and where there are no flush valves on the cold water piping, the demand corresponding to a given number of fixture units is determined from the values given for the flush tank system.

The accuracy of Hunter's curve, however, has come into serious question. Results utilizing the curve have proven to be as much as 100% inflated in some instances. The consistent overdesign, however, should in no way be interpreted as indicating that Hunter's basic research and approach are incorrect.

His method is demonstrably accurate, but it must be remembered that his basic assumptions and criteria were promulgated more than 60 years ago. Many things have changed, and changed drastically, in the interim. Improvements have been made in flush valve design as well as in faucets and fixtures. Social customs and living patterns have changed. The public emphasis on water and energy conservation has altered many basic criteria. It is now necessary to change some of Hunter's basic assumptions (but not his concept).

It has been demonstrated by thousands of projects operating satisfactorily that it is safe to reduce the values obtained by use of Hunter's curve by 40%. It is stressed again that this reduction can be applied only for systems with a large number of fixtures. The opposite is true for water use in toilet facilities where large numbers of people gather, such as sport facilities and auditoriums. In

Chapter 13—Water System Design

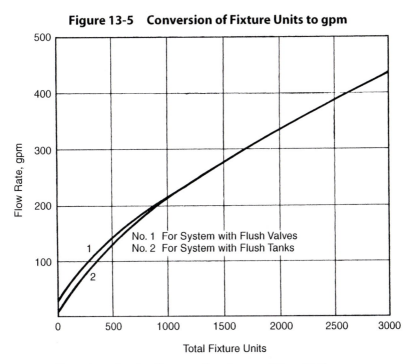

Figure 13-5 Conversion of Fixture Units to gpm

See enlarged scale of lower portion of curves (Figure 13-6).

Figure 13-6 Conversion of Fixture Units to gpm (enlarged scale)

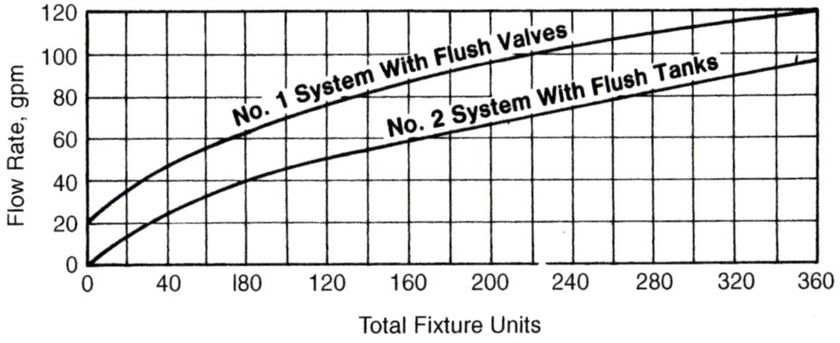

these types of facilities, demand flow rates will exceed those determined by Hunter's curve because many people will use the toilet rooms during breaks in the game or performance.

The student is again warned to use the table of fixture unit values in the code applicable to the locality of the project. The values vary slightly from code to code. The student is also alerted to the fact that water supply fixture units are not the same as drainage fixture unit values. The discharge rates of certain fixtures are entirely different from the rate at which water is supplied, e.g., bathtubs. The loading effect is therefore different on the drainage system than it is on the water supply system for specific fixtures.

Engineered Plumbing Design

For supply outlets that are likely to impose continuous demands, estimate the continuous demand separately from the intermittent demand and add this amount in gallons per minute to the demand of the fixtures in gallons per minute.

It should be kept in mind when calculating maximum probable demands that, except for continuous demands, fixture unit values are always added, never gpm values. For example, if the maximum probable demand for two branches is required and one branch has a load of 1250 FU and the other 1750 FU, it would be wrong to add 240 gpm + 294 gpm to obtain 534 gpm for the total demand. The correct procedure is to add 1250 FU + 1750 FU to obtain a total FU value of 3000 and then from Table 13-4 determine the correct peak demand as 432 gpm. The 432 gpm value reflects the proper application of the theory of probability.

The following example illustrates the procedure for sizing a system.

Example 13-1

Determine the peak demands for hot and cold and total water for an office building that has 60 flush valve water closets, 12 wall hung urinals, 40 lavatories, and 2 hose bibbs and requires 30 gpm for air-conditioning water makeup.

From Table 13-3 determine the FU values:

	Hot Water	Cold Water	Total (Hot & Cold)
60 WC × 10	—	600	600
12 UR × 5	—	60	60
40 Lavs × 2	—	—	80
40 Lavs × 1.5	60	60	—
	60 FU	720 FU	740 FU

From Table 13-4 or Figure 13-5:

60 FU = 32 gpm hot water demand
720 FU = 174 gpm cold water demand
740 FU = 177 gpm total water demand

To the cold water and total water demand must be added the continuous demand:

$$\begin{array}{ll} \text{2 hose bibbs} \times 5 \text{ (from Table 13-2)} & = 10 \text{ gpm} \\ \text{Air-conditioning makeup} & = \underline{30 \text{ gpm}} \\ & \;40 \text{ gpm} \end{array}$$

Then:

Hot water demand:	=	32 gpm
Cold water demand: 174 + 40	=	214 gpm
Total water demand: 177 + 40	=	217 gpm

The conversion of fixture unit loads to equivalent gallons per minute demand was obtained from Table 13-4 using straight line interpolations to obtain intermediate values. Total water demand is required for sizing the water service line for the building and also for the cold water piping inside the building up to the point where the connection is taken off to the hot water heater supply.

Figure 13-7 Conversion of Fixture Units to gpm (Mixed System)

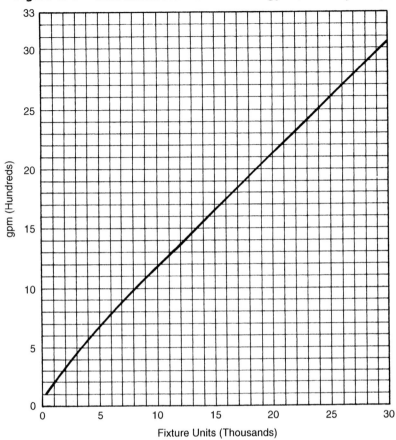

14 Water System Sizing

The water distribution system must always be designed on the basis of the *minimum* pressure available. The source of pressure may be the public water main, gravity tank, hydropneumatic tank, or booster pumps. The normal pressure available from the public main can be obtained from the local utility. As indicated previously, it is good practice to assume the minimum available pressure to be 10 psi less than the stated pressure. When a gravity tank is the source of pressure, the low-water level must be used for design. The low-pressure limit must be used for the hydropneumatic tank system and the delivery pressure of the pressure-regulating valve on the pump discharge of booster pump systems will determine that pressure.

Generally, the minimum pressure to be provided at most fixtures is 8 psi and from 15 to 25 psi for water closets. Remember that these pressures are *flow pressures* and not static pressures.

Friction Head Loss

Water piping must be sized to limit the friction head losses in the piping system so that the highest and most remote water outlet will have the required minimum pressure for adequate flow during periods of peak demand. The maximum friction head loss that can therefore be tolerated in the system during peak demand is the difference between the static pressure at the highest and most remote water outlet at no-flow conditions and the minimum flow pressure required at that outlet.

Pipe friction head loss is directly proportional to the length of run. The longest length of run to the highest outlet should be selected for purposes of sizing the system. If the losses in this run of piping are within the required limits, then every other run of piping will also be within the required friction head losses. The selected longest length of run should be sized in accordance with *uniform friction head loss* distribution throughout its length for ease of calculation. It has been found that this method of uniform friction head loss calculation does not result in any significant variation of pipe sizes throughout the system. The permissible uniform friction head loss, in pounds per square inch per 100 ft can be found by dividing the total friction head loss (in pounds per square inch) permitted by the equivalent length of run of the longest run and multiplying by 100. (Uniform friction head loss = total friction head loss × 100 ÷ equivalent length of run.) Feet of head (ft) and pounds per square inch (psi) are related by, 1 psi = 2.31 feet of head

The equivalent length of piping is its developed length plus the equivalent lengths of pipe corresponding to friction head losses for fittings, valves, strainers, etc. When the size of fittings and valves are known, their equivalent lengths can be obtained from Table 11-5. When a system is sized, the sizes of fittings and valves are not known and the added friction head losses imposed must be approximated. A general rule of thumb that has proven to be surprisingly accurate is to add 50% of the developed length to allow for all fittings and valves. Thus, equivalent length of run (ELR) = [developed length of run (DL) + .50]×DL or ELR = 1.5 × DL

Having established the uniform friction loss, all that is needed now is to employ hydraulic tables to obtain the corresponding rates of flow that will produce that loss for various sizes of pipes of various materials. Tables 14-1 through 14-12 are a compilation of friction head loss and velocity values for various pipe materials and sizes in a form convenient for use by plumbing designers. Other sizing tables may be found in *Cameron Hydraulic Data*, 19th ed. (published by Flowserve, edited by C. C. Heald), *Hydraulic Institute Engineering Data Book* (authored by Maynard Neal), and various catalogs.

For those who would prefer to perform their own calculations, the following formulas (Equations 11-25 and 11-27, respectively) may be used:

Brass and copper piping:

$$q = 60.8 \, d^{2\frac{1}{2}} \left(\frac{p}{L}\right)^{\frac{1}{2}}$$

Galvanized iron or steel piping:

$$q = 43.0 \, d^{2\frac{1}{2}} \left(\frac{p}{L}\right)^{\frac{1}{2}}$$

where q = quantity of flow, gpm
 d = internal diameter of pipe, in.
 p = total permissible friction head loss, psi
 L = equivalent length of longest run, ft

When utilizing the above formulas, it is important to use the actual internal diameter of the pipe rather than its nominal diameter. The actual internal diameter of pipe of various materials is given in Table 13-1 and the values of $d^{2\frac{1}{2}}$ are given in Table 11-4.

Maximum Velocity

The maximum velocity of water flow in the piping during periods of peak demand should always be of prime importance to the designer. As previously discussed, when flow approaches 10 ft/sec in piping, serious problems can develop. High velocities produce noise in the form of whistling, the danger of hydraulic shock and water hammer is increased, the noise of cavitation may occur, and erosion and corrosion are increased. It is for these reasons that piping should be sized so as never to exceed a velocity of flow of 8 ft/sec.

To illustrate how to work with formulas, the following examples are offered.

Example 14-1

For a demand of 100 gpm, a permissible total friction head loss of 16 psi, and an equivalent length of run of 100 ft, what size pipe is required for the following piping materials: brass, type K copper and type L copper?

Using the formula for brass or copper

$$q = 60.8\, d^{2\frac{1}{2}}\, (P/L)^{\frac{1}{2}}$$
$$d^{2\frac{1}{2}} = q/60.8\, (L/p)^{\frac{1}{2}}$$
$$d^{2\frac{1}{2}} = 100/60.8\, (100/16)^{\frac{1}{2}} = 100/60.8\,(6.25)^{\frac{1}{2}}$$
$$d^{2\frac{1}{2}} = 100/60.8\,(2.5) = 100/0.0411$$
$$d^{2\frac{1}{2}} = 4.11$$

Referring to Table 11-4 to determine values of $d^{2\frac{1}{2}}$, we find:
Brass = 2 in.
Type K copper = 2 in.
Type L copper = 2 in.

Example 14-2

Assuming a permissible friction head loss of 36 psi and all other factors the same as in the above example, what are the sizes?

$$d^{2\frac{1}{2}} = 100/60.8\, (100/36)^{\frac{1}{2}}$$
$$d^{2\frac{1}{2}} = 2.74$$

then:
Brass = 1½ in.
Type K copper = 2 in.
Type L copper = 1½ in.

It can be seen from this example that material selection can have an effect on the size of a pipe for a particular system.

Minimum Sizes

Most codes establish minimum sizes for the piping supplying the outlets for the various kinds of fixtures. Table 14-13 lists fixtures and the minimum size of fixture supply pipe. Sizes given in the table are generally such as to maintain velocity of flow below the maximum of 10 ft/sec.

Procedure for Sizing

Before an attempt to size any system, a riser diagram of the complete water distribution system should be drawn. In this riser diagram, the floor-to-floor heights should be shown. It often proves useful also to note the static pressure at each floor. On this drawing, the minimum pressure required at the highest outlet as well as the minimum available pressure should be noted. This can be considered Step 1. Then:

Table 14-13 Minimum Size of Fixture Supply Pipes

Fixture or Device	Pipe Size, in.
Bathtubs	½
Combination sink and tray	½
Drinking fountain	⅜
Dishwasher (domestic)	½
Kitchen sink (residential)	½
Kitchen sink (commercial)	¾
Lavatory	⅜
Laundry tray, 1, 2 or 3 compartments	½
Shower (single head)	½
Sinks (service, slop)	½
Sinks (flushing rim)	¾
Urinal (flush tank)	½
Urinal (¾" flush valve)	¾
Water closet (flush tank)	⅜
Water closet (flush valve)	1
Hose bibb	½
Wall hydrant	½

2. Mark the FU value at every outlet and the sum of fixture units for every section of the system. (It is important to stress that when adding loads

it is mandatory to add fixture unit values. *Never add gallons per minute demands except for continuous demands.*)
3. Convert all FU values to gpm demand and assign the gpm values to continuous demand outlets.
4. Determine the pressure available for friction head loss. Using the longest run to the highest fixture (refer to plans as well as the riser diagram to determine the longest run), establish the uniform friction head loss.
5. Use hydraulic tables to select sizes. The selection will be based on the gpm demand, the uniform friction head loss, and the maximum design velocity selected. If the size indicated by the tables produces a velocity in excess of the selected maximum velocity, then a size must be selected that produces the required velocity.

Many friction head loss tables are based upon the Hazen and Williams formula:

(14-1)
$$f = .2083 \left(\frac{100}{C}\right)^{1.85} \times \frac{q^{1.85}}{d^{4.8655}}$$

where f = friction head in ft of liquid per 100 ft of pipe (if psi is desired, multiply f by .433)
d = inside diameter of pipe, in.
q = flow, gpm
C = constant, reflecting roughness of pipe

Tables 14-1 through 14-12 are a compilation of values based on the Hazen and Williams formula, employing the commonly used value for design purposes of $C = 100$ for steel pipe and $C = 130$ for brass and copper. Maximum design velocity should generally be 8 fps, and a velocity of 10 fps (at which velocity noise becomes a problem) should never be exceeded. Maximum friction head loss (uniform head loss per 100 ft) is determined exclusively on the basis of total pressure available for friction loss and the longest equivalent length of run. The Hazen and Williams formula is most popular with civil engineers because its use is accurate for the flow of 60°F water in pipes larger than 2 in. and smaller than 72 in. Use of the formula for water temperatures much higher or lower than 60°F results in some error. Because the formula uses the hydraulic radius, it can also be used for noncircular sections.

A Hydropneumatic or Booster Pump System

The foregoing procedure works very well when the street pressure is adequate to supply the requirements of the building or where a gravity tank system is installed. Under these conditions the minimum available pressure is already established and thus pressure available for friction head loss can be calculated. It is an entirely different situation if a hydropneumatic tank or booster pump system is selected to provide the pressure for a system. The minimum available pressure is no longer a fixed and unchangeable quantity. The pressure available can now be that selected and determined by the designer, and the economic impact of that decision must be evaluated. If the piping system is designed for a high uniform friction head loss, the pipe sizes will be correspondingly smaller than for a lower uniform pressure loss, but the minimum available pressure must of necessity be higher. This means that pump head

Engineered Plumbing Design

Table 14-1 ½ inch

Flow Gal. Per Min.	Sch. 40 Steel .622" I.D.		Sch. 80 Steel .546" I.D.		Type K Copper .527" I.D. .576" O.D.		Type L Copper .545" I.D. .555" O.D.		Type M Copper .569" I.D. .577" O.D.		Copper or Brass Pipe .625" I.D. .7325" O.D.		Flow Gal. Per Min.
	Vel fps	Head Loss psi/100'	Vel fps	Head Loss psi/100'	Vel fps	Head Loss psi/100'	Vel fps	Head Loss psi/100'	Vel fps	Head Loss psi/100'	Vel fps	Head Loss psi/100'	
.5	.528	.25	.69	.48	.74	.35	.69	.29	.63	.24	.52	.15	.5
1.0	1.06	.91	1.37	1.72	1.47	1.26	1.38	1.07	1.26	.87	1.04	.55	1.0
1.5	1.58	1.92	2.06	3.63	2.20	2.68	2.06	2.28	1.90	1.85	1.57	1.16	1.5
2.0	2.11	3.28	2.74	6.19	2.94	4.55	2.75	3.85	2.53	3.13	2.09	1.98	2.0
2.5	2.64	4.94	3.43	9.35	3.67	6.84	3.44	5.80	3.16	4.72	2.61	2.98	2.5
3.0	3.17	6.93	4.11	13.1	4.40	9.61	4.12	8.14	3.79	6.63	3.13	4.18	3.0
3.5	3.70	9.22	4.80	17.4	5.14	12.77	4.81	10.87	4.42	8.83	3.66	5.59	3.5
4.0	4.23	11.82	5.48	22.3	5.87	16.3	5.50	13.9	5.05	11.3	4.18	7.10	4.0
4.5	4.75	14.68	6.17	27.7	6.61	20.3	6.19	17.2	5.68	13.9	4.70	8.8	4.5
5	5.28	17.84	6.86	33.6	7.35	24.6	6.87	20.9	6.31	17.0	5.22	10.7	5
6	6.34	25.00	8.23	47.2	8.81	34.5	8.25	29.3	7.59	23.8	6.26	15.1	6
7	7.39	32.30	9.60	62.8	10.3	45.9	9.62	38.9	8.84	31.7	7.31	20.0	7
8	8.45	42.60	10.3	71.5	-	-	11.00	50.2	10.10	40.6	8.35	25.7	8
9	9.51	52.8	-	-	-	-	-	-	-	-	9.40	31.8	9
10	10.60	64.5	-	-	-	-	-	-	-	-	10.40	38.7	10

Table 14-2 ¾ inch

Flow Gal. Per Min.	Sch. 40 Steel .824" I.D.		Sch. 80 Steel .742" I.D.		Type K Copper .745" I.D. .810" O.D.		Type L Copper .785" I.D. .830" O.D.		Type M Copper .811" I.D. .843" O.D.		Copper or Brass Pipe .822" I.D .936" O.D.		Flow Gal. Per Min.
	Vel fps	Head Loss psi/100'	Vel fps	Head Loss psi/100'	Vel fps	Head Loss psi/100'	Vel fps	Head Loss psi/100'	Vel fps	Head Loss psi/100'	Vel fps	Head Loss psi/100'	
3	1.81	1.77	2.23	2.94	2.21	1.77	1.99	1.38	1.86	1.17	1.81	1.11	3
4	2.41	3.00	2.97	5.02	2.94	3.01	2.65	2.34	2.48	1.99	2.42	1.89	4
5	3.01	4.55	3.71	7.58	3.67	4.55	3.31	3.53	3.10	3.01	3.02	2.85	5
6	3.61	6.37	4.45	10.60	4.41	6.37	3.98	4.98	3.72	4.22	3.62	4.00	6
8	4.82	10.8	5.94	18.1	5.88	10.9	5.30	8.4	4.96	7.2	4.83	6.8	8
10	6.02	16.4	7.42	27.3	7.35	16.4	6.62	12.7	6.20	10.9	6.04	10.3	10
11	6.62	19.5	8.17	32.5	8.09	19.5	7.29	15.2	6.82	12.9	6.64	12.2	11
12	7.22	22.9	8.91	38.2	8.83	23.0	7.95	17.8	7.44	15.2	7.25	14.4	12
13	7.82	26.6	9.63	44.2	9.56	26.6	8.61	20.7	8.06	17.6	7.85	16.7	13
14	8.43	30.5	10.40	50.7	10.3	30.6	9.27	23.8	8.68	20.2	8.45	19.1	14
15	9.03	34.8	-	-	-	-	9.94	27.0	9.30	23.0	9.05	21.7	15
16	9.63	39.1	-	-	-	-	10.60	30.4	9.92	25.9	9.65	24.5	16
17	-	-	-	-	-	-	-	-	10.55	29.0	10.25	27.5	17

Chapter 14—Water System Sizing

Table 14-3　1 inch

Flow Gal. Per Min.	Sch. 40 Steel 1.049" I.D.		Sch. 80 Steel .957" I.D.		Type K Copper .995" I.D. 1.060" O.D.		Type L Copper 1.025" I.D. 1.075" O.D.		Type M Copper 1.055" I.D. 1.090" O.D.		Copper or Brass Pipe 1.062" I.D 1.1885" O.D.		Flow Gal. Per Min.
	Vel fps	Head Loss psi/100'	Vel fps	Head Loss psi/100'	Vel fps	Head Loss psi/100'	Vel fps	Head Loss psi/100'	Vel fps	Head Loss psi/100'	Vel fps	Head Loss psi/100'	
3	1.11	.55	1.34	.85	1.24	.44	1.17	.38	1.10	.33	1.08	.32	3
4	1.49	.93	1.79	1.42	1.65	.74	1.56	.64	1.47	.56	1.45	.54	4
5	1.86	1.40	2.23	2.20	2.06	1.12	1.95	.97	1.83	.84	1.81	.81	5
6	2.23	1.97	2.68	3.07	2.48	1.57	2.34	1.36	2.20	1.18	2.17	2.99	6
8	2.97	3.4	3.57	5.2	3.30	2.7	3.11	2.3	2.93	2.0	2.89	2.0	8
10	3.71	5.1	4.46	7.9	4.12	4.0	3.89	3.5	3.66	3.0	3.61	2.9	10
12	4.46	7.1	5.36	11.1	4.95	5.6	4.67	4.9	4.40	4.3	4.34	4.1	12
14	5.20	9.4	6.25	14.7	5.77	7.5	5.45	6.5	5.13	5.7	5.05	5.5	14
16	5.94	12.1	7.14	18.9	6.60	9.6	6.22	8.3	5.86	7.3	5.78	7.0	16
18	6.68	15.0	8.03	23.5	7.42	11.9	7.00	10.3	6.60	9.0	6.50	8.7	18
20	7.43	18.2	8.92	28.5	8.24	14.5	7.78	12.6	7.33	10.9	7.22	10.6	20
25	9.28	27.6	-	-	10.30	22.0	9.74	19.0	9.16	16.5	9.03	15.9	25
30	11.10	38.6	-	-	-	-	11.68	26.6	11.00	23.1	10.84	22.3	30

Table 14-4　1¼ inch

Flow Gal. Per Min.	Sch. 40 Steel 1.380" I.D.		Sch. 80 Steel 1.278" I.D.		Type K Copper 1.245" I.D. 1.310" O.D.		Type L Copper 1.265" I.D. 1.320" O.D.		Type M Copper 1.291" I.D. 1.333" O.D.		Copper or Brass Pipe 1.368" I.D 1.514" O.D.		Flow Gal. Per Min.
	Vel fps	Head Loss psi/100'	Vel fps	Head Loss psi/100'	Vel fps	Head Loss psi/100'	Vel fps	Head Loss psi/100'	Vel fps	Head Loss psi/100'	Vel fps	Head Loss psi/100'	
5	1.07	.37	1.25	.54	1.32	.37	1.28	.35	1.22	.31	1.09	.24	5
6	1.29	.52	1.50	.75	1.58	.52	1.53	.49	1.47	.44	1.31	.33	6
7	1.50	.69	1.75	1.00	1.84	.70	1.79	.65	1.71	.59	1.53	.44	7
8	1.72	.9	2.00	1.3	2.11	.9	2.04	.8	1.96	.8	1.75	.6	8
10	2.15	1.3	2.50	1.9	2.63	1.1	2.55	1.3	2.45	1.1	2.18	.9	10
12	2.57	1.9	3.00	2.7	3.16	1.9	3.06	1.8	2.93	1.6	2.62	1.2	12
15	3.22	2.6	3.75	4.1	3.95	2.9	3.83	2.7	3.66	2.4	3.27	1.8	15
20	4.29	4.8	5.00	7.0	5.26	4.9	5.10	4.5	4.89	4.1	4.36	3.1	20
25	5.36	7.3	6.25	10.8	6.58	7.4	6.38	6.8	6.11	6.2	5.46	4.6	25
30	6.43	10.2	7.50	14.8	7.90	10.3	7.65	9.6	7.33	8.6	6.55	6.5	30
35	7.51	13.5	8.75	19.7	9.21	13.7	8.94	12.7	8.55	11.5	7.65	8.7	35
40	8.58	17.3	10.00	25.2	10.5	17.5	10.20	16.3	9.77	14.7	8.74	11.1	40
45	9.16	21.7	-	-	-	-	-	-	11.00	18.4	9.83	13.8	45
50	10.70	26.2	-	-	-	-	-	-	-	-	10.90	16.8	50

Engineered Plumbing Design

Table 14-5 1½ inch

Flow Gal. Per Min.	Sch. 40 Steel 1.610" I.D.		Sch. 80 Steel 1.500" I.D.		Type K Copper 1.481" I.D. 1.553" O.D.		Type L Copper 1.505" I.D. 1.565" O.D.		Type M Copper 1.527" I.D. 1.576" O.D.		Copper or Brass Pipe 1.600" I.D. 1.615" O.D.		Flow Gal. Per Min.
	Vel fps	Head Loss psi/100'	Vel fps	Head Loss psi/100'	Vel fps	Head Loss psi/100'	Vel fps	Head Loss psi/100'	Vel fps	Head Loss psi/100'	Vel fps	Head Loss psi/100'	
10	1.58	.6	1.82	.9	1.86	.6	1.80	.5	1.75	.5	1.59	.4	10
12	1.89	.9	2.18	1.2	2.23	.8	2.16	.8	2.10	.7	1.91	.6	12
15	2.36	1.3	2.72	1.9	2.79	1.2	2.70	1.1	2.63	1.1	2.39	.8	15
20	3.15	2.3	3.63	3.2	3.72	2.1	3.60	1.9	3.50	1.8	3.19	1.4	20
25	3.94	3.4	4.54	5.0	4.65	3.2	4.51	2.9	4.38	2.7	3.98	2.2	25
30	4.73	4.8	5.45	6.8	5.58	4.4	5.41	4.1	5.25	3.8	4.78	3.0	30
35	5.51	5.4	6.36	9.0	6.51	5.9	6.31	5.5	6.13	5.1	5.58	4.1	35
40	6.30	8.2	7.26	11.6	7.44	7.5	7.21	7.0	7.00	6.5	6.37	5.2	40
45	7.09	10.2	8.17	14.4	8.37	9.4	8.11	8.7	7.88	8.1	7.16	6.5	45
50	7.88	12.4	9.08	17.5	9.30	11.4	9.01	10.5	8.76	9.8	7.96	7.8	50
60	9.46	17.3	10.9	24.4	11.2	15.9	10.8	14.8	10.5	13.8	9.56	11.0	60
70	11.00	23.0	-	-	-	-	-	-	-	-	11.20	14.6	70

Table 14-6 2 inch

Flow Gal. Per Min.	Sch. 40 Steel 2.067" I.D.		Sch. 80 Steel 1.939" I.D.		Type K Copper 1.959" I.D. 2.042" O.D.		Type L Copper 1.985" I.D. 2.055" O.D.		Type M Copper 2.009" I.D. 2.067" O.D.		Copper or Brass Pipe 2.062" I.D. 2.2185" O.D.		Flow Gal. Per Min.
	Vel fps	Head Loss psi/100'	Vel fps	Head Loss psi/100'	Vel fps	Head Loss psi/100'	Vel fps	Head Loss psi/100'	Vel fps	Head Loss psi/100'	Vel fps	Head Loss psi/100'	
10	.96	.2	1.09	.3	1.07	.2	1.04	.1	1.01	.1	.96	.1	10
12	1.15	.3	1.30	.4	1.28	.2	1.24	.2	1.21	.2	1.15	.2	12
14	1.34	.4	1.52	.5	1.49	.3	1.45	.3	1.42	.3	1.34	.2	14
16	1.53	.5	1.74	.6	1.70	.4	1.66	.3	1.62	.3	1.53	.3	16
18	1.72	.6	1.96	.8	1.92	.4	1.87	.4	1.82	.4	1.72	.3	1S
20	1.91	.7	2.17	.9	2.13	.5	2.07	.5	2.02	.5	1.92	.4	20
25	2.39	1.0	2.72	1.4	2.66	.8	2.59	.8	2.53	.7	2.39	.6	25
30	2.87	1.4	3.26	1.9	3.19	1.1	3.11	1.1	3.03	1.0	2.87	.9	30
35	3.35	1.9	3.80	2.6	3.73	1.5	3.62	1.4	3.54	1.3	3.35	1.2	35
40	3.82	2.4	4.35	3.3	4.26	1.9	4.14	1.8	4.05	1.7	3.83	1.5	40
45	4.30	3.0	4.89	4.1	4.79	2.4	4.66	2.3	4.55	2.1	4.30	1.9	45
50	4.78	3.7	5.43	5.0	5.32	2.9	5.17	2.7	5.05	2.6	4.80	2.3	50
60	5.74	5.2	6.52	7.0	6.39	4.1	6.21	3.8	6.08	3.6	5.75	3.2	60
70	6.69	6.8	7.61	9.3	7.45	5.5	7.25	5.1	7.07	4.9	6.70	4.2	70
80	7.65	8.8	8.69	12.0	8.52	7.0	8.28	6.5	8.09	6.2	7.65	5.4	80
90	8.61	10.9	9.78	14.9	9.58	8.7	9.31	8.1	9.10	7.7	8.61	6.8	90
100	9.56	13.2	10.90	18.0	10.65	10.6	10.40	9.8	10.10	9.4	9.57	8.2	100
110	10.50	15.8	-	-	-	-	-	-	-	-	10.50	9.7	110

Chapter 14—Water System Sizing

Table 14-7 2½ inch

Flow Gal. Per Min.	Sch. 40 Steel 2.469" I.D.		Sch. 80 Steel 2.323" I.D.		Type K Copper 2.435" I.D. 2.530" O.D.		Type L Copper 2.465" I.D. 2.545" O.D.		Type M Copper 2.495" I.D. 2.560" O.D.		Copper or Brass Pipe 2.500" I.D. 2.6875" O.D.		Flow Gal. Per Min.
	Vel fps	Head Loss psi/100'	Vel fps	Head Loss psi/100'	Vel fps	Head Loss psi/100'	Vel fps	Head Loss psi/100'	Vel fps	Head Loss psi/100'	Vel fps	Head Loss psi/100'	
20	1.34	.3	1.51	.4	1.38	.2	1.34	.2	1.31	.2	1.31	.2	20
25	1.68	.4	1.90	.6	1.72	.3	1.68	.3	1.64	.3	1.63	.3	25
30	2.01	.6	2.27	.8	2.07	.4	2.02	.4	1.97	.4	1.96	.3	30
35	2.35	.8	2.65	1.1	2.41	.5	2.35	.5	2.30	.5	2.29	.5	35
40	2.68	1.0	3.03	1.4	2.76	.7	2.69	.6	2.62	.6	2.61	.6	40
45	3.02	1.3	3.41	1.7	3.10	.8	3.02	.8	2.95	.7	2.94	.7	45
50	3.35	1.5	3.79	2.1	3.45	1.0	3.36	1.0	3.28	.9	3.26	.9	50
60	4.02	2.2	4.54	2.9	4.14	1.4	4.03	1.3	3.93	1.3	3.92	1.2	60
70	4.69	2.9	5.30	3.9	4.82	1.9	4.70	1.8	4.59	1.7	4.57	1.7	70
80	5.36	3.7	6.05	4.9	5.51	2.4	5.37	2.3	5.25	2.2	5.22	2.1	80
90	6.03	4.6	6.81	6.0	6.20	3.0	6.04	2.8	5.90	2.7	5.88	2.6	90
100	6.70	5.5	7.57	7.5	6.89	3.7	6.71	3.4	6.55	3.2	6.53	3.2	100
110	7.37	6.6	8.33	8.9	7.58	4.4	7.38	4.1	7.21	3.9	7.19	3.8	110
120	8.04	7.8	9.08	10.5	8.27	5.1	8.05	4.8	7.86	4.6	7.84	4.5	120
130	8.71	9.1	9.84	12.2	8.96	5.9	8.73	5.6	8.52	5.2	8.49	5.2	130
140	9.38	10.4	10.60	13.9	9.65	6.8	9.40	6.4	9.18	6.0	9.14	5.9	140
150	10.00	11.8	-	-	10.35	7.8	10.10	7.3	9.83	6.8	9.79	6.8	150
160	10.70	13.3	-	-	-	-	-	-	10.50	7.7	10.45	7.6	160

Table 14-8 3 inch

Flow Gal. Per Min.	Sch. 40 Steel 3.068" I.D.		Sch. 80 Steel 2.900" I.D.		Type K Copper 2.907" I.D. 3.016" O.D.		Type L Copper 2.945" I.D. 3.035" O.D.		Type M Copper 2.981" I.D. 3.053" O.D.		Copper or Brass Pipe 3.062" I.D. 3.281" O.D.		Flow Gal. Per Min.
	Vel fps	Head Loss psi/100'	Vel fps	Head Loss psi/100'	Vel fps	Head Loss psi/100'	Vel fps	Head Loss psi/100'	Vel fps	Head Loss psi/100'	Vel fps	Head Loss psi/100'	
20	.87	.1	1.04	.1	.96	.1	.94	.1	.92	.1	.87	.1	20
30	1.30	.2	1.56	.3	1.45	.2	1.41	.2	1.37	.2	1.30	.1	30
40	1.74	.4	2.08	.5	1.93	.3	1.88	.3	1.83	.3	1.74	.2	40
50	2.17	.5	2.60	.7	2.41	.4	2.35	.4	2.29	.4	2.17	.3	50
60	2.60	.8	3.12	1.0	2.89	.6	2.82	.6	2.75	.5	2.61	.5	60
70	3.04	1.0	3.64	1.3	3.38	.8	3.29	.8	3.20	.7	3.04	.6	70
80	3.47	1.3	4.16	1.7	3.86	1.0	3.76	1.0	3.66	.9	3.48	.8	80
90	3.91	1.6	4.68	2.1	4.34	1.3	4.23	1.2	4.12	1.1	3.91	1.0	90
100	4.34	1.9	5.21	2.5	4.82	1.6	4.70	1.5	4.59	1.4	4.35	1.2	100
110	4.77	2.3	5.73	3.0	5.30	1.9	5.17	1.7	5.05	1.6	4.79	1.4	110
120	521	2.7	6.25	3.6	5.79	2.2	5.64	2.0	5.50	1.9	5.21	1.7	120
130	5.64	3.1	6.77	4.1	6.27	2.5	6.11	2.4	5.95	2.2	5.65	2.0	130
140	6.08	3.6	7.29	4.7	6.75	2.9	6.58	2.7	6.41	2.6	6.09	2.2	140
150	6.51	4.1	7.81	5.4	7.24	3.3	7.05	3.1	6.87	2.9	6.52	2.6	150
160	6.94	4.6	8.33	6.1	7.72	3.7	7.52	3.5	7.34	3.3	6.95	2.9	160.
170	7.37	5.1	8.85	6.8	8.20	4.1	7.99	3.9	7.79	3.7	7.39	3.2	170
180	7.81	5.7	9.37	7.5	8.69	4.6	8.46	4.3	8.25	4.1	7.82	3.6	180
190	8.24	6.3	9.89	8.4	9.16	5.1	8.93	4.8	8.70	4.5	8.25	3.9	190
200	8.68	7.0	10.40	9.2	9.64	5.6	9.40	5.2	9.16	4.9	8.70	4.3	200
220	9.55	8.3	-	-	10.60	6.6	10.30	6.2	10.10	5.9	9.56	5.2	220
240	10.40	9.8	-	-	-	-	-	-	-	-	10.40	6.1	240

Engineered Plumbing Design

Table 14-9   4 inch

Flow Gal. Per Min.	Sch. 40 Steel 4.026" I.D.		Sch. 80 Steel 3.826" I.D.		Type K Copper 3.857" I.D. 3.991" O.D.		Type L Copper 3.905" I.D. 4.015" O.D.		Type M Copper 3.935" I.D. 4.030" O.D.		Copper or Brass Pipe 4.000" I.D. 4.250" O.D.		Flow Gal. Per Min.
	Vel fps	Head Loss psi/100'	Vel fps	Head Loss psi/100'	Vel fps	Head Loss psi/100'	Vel fps	Head Loss psi/100'	Vel fps	Head Loss psi/100'	Vel fps	Head Loss psi/100'	
100	2.52	.5	2.79	.7	2.74	.4	2.68	.4	2.64	.4	2.55	.3	100
110	2.77	.6	3.07	.8	3.02	.5	2.94	.4	2.90	.4	2.81	.4	110
120	3.02	.7	3.35	.9	3.29	.6	3.21	.5	3.16	.5	3.06	.5	120
130	3.28	.8	3.63	1.1	3.57	.6	3.48	.6	3.42	.6	3.31	.5	130
140	3.53	1.0	3.91	1.2	3.84	.7	3.74	.7	3.69	.7	3.57	.6	140
150	3.78	1.1	4.19	1.4	4.11	.8	4.01	.8	3.95	.8	3.83	.7	150
160	4.03	1.2	4.47	1.6	4.39	.9	4.28	.9	4.21	.8	4.08	.8	160
170	4.29	1.4	4.75	1.8	4.66	1.0	4.55	1.0	4.48	.9	4.33	.9	170
180	4.54	1.5	5.02	2.0	4.94	1.2	4.81	1.1	4.74	1.1	4.58	1.0	180
190	4.79	1.7	5.30	2.2	5.21	1.3	5.08	1.2	5.00	1.2	4.84	1.1	190
200	5.05	1.9	5.58	2.4	5.49	1.4	5.35	1.3	5.27	1.3	5.10	1.2	200
220	5.55	2.2	6.14	2.8	6.04	1.7	5.89	1.6	5.80	1.5	5.61	1.4	220
240	6.05	2.6	6.70	3.3	6.59	2.0	6.42	1.9	6.32	1.8	6.12	1.7	240
260	6.55	3.0	7.26	3.9	7.14	2.3	6.95	2.1	6.85	2.1	6.63	1.9	260
280	7.06	3.5	7.82	4.4	7.69	2.6	7.49	2.5	7.38	2.4	7.14	2.2	280
300	7.57	3.9	8.38	5.0	8.24	3.0	8.02	2.8	7.90	2.7	7.65	2.5	300
350	8.83	5.2	9.80	6.7	9.60	4.0	9.36	3.7	9.22	3.6	8.92	3.3	350
400	10.10	6.7	11.20	8.6	11.00	5.1	10.70	4.8	10.50	4.6	10.20	4.3	400

Table 14-10   5 inch

Flow Gal. Per Min.	Sch. 40 Steel 4.026" I.D.		Sch. 80 Steel 3.826" I.D.		Type K Copper 3.857" I.D. 3.991" O.D.		Type L Copper 3.905" I.D. 4.015" O.D.		Type M Copper 3.935" I.D. 4.030" O.D.		Copper or Brass Pipe 4.000" I.D. 4.250" O.D.		Flow Gal. Per Min.
	Vel fps	Head Loss psi/100'	Vel fps	Head Loss psi/100'	Vel fps	Head Loss psi/100'	Vel fps	Head Loss psi/100'	Vel fps	Head Loss psi/100'	Vel fps	Head Loss psi/100'	
150	2.40	.4	2.65	.5	2.64	.3	2.58	.3	2.53	.3	2.38	.2	150
160	2.56	.4	2.82	.5	2.82	.3	2.75	.3	2.70	.3	2.54	.3	160
170	2.72	.5	3.00	.5	3.00	.4	2.92	.3	2.87	.3	2.70	.3	170
180	2.88	.5	3.17	.6	3.17	.4	3.09	.4	3.04	.4	2.86	.3	180
190	3.04	.6	3.35	.7	3.35	.4	3.26	.4	3.21	.4	3.02	.3	190
200	3.20	.6	3.52	.8	3.53	.5	3.44	.5	3.38	.4	3.18	.4	200
220	3.52	.7	3.88	.9	3.88	.6	3.78	.6	3.72	.5	3.50	.5	220
240	3.85	.9	4.23	1.1	4.24	.7	4.12	.7	4.05	.6	3.81	.5	240
260	4.17	1.0	4.58	1.3	4.59	.8	4.46	.8	4.39	.7	4.14	.6	260
280	4.49	1.2	4.94	1.5	4.94	.9	4.81	.9	4.73	.8	4.45	.7	280
300	4.81	1.3	5.29	1.7	5.29	1.0	5.15	1.0	5.07	.9	4.76	.8	300
350	5.61	1.7	6.17	2.1	6.17	1.4	6.01	1.3	5.91	1.2	5.56	1.1	350
400	6.41	2.2	7.05	2.8	7.05	1.8	6.87	1.7	6.75	1.6	6.35	1.3	400
450	7.22	2.8	7.94	3.6	7.94	2.2	7.73	2.1	7.60	2.0	7.15	1.7	450
500	8.02	3.4	8.82	4.3	8.81	2.6	8.59	2.5	8.45	2.4	7.95	2.0	500
550	8.82	4.0	9.70	5.1	9.70	3.2	9.45	3.0	9.29	2.8	8.75	2.4	550
600	9.62	4.7	10.60	5.9	10.60	3.7	10.30	3.6	10.10	3.3	9.54	2.9	600
650	10.40	5.5	-	-	-	-	-	-	-	-	10.30	3.4	650

Table 14-11 6 inch

Flow Gal. Per Min.	Sch. 40 Steel 6.065" I.D. Vel fps	Sch. 40 Steel Head Loss psi/100'	Sch. 80 Steel 5.761" I.D. Vel fps	Sch. 80 Steel Head Loss psi/100'	Type K Copper 5.741" I.D. 5.933" O.D. Vel fps	Type K Copper Head Loss psi/100'	Type L Copper 5.845" I.D. 5.985" O.D. Vel fps	Type L Copper Head Loss psi/100'	Type M Copper 5.881" I.D. 6.003" O.D. Vel fps	Type M Copper Head Loss psi/100'	Copper or Brass Pipe 6.125" I.D. 6.375" O.D. Vel fps	Copper or Brass Pipe Head Loss psi/100'	Flow Gal. Per Min.
240	2.67	.4	2.96	.5	2.98	.3	2.87	.3	2.84	.3	2.61	.2	240
260	2.89	.4	3.20	.5	3.22	.3	3.11	.3	3.07	.3	2.83	.2	260
280	3.11	.5	3.45	.6	3.48	.4	3.35	.4	3.31	.3	3.05	.3	280
300	3.33	.5	3.69	.7	3.72	.4	3.58	.4	3.54	.4	3.26	.3	300
350	3.89	.7	4.31	.9	4.35	.6	4.19	.5	4.14	.5	3.81	.4	350
400	4.44	.9	4.93	1.2	4.97	.7	4.79	.7	4.72	.7	4.35	.5	400
450	5.00	1.1	5.54	1.5	5.59	.9	5.38	.8	5.31	.8	4.90	.7	450
500	5.56	1.4	6.16	1.8	6.20	1.1	5.98	1.0	5.90	1.0	5.44	.8	500
550	6.11	1.7	6.77	2.1	6.82	1.3	6.57	1.2	6.50	1.2	5.98	1.0	550
600	6.66	1.9	7.39	2.5	7.45	1.6	7.17	1.4	7.10	1.4	6.53	1.1	800
650	7.22	2.2	8.00	2.9	8.07	1.8	7.76	1.7	7.68	1.6	7.07	1.3	650
700	7.78	2.6	8.63	3.3	8.69	2.1	8.36	1.9	8.27	1.8	7.61	1.5	700
750	8.34	2.9	9.24	3.8	9.31	2.4	8.96	2.2	8.86	2.1	8.15	1.7	750
800	8.90	3.3	8.95	4.2	9.93	2.6	9.56	2.4	9.45	2.4	8.70	1.9	800
850	9.45	3.7	10.50	4.7	10.60	3.0	10.20	2.7	10.00	2.6	9.25	2.1	850
900	10.00	4.1	-	-	-	-	-	-	-	-	9.79	2.4	900

Table 14-12 8 inch

Flow Gal. Per Min.	Sch. 40 Steel 7.981" I.D. Vel fps	Sch. 40 Steel Head Loss psi/100'	Sch. 80 Steel 7.625" I.D. Vel fps	Sch. 80 Steel Head Loss psi/100'	Type K Copper 7.583" I.D. 7.854" O.D. Vel fps	Type K Copper Head Loss psi/100'	Type L Copper 7.725" I.D. 7.925" O.D. Vel fps	Type L Copper Head Loss psi/100'	Type M Copper 7.785" I.D. 7.955" O.D. Vel fps	Type M Copper Head Loss psi/100'	Copper or Brass Pipe 8.000" I.D. 8.3125" O.D. Vel fps	Copper or Brass Pipe Head Loss psi/100'	Flow Gal. Per Min.
500	3.20	.4	3.51	.5	3.55	.3	3.42	.3	3.37	.3	3.19	.2	500
550	3.52	.4	3.86	.5	3.91	.3	3.76	.3	3.71	.3	3.51	.3	550
600	3.85	.5	4.22	.6	4.26	.4	4.10	.4	4.05	.4	3.83	.3	600
650	4.17	.6	4.57	.7	4.61	.5	4.44	.4	4.39	.4	4.15	.4	650
700	4.49	.7	4.92	.8	4.97	.5	4.78	.5	4.72	.5	4.46	.4	700
750	4.81	.8	5.27	1.0	5.32	.6	5.12	.6	5.06	.5	4.79	.5	750
800	5.13	.9	5.62	1.1	5.68	.7	5.46	.6	5.40	.6	5.10	.5	800
850	5.45	1.0	5.97	1.2	6.04	.8	5.80	.7	5.73	.7	5.42	.6	850
900	5.77	1.1	6.32	1.3	6.39	.8	6.15	.8	6.06	.8	5.74	.7	900
950	6.09	1.2	6.67	1.5	6.75	.9	6.49	.9	6.40	.8	6.05	.7	950
1000	6.41	1.3	7.03	1.6	7.10	1.0	6.84	.9	6.74	.9	6.38	.8	1000
1100	7.05	1.6	7.83	1.9	7.81	1.2	7.52	1.1	7.42	1.1	7.01	1.0	1100
1200	7.69	1.8	8.43	2.3	8.52	1.4	8.20	1.3	8.10	1.3	7.65	1.1	1200
1300	8.33	2.1	9.13	2.7	9.24	1.7	8.88	1.5	8.76	1.5	8.30	1.3	1300
1400	8.97	2.4	9.83	3.0	9.95	1.9	9.56	1.8	9.44	1.7	8.93	1.5	1400
1500	9.61	2.8	10.50	3.5	10.70	2.2	10.30	2.0	10.10	1.9	9.56	1.7	1500
1600	10.30	3.1	-	-	-	-	-	-	-	-	10.20	1.9	1600

and horsepower requirements must be higher. The designer is therefore faced with a comparison of initial savings in piping and insulation versus increased initial costs for pumps as well as increased operating costs. A design criterion of approximately 5 psi per 100 ft uniform friction head loss generally results in an economically designed system. There are, however, many specific installations where it is far more advantageous to design for much lower or higher pressure drops. Each system must be analyzed and the parameters set in accordance with that analysis.

The following example illustrates the procedure for sizing a system with booster pumps as the source of pressure:

Example 14-3

Assume there is a 16-story building where the street pressure is reported as 45 psi. The highest fixture outlet is 180 ft above the level of the pumps and the pumps are at the same level as the street main. The building has flush valve water closets and the pressure required at the highest fixture is 20 psi.. The total combined cold and hot water fixture count for the building is 4,840 FU. The length of run from the pumps to the furthest and highest fixture is 350 ft. Material for the water service into the building will be type K copper tubing and the water distribution system within the building will be type L copper tubing. Fixture unit load for each riser is shown in Tables 14-14 and 14-15 and Figure 14-1.

The procedure for calculation is as follows:
1. Static pressure for highest outlet = $180 \times .433 =$ 77.94 psi
2. Friction head loss (it is decided to size at a uniform friction head loss of 10 psi/100ft)
   = $1.5 \times 350 = 525$ ft
   ELR ($^{10}/_{100}$) $\times 525 =$ 52.50
3. Required outlet pressure =   20.00
4. Head loss through PRV at pump =   5.00
   Total required initial pressure =   155.44 psi
5. Total FU load = 4,840 FU. Interpolating in Table 13-4: 4,840 FU = 582 gpm. The pump capacity and head can now be selected.
6. Head on pump = 155.44 psi.
7. Suction pressure: Since street pressure is 45 psi, by allowing for lower pressure during heavy demand, we can use 35 psi. Then 155.44 - 35 = 120.44 psi head.
8. Select pump for 582 gpm at a head of 120 psi.

Table 14-14 Cold Water

Riser								
5:	1376	FU	=	259	gpm	=	4"	(6.95 fps, 2.1 psi/100')
3:	430	FU	=	131	gpm	=	2½"	(8.73 fps, 5.6 psi/100')
4:	688	FU	=	169	gpm	=	3"	(7.99 fps, 3.9 psi/100')
3+4:	1118	FU	=	223	gpm	=	4"	(5.89 fps, 1.6 psi/100')
3+4+5:	2494	FU	=	375	gpm	=	5"	(6.5 fps, 1.7 psi/100')
2:	860	FU	=	191	gpm	=	3"	(8.93 fps, 4.8 psi/100')
3+4+5+2:	3354	FU	=	465	gpm	=	5"	(8.0 fps, 2.2 psi/100')
1:	1376	FU	=	259	gpm	=	4"	(6.95 fps, 2.1 psi/100')
3+4+5+2+1:	4730	FU	=	575	gpm	=	6"	(6.87 fps, 1.3 psi/100')

Table 14-15 Hot Water

Riser								
	5:	96	FU	=	46	gpm	=	1½"	(8.1 fps, 8.7 psi/100')
	3:	30	FU	=	20	gpm	=	1¼"	(5.1 fps, 4.5 psi/100')
	4:	48	FU	=	28	gpm	=	1¼"	(7.65 fps, 9.6 psi/100')
	3+4:	78	FU	=	37	gpm	=	1½"	(6.7 fps, 6.2 psi/100')
	3+4+5:	174	FU	=	60	gpm	=	2"	(6.21 fps, 3.8 psi/100')
	2:	60	FU	=	32	gpm	=	1¼"	(8.2 fps, 11.1 psi/100')
	3+4+5+2:	234	FU	=	72	gpm	=	2"	(7.3 fps, 5.3 psi/100')
	1:	96	FU	=	46	gpm	=	1½"	(8.1 fps, 8.7 psi/100')
	3+4+5+2+1:	330	FU	=	91	gpm	=	2½"	(6.04 fps, 2.8 psi/100')

The water service must supply 582 gpm. From the friction head loss tables, a 5-in. type K copper tubing line will deliver 600 gpm at a velocity of 10.6 ft/sec and a 3.7 psi/100 ft pressure loss. The service is buried and thus the high velocity should not create a noise problem. From the pump discharge to the takeoff for the hot water heater, the piping will be sized for the combined total hot and cold water load, and from that point, the rest of the system will be sized for the cold water load and the hot water load, respectively.

Friction head loss tables for type L copper tubing will be utilized. The tables indicate that a 5-in. size for a flow of 582 gpm from the pumps would produce a velocity of 10 fps and a pressure loss of 3.6 psi/100 ft. This velocity would probably produce noise within the building and thus it would be advisable to select a 6-in. size that delivers 600 gpm at 7.17 ft/sec and 1.4 psi/100 ft pressure loss.

See Figure 14-1 for the schematic layout of the distribution mains in the basement. Starting at the most remote riser, work toward the source of supply. Use Table 13-4 for converting FU to gpm and the friction loss tables for selecting pipe sizes for the corresponding gpm flow.

Figure 14-1 Sizing of Distribution System

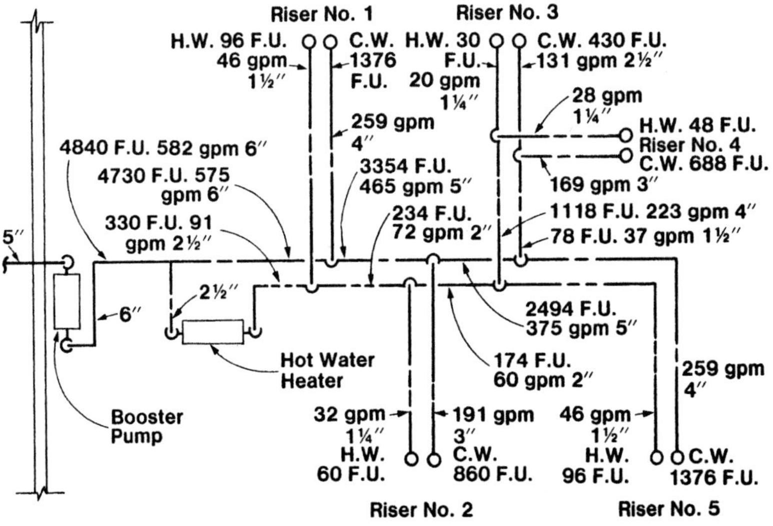

From the foregoing sizes required to maintain velocities at or below 8 fps, it can be seen that the corresponding pressure losses are significantly below the selected criterion of 10 psi per 100 ft. A much more realistic and economical criterion of 3 psi/100 ft would not result in any significant changes in pipe sizes. As a result, the head on the pump would be appreciably decreased. The total friction head loss would now be 3 psi/100 ft. × 525 = 15.75 psi and the pump head would now be decreased by 52.50 - 15.75 = 36.75 psi.

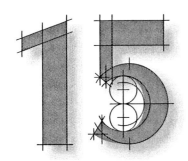

Water System Components

The physical and chemical characteristics of water to be used for human consumption are usually clearly defined by the public health department of the municipality or state and by most codes. Rules and regulations have been promulgated and enforced to ensure that only potable water is distributed in the public water mains. It is mandatory that building systems connect to available public mains as a positive measure for the protection of the public's health. Public water supply systems offer reliability, capacity, convenience, and, above all, high-quality potable water.

In areas where public water mains are not available, it is permissible to connect to an approved private source of potable water. When the source is a private utility, that utility is subject to the same criteria established for the public systems. Wells are another acceptable source of supply when public water is not available. Under no circumstances should water from any well be used until a recognized water-testing laboratory has analyzed the water and certified to be in conformance with the criteria established by the health authority having jurisdiction.

Potable water should be used for all drinking, food preparation, bottling, canning, washing of dishes, glasses, kitchen utensils, and similar purposes. When a nonpotable water supply is provided in the same building as a potable supply, extreme precautions must be taken. All nonpotable water faucets and outlets and all nonpotable piping must be adequately identified by conspicuous markings, using signs, colors, or other symbols required by the authority having jurisdiction.

Whenever there is nonpotable piping in a building there is always the real danger of a cross-connection being made to the potable piping, resulting in contamination and the likelihood of disease and death.

Protection of the Potable Water Supply

The following rules should be scrupulously followed in the design of all water supply distribution systems:

1. No materials or chemicals that can produce toxic effects should be used.
2. There should never be a cross-connection between a private and a public water supply system.
3. Water supply piping must never be directly connected to the drainage system.
4. Water supply piping must never be directly connected to embalming, mortuary, operating, or dissection tables.
5. There must be no direct connection for pump priming.

6. There should be no direct connections to sterilizers, aspirators, or similar equipment.
7. Water that is used for cooling, heating, or processing must not be reintroduced to the water supply system.
8. An air gap must be provided between the overflow level of the fixture and the water supply outlet.
9. Below-the-rim water supply connections must never be made except where the configuration of the fixture makes this impossible. The connection is permitted if special precautions are taken.

Protective Methods for Below-the-Rim Supply

Where below-the-rim connections must be made, vacuum breakers located at least 4 in. above the overflow level of the fixture have proven to give satisfactory results. When back-pressure exists in the supply connection, a check valve should be installed between the fixture and the vacuum breaker. The danger posed by submerged inlets is so great that presentation of the following rules covering specific conditions is warranted. The rules are not original with the author but have been excerpted from various codes.

Rules Relative to Submerged Inlets

1. **Vacuum Breakers.** All vacuum breakers shall be set at least 4 in. above the flood level rim of fixtures. Each fixture with submerged inlets shall be independently protected by an approved vacuum breaker or breakers the full size of supply pipe. All hose coupling outlets and serrated tip outlets for hose connections are deemed to be submerged inlets, and shall be independently protected with an approved vacuum breaker. Drain cocks for hot water and steam boilers, and hose connections on firefighting equipment are excepted. Every underground lawn and/or garden watering system shall be protected by an approved vacuum breaker placed at least 12 in. above the highest elevation of the sprinkling and/or spraying discharge point.
2. **Submerged Inlets.** Water supply inlets to fixtures shall be located at least 1 in. above the flood level rim of the fixture, except when submerged inlets are absolutely essential for the proper functioning of a fixture. Water supply lines to water preheating apparatus utilizing waste water shall be equipped with an approved vacuum breaker located at least 4 in. above the highest elevation of the preheating apparatus or coil with a check valve between the vacuum breaker and the preheating apparatus. Any hot water boiler supplied through such a preheating device and having an independent cold water supply line shall have the cold water supply line equipped with a vacuum breaker and check valve located at least 4 in. above the highest elevation of the boiler.
3. **Flush Valve Fixtures.** Flush valve controlled fixtures, with submerged inlets supplied directly or indirectly from the city water supply system, shall be equipped with an approved vacuum breaker in the supply line not less than 4 in. above the flood level rim of the fixture.
4. **Ballcocks.** All flush tanks operated by ballcocks shall have an approved vacuum breaker located not less than ¾ in. above the overflow outlet

of the flush tank. Ballcocks controlling water supply to suction, roof, or other intermediate tanks shall be located at least 1 ft above the flood level rim of the tank, except that in roof tanks equipped with an overflow of a size at least one commercial diameter larger than the water supply pipe the ballcock may be located 2 in. above the highest elevation of the overflow pipe. The overflow and emptying pipes of roof, suction, and intermediate tanks shall not be directly connected to a drainage system.

5. **Sterilizers.** When approved sterilizers are used, the waste piping from sterilizers shall not connect directly with any drainage system.
6. **Aspirators—Water Syphons.** The waste pipe for this type of apparatus shall not connect directly with any drainage system.
7. **Bedpan Washers.** Water supply connections to bedpan washers or similar apparatus shall be equipped with an approved vacuum breaker and check valve, the latter located between the fixture and the vacuum breaker. Water supply to bedpan washers equipped with an approved flush valve shall be governed by Rule 3.
8. **Sump or Well Pumps.** Direct water supply connections for priming purposes to sump, well, or similar type pumps, when permitted, shall be connected to the inlet side of the pump and equipped with an approved vacuum breaker and check valve. The check valve is to be located between the pump and the vacuum breaker.
9. **Condensers.** Direct water connection to a refrigeration unit for cooling purposes shall be equipped with an approved check valve to prevent possible backflow of ammonia or other refrigerant from defective condenser coils or jackets, except in such installations where the water supply piping is entirely outside of the piping or tank containing the refrigerant and two independent wall thicknesses of metal separate the refrigerant from the city water supply. Refrigeration units containing more than 20 lb of refrigerant shall be provided with an additional safeguard in the form of an approved relief valve installed at the outlet side of the check valve, such relief valve being set at 5 lb above the maximum water pressure at the point of installation.
10. **Mortuary, Dissection, Operating, and Embalming Tables or Equipment.** Mortuary, dissection, operating, and embalming tables or equipment shall have no direct water supply connection. Hose used with such equipment shall terminate at least 12 in. at any point from the table or attachments.
11. **Dishwashing and Laundry Machines.** Direct water supply to dishwashing and laundry machines shall be equipped with an approved vacuum breaker and a check valve located between the vacuum breaker and the fixture. The vacuum breaker shall be located at least 4 in. above the highest elevation of the machine.
12. **Chemical Solution Tanks or Apparatus.** Direct water supply connections to any tank or apparatus containing any chemical, when permitted, shall be equipped with an approved vacuum breaker and a check valve located between the vacuum breaker and the tank or apparatus. All hose,

Engineered Plumbing Design

piping, or portable equipment when used for the application of chemicals or other solutions shall be equipped with an approved vacuum breaker.

13. **Steam Tables.** Steam tables used exclusively for the warming of foodstuffs and provided with a water supply inlet elevation at least 1 in. above the overflow outlet, the area of which is at least four times that of the water supply piping, shall be equipped with an approved horizontal swing check valve in the water supply piping. Where elevation of the water supply piping inlet is less than 1 in. above the overflow outlet, or when the area of the overflow outlet is less than four times that of water supply piping, the water supply piping shall be equipped with an approved vacuum breaker located at least 4 in. above the maximum overflow level of the fixture.

Water Meters

A separate water service is generally provided for each building. When several buildings are constructed in one complex and are to remain permanently under a single ownership, it may be desirable to provide one service to supply water for all buildings. Provisions should be made so that each building can be isolated from the others by proper valving.

Water meters should be installed only when it is required that the amount of water supplied to a building be measured. Refer to Figure 15-1 for typical meter setting. There are three principal types of water meters used for measuring the quantity of flow of water used by residences, commercial buildings, industrial plants, restaurants, etc. They are

1. Displacement meters
2. Current meters
3. Proportional meters

A combination of the displacement and current meters in a common housing is called a *compound meter*.

Figure 15-1   Typical Meter Setting Including Bypass

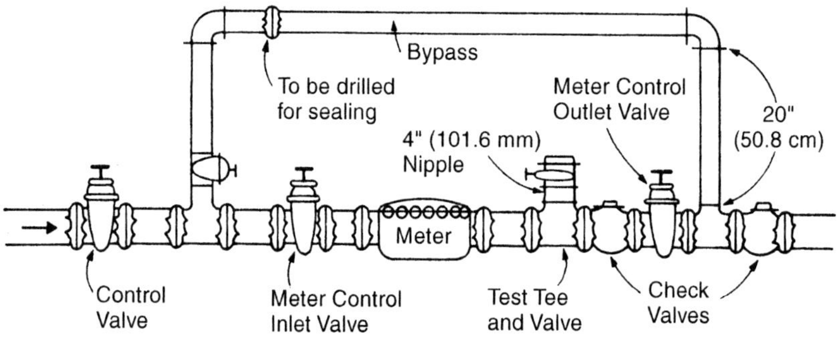

NOTE: Bypass to be of sufficient height so that top case and interior of meter can be removed for repairs and replacement, or install bypass in horizontal plane. Bypasses may not be allowed by local codes or ordinances, especially on 2" and smaller meters.

Figure 15-2 Typical Internal View of a Disc Meter

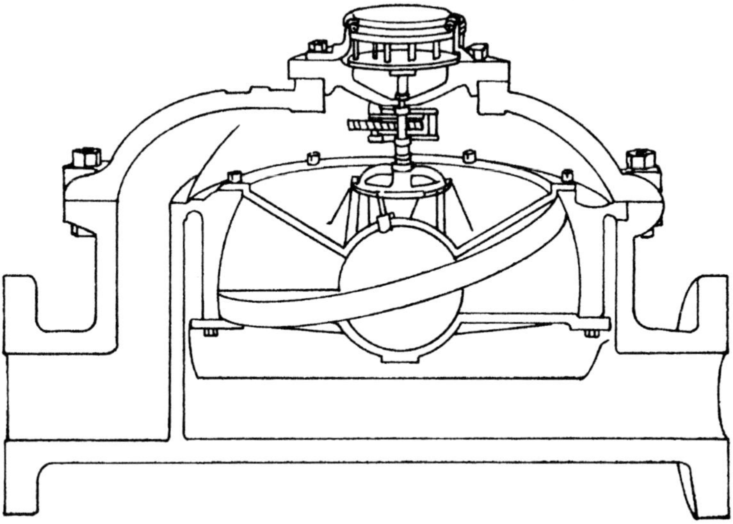

Displacement Meters

Displacement meters are positive measuring devices. A chamber fills and empties, recording the volume of water flowing through it. The actual displacement of the water is measured. The *disc meter* is the most common meter of this type. The disc meter operates by means of the nutating motion of a disc, which is closely fitted into a measuring chamber. (See Figure 15-2). The nutating motion of the disc causes the end of the spindle, attached to its center, to swing about a fixed point where it is in contact with a cam on a shaft, which is connected to a gear train. The shaft is rotated by the nutating motion of the disc and this rotation is translated into meter readings by the gear train. Registration can be calibrated to read gallons, cubic feet, or pounds of water. Meters are now available that have a magnetic coupling instead of the physical connection to the gear train.

The disc meter finds its greatest application in domestic service and medium industrial service. The distinctive characteristics of the meter are its large capacity, extreme sensitivity, and consistent accuracy. As the maximum limit of flow is approached, for each specific size of meter, the pressure loss through the meter becomes prohibitive. These meters are therefore not suitable for use where large flows with minimum pressure loss are required. The pressure loss in this meter, however, is low in comparison to the losses that occur through other displacement meters of the same capacity.

Table 15-1 Disc Meter

Meter Size, inches	Safe Maximum Operating Capacity, gpm	Maximum Pressure Loss, psi
⅝	20	15
¾	30	15
1	50	15
1½	100	20
2	160	20
3	300	20
4	500	20
6	1000	20

Engineered Plumbing Design

Other types of displacement meters are the *rotary piston* and the *plunger duplex piston*. An extremely high pressure drop occurs through the plunger duplex piston meter. This type should be used only where absolute accuracy is required, such as in checking feed water to boilers, or where high initial pressures are available.

Where the location of the meter exposes it to freezing temperatures, the disc meter can be furnished with a cast-iron bottom or cast-iron bolts. If the water freezes, the cast-iron bottom or bolts will break as pressure builds up within the meter and thus prevent damage to the more expensive parts. Disc meters are available in sizes ranging from 5/8 in. through 6 in. Table 15-1 tabulates the maximum operating capacity and maximum pressure drop for the various sizes.

The maximum capacities as given are the maximum rates of flow that can safely pass through the meter for *short periods of time*. For continuous flow, the safe maximum is 1/5 of the capacity shown in the table.

Current (or Velocity) Meters

Current meters are of the inferential type. The velocity of the water passing through is measured, and from this data the volume of water is inferred. The most common type is the *turbine meter*. The water in flowing through this meter passes through a turbine, which is rotated by the flow. The greater the velocity of flow, the faster the rotation of the turbine. The turbine is attached to a spindle, which rotates and transmits the motion to the register for recording.

The Current or velocity type meters find their greatest application where large volumes of water flowing at continuously high rates are required with a minimum pressure drop through the meter. Large industrial or process plants fall within this category. These meters are available in sizes ranging from 2 in. through 20 in.

Torrent and *venturi meters* are two other types classified as current meters.

Proportional Meters

Proportional meters derive their name from the fact that only a small portion of the water they record passes through the recording section of the meter. The greater portion of the water flows through a waterway containing a friction ring, which creates a drop in pressure on the downstream side, causing a portion of the water to bypass through the recording meter. This meter is calibrated to record the quantity of flow passing through both the waterway and the recording meter. A displacement meter is usually employed as the recording meter. Proportional meters, similar to current or velocity meters, find their application where large volumes of flow at continuously high rates are required with a minimum pressure drop through the meter. Such applications are for recording flows in street mains, pumping stations, or reservoirs.

Compound Meters

A compound meter is a combination of a displacement meter and a current meter in a single housing. The most common type is composed of a *disc meter* for measuring small flows, and a *turbine meter* for measuring large flows. A differential check valve is located on the outlet port of the turbine meter. When small flows occur, the flow is not enough to raise the check off its seat and the water is diverted to the disc meter. As flow increases, to the capacity of the disc meter, it is capable of opening the differential check. This action closes the port from the disc meter and all flow is then through the current meter.

Compound meters find their greatest application where large fluctuations in water flow do not permit the accurate measurement of the various flows within the limits of a single type of meter. In addition to its application to large facilities having fluctuating demands, it is ideal for use on fire protection systems. In the case of fire service, the compound meter prevents the unauthorized use of water for any purpose other than standpipe or sprinkler protection because the smallest water use will be detected by the disc meter. In use, small flow for the fire system is accurately measured as well as the flow when maximum demand occurs. Any meter used for fire service must be of the type approved by the National Fire Protection Association (NFPA) and contain the necessary integral check valves. These meters are available in sizes ranging from 3 in. through 12 in.

Water Meter Rules
1. A water meter should be restricted to a size and type that will ensure accurate registration on the basis of the water requirements of the premises to be metered.
2. A meter should never be larger than the pipe supplying the meter.
3. The meter should be set level and properly supported. The recording dials should not be more than 3 ft above the floor. (This is necessary for ease of reading.)
4. Disc meters should be set within 1 ft of the inlet meter control valve.
5. Current and compound meters should be set with a straight section of pipe, at least eight times the diameter of the size of the meter, installed between the inlet meter control valve and the meter connections. (If this length of pipe cannot be installed, the meter must be calibrated in place to guarantee accuracy.) Fish traps (strainers) should be installed immediately after the inlet meter control valve.
6. Meters should be set within 3 ft of the building wall at the point of entry of the service pipe. When conditions exist that prevent the setting of the meter at the point of entry, the meter may be set outside of the building in a proper waterproof and frost proof pit.

Piping Installation
Water piping should always be installed in alignment and parallel to the walls of the building. The piping should be arranged so that the entire system can be drained. There should be no sags where sediment could collect or high points where air pockets might be created.

Piping should be so routed that it does not pass over or within 2 ft of electrical switchgear, transformers, panel boards, control boards, motors, telephone equipment, etc. Where it is impossible to comply with the foregoing, provide a continuous pan, below the piping, that is adequately supported and braced, rimmed, pitched, and drained by a ¾-in. line piped to the nearest floor drain or slop sink.

Piping should be protected where there is danger of external corrosion (such as when buried in floor fill or concrete) by applying a heavy coating of black asphaltum paint and covered by 16-gauge black iron U-covers, mitered on the corners and fastened to the floor arch. The U-cover should be large enough to enclose the insulation on the piping.

Mains, risers, and branch connections to risers should be arranged to permit expansion and contraction without strain by means of elbow swings or expansion joints.

All horizontal and vertical piping shall be properly supported by means of hangers, anchors, and guides. Supports should be so arranged as to prevent excessive deflection and avoid excessive bending stresses between supports. Anchored points should be located and constructed to permit the piping to expand and contract freely in opposite directions away from the anchored points. Guide points should be located and constructed at each side of an expansion joint or loop so that only free axial movement occurs without lateral displacement.

The maximum distances between supports for piping of various materials are shown in Table 15-2.

All screwed joints should be made with the best quality approved pipe compound, carefully applied on the male threads only. If the compound is applied to the fitting threads, it will be forced into the piping and impart a distinctive taste to the water.

Table 15-2 Maximum Support Distance

Piping Material	Horizontal	Vertical
Screwed pipe	12 ft.	Alternate floors
Threadless copper & brass (TP)	12 ft	Alternate floors
Copper (types K & L)	10 ft. (2 in. and up)	Every floor
	6 ft (1½ in. and less)	Every floor

All cut and threaded pipe should have the cutting burrs and sharp edges reamed out so as not to impose additional frictional losses in the system. Burrs are also a source of noise propagation, as water vibrates them in passing.

All ferrous to nonferrous pipe connections should be made with dielectric isolating joints to prevent electrolytic action between dissimilar metals.

All copper tubing should be cut square and reamed to remove all burrs. Outsides and insides of the fittings and outsides of the tubing at each end must be well cleaned with steel wool before soldering to remove all traces of oxidation regardless of how clean the surfaces of the pipe and fittings may appear.

Provide unions at connections to each piece of equipment for easy dismantling and at other selected points to facilitate installation. All fittings, unions, and connections at pumps, tanks, or other major equipment 3 in. and over in size should be assembled with flanged joints and gaskets.

Underground piping entering or passing through rigid structures such as building walls, retaining walls, pit walls, etc., should be sleeved to provide not less than 1-in. clearance around the pipe. The opening between the pipe and the pipe sleeve should be tightly packed with oakum and caulked.

Valve Types

Water control valves within the building may be of the gate, globe, check, butterfly, or ball types (see Chapter 22). Standard or extra heavy weight valves should be selected on the basis of the system pressure at the location of the installation.

All valves 2½ in. or smaller should be bronze with solder or threaded ends to match the system in which they are installed. Valves 3 in. and larger should be of cast-iron body with bronze mountings and flanged ends as required by the system in which they are installed. Valves 3 in. and over located at pumps, tanks, and major equipment should be of the indicating type such as—outside screw and yoke type, flanged. Plug cocks or ball valves should be used for water balancing purposes in the hot water circulating piping system. Check valves should be of the horizontal swing type, except in the discharge piping of pumps, where they should be the center-guided, spring-loaded silent check type of the required pressure rating.

Water service lines should be equipped with a gate valve or ground key stopcock near the curb line between the property line and the curb line. A curb box frame and cover, including extension enclosure, or box of required depth should be provided to enclose and protect the water service valve operating mechanism. The type of valve, curb box, and location should always be coordinated with the local water company or municipal department. The curb valve and valve box should never be located under a driveway, where they may be subjected to heavy concentrated loads, which could result in damage. In addition to the curb valve, a valve should be installed in the line inside the building as close to the point of entry as possible.

A riser control valve should be provided for each riser. In addition, an all-brass drain valve should be installed on each riser and located upstream of the riser control valve to provide means for draining the riser. The drain valve should be at least ¾ in. in size. Drain valves should also be provided at all low points of the piping system.

Each zone or section of the distribution system should be provided with a shut-off valve, as well as each group of fixtures or every fixture.

Whether a valve should be installed in any particular location should be evaluated on the basis of ease of maintenance of the water system and the costs for maintenance if the valve is not installed in the particular location being considered. Being overly frugal in the installation of valves is often false economy.

All valves, check valves, pressure-reducing valves, shock absorbers, tempering valves, etc., should always be easily accessible for maintenance or removal. They should always be exposed where possible; if concealed, an access door of adequate size should be provided.

Strainers should always be provided in the inlet lines to all temperature-regulating, pressure-regulating, automatic modulating, or open-and-shut control valves. The strainers should generally be of the "Y" type, full pipe size, and fitted with a blow-off valve.

Hot Water System Design

Proper design of the domestic hot water supply system for any building is extremely important. Careful planning on the basis of all available data will ensure an adequate supply of water at the desired temperature to each fixture at all times. A properly designed system must of course conform with all the regulations of the authorities having jurisdiction. For a complete resource on hot water system design, refer to *Domestic Water Heating Design Manual* (American Society of Plumbing Engineers, Chicago, IL, 2003).

Objectives

The objectives for the design of an efficient hot water distribution system include
1. Providing adequate amounts of water at the prescribed temperature to all fixtures and equipment at all times
2. A system that will perform its function safely
3. The utilization of an economical heat source
4. A cost-effective and durable installation
5. An economical operating system with reasonable maintenance costs.

A brief discussion of each of these objectives is warranted. Any well-designed system should deliver the prescribed temperature at the outlet almost instantaneously to avoid the wasteful running of water until the desired temperature is achieved. The hot water should be available at any time of the day or night and during low-demand periods as well as during peak flows.

Safety must be built into any hot water system and the safety features must operate automatically. The two paramount dangers to be guarded against are excessive pressures and temperatures. Exploding hot water heaters and scalding water at fixtures must be avoided in the design stage.

An economic heat source is of prime importance in conserving energy. Various sources include coal, gas, oil, steam, condensate, and waste hot water. Availability and cost of any of these sources or combinations will dictate selection. If an especially economical source is not adequate to satisfy the total demand, then it can be used to preheat the cold water supply to the heater.

An economical and durable installation can be achieved by judicious selection of the proper materials and equipment. The piping layout will also have a marked effect on this criterion and will later determine ease of replacement and repair.

Following installation, cost-effective operation and maintenance also depend upon the proper preselection of materials and equipment. The choice of instantaneous, semi-instantaneous, or storage type heaters; the selection of insulation on heaters and piping; the location of piping (avoiding cold, unheated areas); the ease of circulation (by avoiding drops and rises in piping); bypasses around pumps and tanks; and adequate valving accessibility, etc., are all items that will affect the operation and maintenance of a system.

Safety Devices

There have been many reports of exploding or bursting tanks, damage to property, and scalding and injury of persons because of hazardous pressures and temperatures. Standard plumbing equipment, including hot water heaters and storage tanks, is designated for a working pressure of 125 psi. Any pressure in excess of this 125-psi limit can therefore be considered hazardous.

Water expands as it is heated. The increase in its volume for a 100° F temperature rise can be calculated to be 1.68% of its original volume. Since water is considered to be incompressible, the increased volume results in a buildup of pressure. It is absolutely necessary to provide a positive means of relieving this excess pressure. A pressure-relief valve or a combined pressure and temperature-relief valve satisfactorily meets this requirement.

A pressure-relief value, or the pressure-relief feature of a combined temperature and pressure-relief valve, must have adequate capacity to prevent excessive pressure in the system when the water-heating source is delivering the maximum rate of heat input.

Many contemporary valves of the pressure-only type and the pressure element of a combined temperature and pressure-relief (T&P) valve are designed just for thermal expansion relief. The combined valve of this type will show only the temperature rating. The thermal-relief capacity of the valve can generally be considered equal to or in excess of the temperature rating; in other words, it has the capacity of relieving thermal expansion in the system so long as the heat input does not exceed the rating of the valve.

The failure of hot water storage tanks and heaters was originally thought to be exclusively because of excessive pressures. The installation of a pressure-relief valve has therefore become a mandatory requirement of most codes. However, even where pressure-relief valves were installed, failures still occurred, which indicated that more extensive protection was required for the safe operation of hot water heating equipment. The opinion was advanced that overheating was the direct cause of explosions and protection against excessive temperature as well as pressure was needed. The most extensive studies were sponsored by the former American Gas Association (AGA) and the results published in their Research Reports 1151 A and 1151 B. The reports supported the premise that emergency protection against excessive temperature as well as pressure is essential. (The AGA is now part of the Canadian Standards Association [CSA].)

The temperature-relief valve, or the temperature-relief element of a combined valve, must have a relieving capacity not less than the heat input operating at maximum capacity. The name plate rating on the temperature-relief or combined T&P-relief valve shows the temperature rating in terms of the maximum heater input on which the valve can be used.

Separate temperature and pressure-relief valves may be used, but usually a combined T&P-relief valve is preferred because it offers a more economical and yet effective protective procedure.

A relief valve on a water supply system is exposed to many elements that can affect its performance, such as corrosive water that attacks materials and deposits of lime that close up waterways and flow passages. Products of corrosion and lime deposits can cause valves to become inoperative or reduce the valve capacity below that of the heater. For these reasons, the minimum size of a valve should be ¾ in. for inlet and outlet connections with the waterways within the valve of an area not less than the area of the inlet connection.

Relief valves must be installed so there is always free passage between the heater or tank and the relief valve. There should never be any valves or check valves installed between the relief valve and the equipment it is protecting.

A pressure-only relief valve can be installed in the cold supply to the tank, in the hot supply from tank, or directly in the heater. The temperature-relief valve or combined T-&P-relief valve must always be installed in a position in which the hottest water in the system can come in contact with the temperature actuator or valve thermostat. If this is not done, the valve becomes an ineffective protective device. Wherever possible, the valve should be installed in a tapping in the tank or heater.

Every valve should have a discharge pipe connected to its outlet and terminated at a point where the discharge will cause no damage to property or injury to persons. The discharge pipe size shall be at least the size of the valve discharge outlet, shall be as short as possible, and shall run down to its terminal without sags or traps.

The pressure setting of the valve should be 25 to 30 lb higher than the system operating pressure to avoid false opening and dripping due to normal pressure surges in the system. The pressure setting, however, must always be less than the maximum working pressure of the material in the system.

An aquastat in the tank or heater (which will shut off the heat source when excessive temperatures develop) is a safety precaution that is strongly recommended and often required by code; it is referred to a high-temperature energy cutoff.

Water Heaters

The brief outline of some of the more common sources of hot water supply that follows sets forth their main characteristics and limitations.

Directly Heated Automatic Storage Heaters

In this category are placed the simple gas, propane-fired, or electrically heated storage tank heaters used so universally in homes, apartments, smaller institutions, and establishments. These water heaters are generally low-demand heaters, with low Btu input so that the heating of the water is spread over several hours. This reduces the rate at which energy is consumed; therefore, the size of the gas or electric service for the heater is also reduced.

These automatic storage heaters are highly efficient if sized properly. They are simple, inexpensive in installation, piping, and controls, and exceptionally trouble-free. They serve well with all kinds of water—hard or soft, alkaline or acid. They can be drained and cleaned readily, especially if flush-out hand holes are provided. They can be procured with corrosion-resistant metal, glass-lined, epoxy-lined or cement-lined tanks.

Instantaneous Heaters

Instantaneous types of water heaters must have sufficient input capacity to meet all demands simultaneously. (See Figures 16-1 and 16—2.) These heaters have no built-in storage volume because they are designed to supply the full load instantly and continuously. The flow of the fuel, gas or liquid, is automatically controlled by the flow of water in an on-off or a modulated system. The water flow rate, temperature, and heat input are fixed at the factory. Heat input must be high in order to ensure the required flow of hot water. Because of these high flow rates and typical on-off operation, efficiencies are lower than they are with the storage types.

Figure 16-1   Conventional U-Tube Instantaneous Water Heater

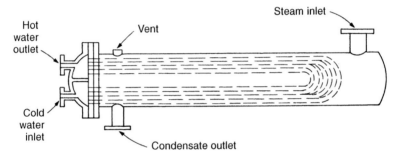

Figure 16-2   Straight Tube, Floating Head Instantaneous Water Heater

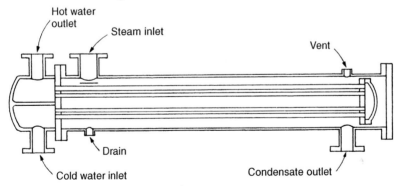

The instantaneous heater finds its best application where water heating demands are level and constant such as they are for swimming pools, certain dishwasher booster requirements, and industrial processes.

The most common form of the instantaneous heater is the U-tube, removable bundle exchanger with steam or boiler water in the shell and domestic water flowing through the tubes. Where the characteristics of the water are such as to cause rapid scale deposits, it is recommended that straight-tube floating head exchangers be used in lieu of the U-tube heater. The water is heated by conduction and the heat transfer rate increases as the water velocity increases through the tubes. A minimum velocity of 4 fps should be maintained through the tubes at full capacity conditions to minimize the rate of scale formation in the tubes.

The greatest problem with instantaneous heaters is adequate control of the temperature of the outlet water within acceptable limits during fluctuating demand. The best temperature-control system cannot respond quickly enough to maintain satisfactory constant outlet water temperature under rapidly changing flow conditions. (This undesirable fluctuation is called overshooting or undershooting.)

Booster Heaters

The term "booster" is applied to describe the function or purpose of the water heater. Heaters may be the ordinary, standard instantaneous types, but they serve to raise the temperature of the regular hot water supply to some higher temperature needed to perform special functions. The advantages in the use of booster heaters are
1. Only as much water need be heated to above normal system temperatures as is required for the specific job. The larger, normal uses throughout the building can be average hot water at lower temperatures.
2. Savings in investment, maintenance and operating costs are derived from the limited use of very hot water.
3. Small boosters can be located near their job, with simple control, minimum waste, and smooth operation.

Semi-Instantaneous Heaters

To overcome the shortcomings of the instantaneous heater, manufacturers have developed a limited storage-type heater known as "semi-instantaneous." This type of heater contains between 10 and 20 sec of domestic water storage according to its rated heating capacity. A 60-gpm rated heater will have approximately 20 gal of water in its shell. This small quantity of water is adequate to allow the temperature-control system to react to sudden fluctuations in water flow and to maintain the outlet water temperature within ± 5°F. The temperature-control system is almost always included with the heater as a package. Figure 16-3 shows a diagram of a shell and coil limited storage type instantaneous heater.

A hot water tempering valve should always be provided for mixing cold water with hot water from an instantaneous heater to prevent scalding water temperatures from entering the distribution system. The semi-instantaneous heater finds its application for apartments, offices, certain institutional structures, or any building in which the peak demand is spread over several hours and where the peak draws are not severe. The designer is referred to the catalogs of manufacturers for a complete and in-depth treatment of the construction and operation of the semi-instantaneous heater.

Storage Water Heaters

The primary reason for using the storage-type water heater instead of the instantaneous or semi-instantaneous one is to smooth out the peak demands on heating systems where there are large volume changes in the hot water demand, such as there are in gymnasiums (showers), laundries, kitchens, and industrial washrooms. When the correct storage capacity is combined with the correct recovery capacity and the proper size heating-medium control valve is selected, a substantial reduction in the peak heat fluid demand can be realized. This results in a smaller boiler installation and less heating-medium piping.

The following example illustrates the above point:

Example 16-1

A large high school gym class will draw 1000 gal of 140°F water in 10 min, none for the next 50 min, then another 1000 gal for 10 min. This is repeated throughout the school day. An ideal storage water selection for this application would have 1500 gallon storage and 1000 gallons per hour recovery.

The larger size storage is selected because only two-thirds of the water stored is usable due to the cooling down of the stored water by the entering cold water. If the heating medium is steam at 10 psi, the steam demand rate will not exceed 1000 lb/hr (30 boiler HP). Based on 10-psi steam at the control valve and 100 ft of steam pipe, the correct size of pipe would be 2½ in. and the control valve size 1½ in. If an instantaneous heater is used, it must heat 1000 gal of water in 10 min, or at a 100 gpm instantaneous rate. This equates to a steam demand rate of 5000 lb/hr (150 boiler HP). This is five times the peak steam demand required by the storage heater. The steam line size would be 6 in. and the control valve 4 in. compared to the 2½-in. line and 1½-in. valve.

Although the storage heater is more costly to install and requires more

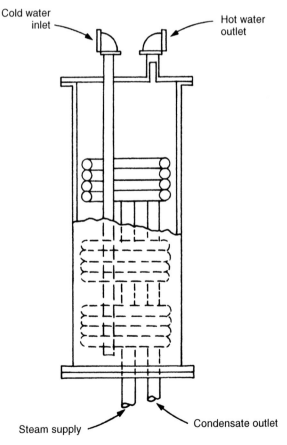

Figure 16-3 Shell and Coil Limited Storage Type Instantaneous Heater

space, the cost and space requirements for 30 boiler HP compared with 150 boiler HP—plus the smaller steam supply pipe size, control system, condensate return system, fittings, etc.—make the storage heater a far more economical selection for this particular application.

A conventional storage water heater generally consists of a removable U-tube, copper-tube bundle installed in the lower half of a horizontal or vertical cylindrical tank. The water to be heated enters the tank below the tube bundle and leaves at the top. The heating fluid—steam, high or medium-temperature water, or other heat transfer fluid—flows through the tubes. The water is heated almost completely by convection flow across the tubes. Temperature control is accomplished by a thermostatic bulb or similar device located to sense the temperature of the water and set to shut the control valve of the heating medium when the selected temperature is reached. (See Figure 16-4 for proper location.)

Selection of the instantaneous heater is a relatively simple and straightforward procedure. The most acceptable method is probably the one based upon the fixture unit method. This has already been covered in the design

Figure 16-4 Conventional Storage Water Heaters—Note Temperature Regulator Location

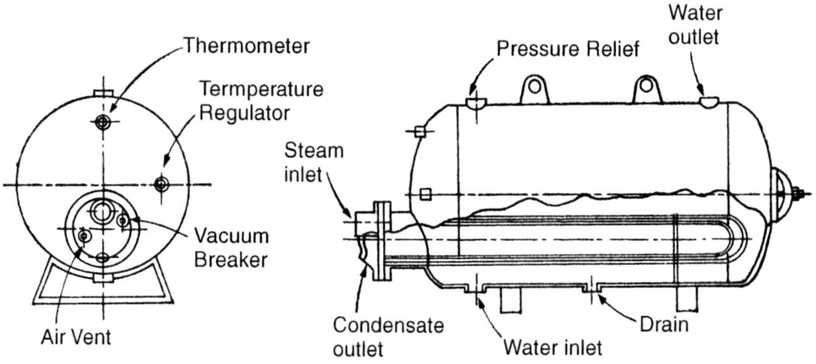

procedures for sizing domestic water systems. (See Chapter 13.) It has been found, however, by more than 15 years of field tests that peak demand for hot water determined by this method is two to three times the actual demand. It is therefore recommended that the following usage factors be applied to select the most economically sized instantaneous water heater. Use 0.5 for hospitals and hotels, 0.33 for residential, and 0.25 for office buildings. When the peak demand has been determined, a heater is selected that will deliver that demand.

Example 16-2

The fixture unit method indicates a peak demand for a hotel of 125 gpm. Assume the water is to be distributed at 120°F with inlet water at 40°F. Then the heater must deliver 125 × 0.5 or 62.5 gal of 120°F water per minute. The Btu per hour required are calculated by multiplying GPM × 500 × (120-40) × usage factor. Thus:

$$\text{Btu/hr} = 125 \times 500 \times 80 \times 0.5 = 2{,}500{,}000$$
$$\text{where } 500 = 60 \text{ min} \times 8.3 \text{ lb/gal}$$

Sizing Storage-Type Heaters

The selection of storage-type heaters is far more complex and, because of the many proposed methods, can be extremely confusing. It would be an exercise in futility to analyze all the proposed methods. The plumbing designer is fortunate that the results of many years of research have been made available. The reports have been very well received by the engineering community, and the older empirical procedures are rapidly being discarded.

In 1969, the Edison Electric Institute published the results of a research project they sponsored to determine the hot water usage for apartments, office buildings, nursing homes, motels, food service establishments, schools, and dormitories. The American Society of Mechanical Engineers (ASME) collected similar data for cold water usage. The Edison report substantiated claims by knowledgeable design professionals that hot water usage figures were grossly inflated. As a result, a minor revolution was created in the design of hot water systems.

Much of the following information consists of recently developed sizing recommendations for hot water storage tank systems. The water usage data and sizing curves are based on the Edison Electric Institute–sponsored research project.

Table 16-1 Hot Water Demands and Use for Various Types of Buildings

Type of Building	Maximum Hour	Maximum Day	Average Day
Men's Dormitories	3.8 gal/student	22.0 gal/student	13.1 gal/student
Women's Dormitories	5.0 gal/student	26.5 gal/student	12.3 gal/student
Motels: No. of units[a]			
20 or less	6.0 gal/unit	35.0 gal/unit	20.0 gal/unit
60	5.0 gal/unit	25.0 gal/unit	14.0 gal/unit
100 or more	4.0 gal/unit	15.0 gal/unit	10.0 gal/unit
Nursing homes	4.5 gal/bed	30.0 gal/bed	18.4 gal/bed
Office buildings	0.4 gal/person	2.0 gal/person	1.0 gal/person
Food service establishments:			
Type A—Full-meal restaurants and cafeterias	1.5 gal/max meals/hr	11.0 gal/max meals/hr	2.4 gal/avg meals/day[b]
Type B—Drive-ins, grilles, luncheonettes, sandwich and snack shops	0.7 gal/max meals/hr	6.0 gal/max meals/hr	0.7 gal/avg meals/day[b]
Apartment houses: No. of apartments			
20 or less	12.0 gal/apt.	80.0 gal/apt.	42.0 gal/apt.
50	10.0 gal/apt.	73.0 gal/apt.	40.0 gal/apt.
75	8.5 gal/apt.	66.0 gal/apt.	38.0 gal/apt.
100	7.0 gal/apt.	60.0 gal/apt.	37.0 gal/apt.
200 or more	5.0 gal/apt.	50.0 gal/apt.	35.0 gal/apt.
Elementary schools	0.6 gal/student	1.5 gal/student	0.6 gal/student[b]
Junior and senior high schools	1.0 gal/student	3.6 gal/student	1.8 gal/student[b]

a Interpolate for intermediate values.
b Per day of operation

Engineered Plumbing Design

Table 16-2 Estimated Hot Water Demand Characteristics for Various Types of Buildings

Type of Building	Hot Water Required per Person	Max. Hourly Demand in Relation to Day's Use	Duration of Peak Load Hours	Storage Capacity in Relation to Day's Use	Heating Capacity in Relation to Day's Use
Residences, apartments, hotels, etc.[a,b]	20–40 gal/day[c]	1/7	4	1/5	1/7
Office buildings	2–3 gal/day[c]	1/5	2	1/5	1/6
Factory buildings	5 gal/day[c]	1/3	1	2/5	1/8
Restaurants			(See text)		

a Daily hot water requirements and demand characteristics vary with the type of hotel. The better class hotel has a relatively high daily consumption with a low peak load. The commercial hotel has a lower daily consumption but a high peak load.
b The increasing use of dishwashers and laundry machines in residences and apartments requires additional allowances of 15 gal/dishwasher and 40 gal/laundry washer.
c At 140°F.

There are two basic determinations that must be made in order to design a hot water storage tank system: (1) the hourly peak demand and (2) the duration of that peak demand. Table 16-1 gives the necessary data to determine the hourly peak demand for various types of buildings as well as other pertinent information. The duration of the peak demand is given in Table 16-2 as well as an alternate method of determining the hourly peak demand.

The total capacity of a hot water storage tank is not *usable* capacity. As hot water is drawn from the tank, cold water enters and due to unavoidable mixture with the hot water in the tank, the temperature of the water leaving is lowered. When 60 to 80% of the total tank capacity has been withdrawn, the temperature of the water leaving is at an unacceptably low level. It is recommended that 70% of the total tank capacity be considered *usable*.

Figures 16-5 through 16-12 show the relationship between recovery and storage capacity for various types of buildings. The required recovery capacity is the difference between the peak hourly demand and the gph available from the tank. As an illustration: If we have a 1000-gal tank and a peak demand of 450 gph, which lasts for two hours, then:

$$1000 \times .70 = 700 \text{ gal usable hot water}$$
$$700/2 = 350 \text{ gph available}$$
$$450 - 350 = 100 \text{ gph recovery capacity required}$$

When the peak demand and the duration of that demand are known, it is possible to select any number of combinations of storage and recovery capacities to satisfy the requirements. Any combination of storage and recovery that falls on the curve for the particular type of building shown in Figures 16-5 through 16-12 will satisfy the requirements of the building. Selection of the minimum recovery capacity and the maximum storage capacity on the curves will yield the smallest hot water capacity capable of satisfying the building demand. Minimizing the recovery capacities will place less demand on the heat source.

The recovery capacities shown are for the actual hot water flow required and do not reflect the system heat losses. Heat losses from storage tanks and

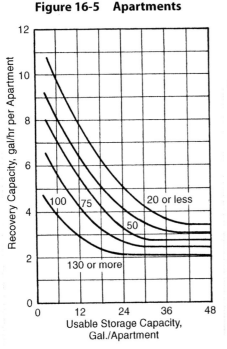

Figure 16-5　Apartments

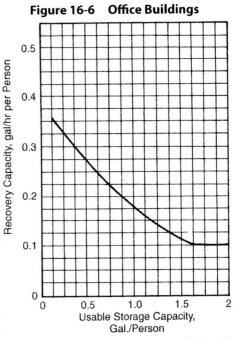

Figure 16-6　Office Buildings

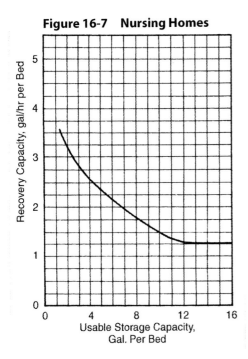

Figure 16-7　Nursing Homes

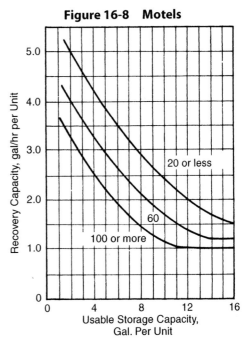

Figure 16-8　Motels

Engineered Plumbing Design

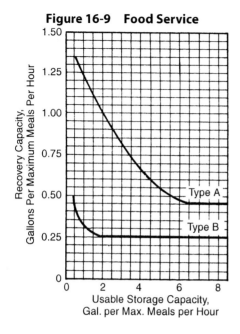

Figure 16-9 Food Service

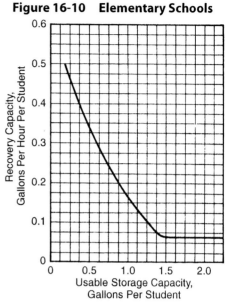

Figure 16-10 Elementary Schools

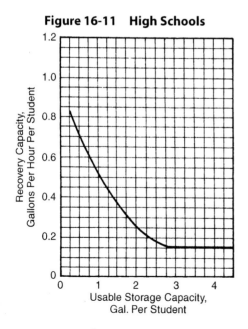

Figure 16-11 High Schools

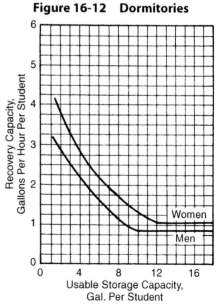

Figure 16-12 Dormitories

Chapter 16—Hot Water System Design

recirculating hot water piping must be calculated and added to the recovery capacities shown. The following examples, taken from the *ASHRAE Handbook*[1] and the American Society of Plumbing Engineers *Plumbing Engineering Design Handbook*, clearly illustrate the use of the tables and curves for selecting storage and recovery capacities.

Example 16-3

Determine the required water heater size for a 300-student women's dormitory using the following criteria:
1. Storage system with minimum recovery rate.
2. Storage system with recovery rate of 2.5 gph per student.
3. With the additional requirement for a cafeteria to serve a maximum of 300 meals per hour for minimum recovery rate combined with item 1, and for a recovery rate of 1.0 gpm per maximum meal per hour combined with item 2.

Solution

1. The minimum recovery rate from Figure 16-12 for women's dormitories is 1.1 gph per student, or a total of 330 gph recovery is required. The storage required is 12 gal per student, or 3600 gal storage. On a 70% net usable basis, the necessary tank size is 1.43 × 3600 = 5150 gal.
2. The same curve also shows 5 gal storage per student at 2.5 gph recovery, or 300 × 5 = 1500 gal storage with recovery of 300 × 2.5 = 750 gph The tank size will be 1.43 × 1500 = 2150 gal.
3. The additional requirement for a cafeteria can be determined from the recommendations for food service establishments (Table 16-1) with the storage and recovery capacity added to that for the dormitory.

For the case of minimum recovery rate, the cafeteria requires 300 × 0.45 = 135 gph recovery rate and 300 × 7 × 1.43 = 3000 gal of additional storage. The entire building then requires 330 + 135 = 465 gph recovery and 5150 + 3000 = 8150 gal of storage.

With 1 gal recovery per maximum meal per hour, the recovery required is 300 gph, with 300 × 2 × 1.43 = 860 gal of additional storage. Combining this with item 2, the entire building requires 750 + 300 = 1050 gph recovery and 2150 + 860 = 3010 gal of storage.

Note: Recovery capacities shown are for heating water only. To these capacities, additional capacity must be added to offset the system heat losses.

Example 16-4

Determine the water heater size and monthly hot water consumption for an office building to be occupied by 300 people:

1. Storage system with minimum recovery rate.
2. Storage system with 1.0 gal per person storage.
3. Additional minimum recovery rate requirement for a luncheonette, open five days a week, seeing a maximum of 100 meals in 1 hr, and an average of 200 meals per day.

Solution

---
[1] American Society of Heating, Refrigeration, and Air-Conditioning Engineers. 2003. *Applications Handbook.*

1. With minimum recovery rate of 0.10 gph per person from Figure 16-6 for office buildings, 30 gph recovery is required, while the storage is 1.6 gal per person, or 300 × 1.6 = 480 gal storage. The tank size will be 1.43 × 480 = 690 gal.
2. The curve also shows 1-gal storage per person at 0.175 gph per person recovery, or 300 × 1 = 300 gal storage with recovery of 300 × 0.175 = 52.5 gph. The tank size will be 1.43 × 300 = 430 gal.
3. The hot water requirements for a luncheonette are contained in the recommendations for food service establishments (Table 16-1).
   The recovery vs. storage curve shows that with minimum recovery capacity of 0.25 gph per maximum meals per hour, 100 meals would require 25 gph recovery, while the storage would be 2 gal per meal, or 100 × 25 × 1.43 = 386-gal storage. The combined requirement with item 1 would then be 55-gph recovery and 1076-gal storage.
   Combined with item 2, the requirement will be 77.5-gph recovery and 816-gal storage.
4. From Table 16-1 the office building will consume an average of 1 gal per person per day × 30 days × 300 people = 9000 gal per month, while the luncheonette will consume 0.7 gal per meal × 200 meals per day × 22 days per month = 3080 gal per month, for a total of 12,080 gal per month.

Note: Recovery capacities shown are for heating water only. To these capacities, additional capacity must be added to offset the system heat losses.

Example 16-5

Determine the water heater size for a 200-unit apartment house:
1. Storage system with minimum recovery rate, with a single tank.
2. Storage system with 4 gph per apartment recovery rate, with a single tank.
3. Storage system for each of two 100-unit wings.
   A. Minimum recovery rate.
   B. Recovery rate of 4 gph per apartment.

Solution

1. The minimum recovery rate, from Figure 16-5, for apartment buildings with 200 apartments is 2.1 gph per apartment, or a total of 420 gph recovery is required. The storage required is 24 gal per apartment, or 4800 gal. Based on 70% usable capacity in the storage tank, the necessary tank size is 1.43 × 4800 = 6860 gal.
2. The same curve also shows 5 gal storage per apartment at 4 gph recovery, or 200 × 4 = 800 gph. The tank size will be 1.43 × 1000 = 1430 gal.
3. Alternate solution for a 200-unit apartment house with two wings, each with its own hot water system. In this instance, the solution for each 100-unit wing would be:
   A. With minimum recovery rate of 2.5 gph per apartment, from the curve, a 250-gph recovery is required, while the necessary storage is

Chapter 16—Hot Water System Design

28 gal per apartment, or $100 \times 28 = 2800$ gal. The required tank size is $1.43 \times 2800 = 4000$ gal for each wing.

B. The curve also shows that for a recovery rate of 4 gph per apartment, the storage would be 14 gal or $100 \times 14 = 1400$ gal, with recovery of $100 \times 4 = 400$ gph. The necessary tank size is $1.43 \times 1400 = 2000$ gal in each wing.

Note: Recovery capacities shown are for heating water only. To these capacities, additional capacity must be added to offset the system heat losses.

Example 16-6

Determine the water heater size and monthly hot water consumption for a 2000-student high school.
1. Storage system with minimum recovery rate.
2. Storage system with 4000-gal maximum storage capacity.

Solution
1. With the minimum recovery rate of 0.15 gph per student from Figure 16-11 for high schools, 300-gph recovery is required. The storage required is 3 gal per student, or $2000 \times 3 = 6000$-gal storage. The tank size is $1.43 \times 6000 = 8600$ gal.
2. The net storage capacity will be $0.7 \times 4000 = 2800$ gal or 1.4 gal per student. From the curve 0.37 gph per student recovery capacity is required, or $0.37 \times 2000 = 740$ gph.
3. From Table 16-1 the monthly hot water consumption will be 2000 students $\times$ 1.8 gal per student per day $\times$ 22 days = 79,200 gal.

Note: Recovery capacities shown are for heating water only. To these capacities, additional capacity must be added to offset the system heat losses.

Example 16-7

Determine the required heater capacity for an apartment building housing 200 people, if the storage tank has a capacity of 1000 gal. What heater capacity will be required if the storage tank is changed to 2500-gal capacity?

Solution

Assume an apartment building housing 200 people.

From the data in Table 16-2: Daily requirements = $200 \times 40 = 8000$ gal. Maximum hours demand = $8000 \times \frac{1}{7} = 1140$ gal. Duration of peak load = 4 hr. Water required for 4-hr peak = $4 \times 1140 = 4560$.

If a 1000-gal storage tank is used, hot water available from the tank = $1000 \times 0.70 = 700$. Water to be heated in 4 hr = $4560 - 700 = 3860$ gal. Heating capacity per hour = $3860/4 = 965$ gal.

If instead of a 1000-gal tank, a 2500-gal tank had been installed, the required heating capacity per hour would be $[4560 - (2500 \times 0.70)]/4 = 702$ gal.

Table 16-3 may be used to determine the size of water heating equipment from the number of fixtures. To obtain the *probable maximum demand*, multiply the total quantity for the fixtures by the *demand factor* in line 19. The heater or coil should have a water heating capacity equal to this *probable maximum demand*. The storage tank should have a capacity equal to the *probable maximum demand* multiplied by the storage capacity factor in line 20. Example 16-8 illustrates the procedure.

Example 16-8

Determine the heater and storage tank size for an apartment building from number of fixtures.

60 lavatories	×2	= 120 gph
30 bathtubs	×20	= 600 gph
30 showers	×30	= 900 gph
60 kitchen sinks	×10	= 600 gph
15 laundry tubs	×20	= 300 gph
Possible maximum demand		= 2520 gph
Probable maximum demand		= 2520 × 0.30 = 756 gph
Heater or coil capacity		= 756 gph
Storage tank capacity		= 756 × 1.25 = 945 gal

Showers. In many mass housing installations such as motels, hotels, military barracks, and dormitories, the peak hot water load usually results from the use of showers.

Hotels will nominally have a 3 to 4-hr peak shower load, and Table 16-3 indicates the probable hourly hot water demand and the recommended demand and storage capacity factors to be applied. In motels, similar volumes of hot water are required but the peak demand may last for only a 2-hr period.

In some types of housing, such as barracks, fraternities, or dormitories, all occupants may, on some occasions, take showers within a short period of time. In this case, it is best to determine the peak load by determining the number of shower heads and the rate of flow per head, and estimate the length of time that the showers will be on.

The rate of flow from a shower head will vary depending on the type, size, water pressure and flow control. At 40-psi water pressure and without flow control, available shower heads have nominal flow rates from about 2.5 to 10 gpm. In multiple shower installations flow-control valves are recommended since they reduce the flow rate and maintain it regardless of fluctuations in water pressure. Regulations now require that water flow rates for showers and other fixtures be limited; see page 132. Manufacturers are required to provide shower heads with maximum of 2.5 gpm.

Table 16-4 shows the *number of gallons of hot water that, when mixed with 1 gal of cold water, produces a given mixture temperature.* For example, if hot water at 200°F and cold water at 40°F are to be mixed to a temperature of 170°F, the table shows that 4.3 gal of hot water are required for every gallon of cold water. It is a solution of the equation:

Chapter 16—Hot Water System Design

Table 16-3 Hot Water Demand per Fixture for Various Types of Buildings
(Gallons of water per hour per fixture, calculated at a final temperature of 140°F)

	Apartment House	Club	Gymnasium	Hospital	Hotel	Industrial Plant	Office Building	Private Residence	School	Y.M.C.A.
1. Basins, private lavatory	2	2	2	2	2	2	2	2	2	2
2. Basins, public lavatory	4	6	8	6	8	12	6	-	15	8
3. Bathtubs	20	20	30	20	20	-	-	20	-	30
4. Dishwashers	15	50-150	-	50-150	50-200	20-100	-	15	20-100	20-100
5. Foot basins	3	3	12	3	3	12	-	3	3	12
6. Kitchen sink	10	20	-	20	30	20	20	10	20	20
7. Laundry, stationary tubs	20	28	-	28	28	-	-	20	-	28
8. Pantry sink	5	10	-	10	10	-	10	5	10	10
9. Showers	30	150	225	75	75	225	30	30	225	225
10. Slop sink	20	20	-	20	30	20	20	15	20	20
11. Hydro-therapeutic showers				400						
12. Hubbard baths				600						
13. Lap baths				100						
14. Arm baths				35						
15. Sitz baths				30						
16. Continuous-flow baths				165						
17. Circular wash sinks				20	20	30	20		30	
18. Semi-circular wash sinks				10	10	15	10		15	
19. Demand factor	0.30	0.30	0.40	0.25	0.25	0.40	0.30	0.30	0.40	0.40
20. Storage capacity factor*	1.25	0.90	1.00	0.60	0.80	1.00	2.00	0.70	1.00	1.00

*Ratio of storage tank capacity to probable maximum demand per hour. Storage capacity may be reduced where an unlimited supply of steam is available from a central street steam system or large boiler plant.

Table 16-4 Mixing of Hot and Cold Water (Ratio of Gallons)

Temperature of Mixture, °F	Temperature of Cold Water, °F															
	40	50	60	70	80	90	40	50	60	70	80	90	40	50	60	70	80	90
	210° Hot Water						200° Hot Water						190° Hot Water					
200	16.0	15.0	14.0	13.0	12.0	11.0	–	–	–	–	–	–	–	–	–	–	–	–
190	7.5	7.0	6.5	6.0	5.5	5.0	15.0	14.0	13.0	12.0	11.0	10.0	–	–	–	–	–	–
180	4.7	4.3	4.0	3.7	3.3	3.0	7.0	6.5	6.0	5.5	5.0	4.5	14.0	13.0	12.0	11.0	10.0	9.0
170	3.3	3.0	2.8	2.5	2.3	2.0	4.3	4.0	3.7	3.3	3.0	2.7	6.5	6.0	5.5	5.0	4.5	4.0
160	2.4	2.2	2.0	1.8	1.6	1.4	3.0	2.8	2.5	2.3	2.0	1.8	4.0	3.7	3.3	3.0	2.7	2.3
150	1.8	1.7	1.5	1.3	1.2	1.0	2.2	2.0	1.8	1.6	1.4	1.2	2.8	2.5	2.3	2.0	1.8	1.5
140	1.4	1.3	1.1	1.0	0.9	0.7	1.7	1.5	1.3	1.2	1.0	0.8	2.0	1.8	1.6	1.4	1.2	1.0
130	1.1	1.0	0.9	0.8	0.6	0.5	1.3	1.1	1.0	0.9	0.7	0.6	1.5	1.3	1.2	1.0	0.8	0.7
120	0.9	0.8	0.7	0.6	0.4	0.3	1.0	0.9	0.8	0.6	0.5	0.4	1.1	1.0	0.9	0.7	0.6	0.4
110	0.7	0.6	0.5	0.4	0.3	0.2	0.7	0.8	0.6	0.4	0.3	0.2	0.9	0.8	0.6	0.5	0.4	0.3
100	0.6	0.5	0.4	0.3	0.2	0.1	0.6	0.5	0.4	0.3	0.2	0.1	0.7	0.6	0.4	0.3	0.2	0.1
	180° Hot Water						170° Hot Water						160° Hot Water					
170	13.0	12.0	11.0	10.0	9.0	8.0	–	–	–	–	–	–	–	–	–	–	–	–
160	6.0	5.5	5.0	4.5	4.0	3.5	12.0	11.0	10.0	9.0	8.0	7.0	–	–	–	–	–	–
150	3.7	3.3	3.0	2.7	2.3	2.0	5.5	5.0	4.5	4.0	3.5	3.0	11.0	10.0	9.0	8.0	7.0	6.0
140	2.5	2.3	2.0	1.8	1.5	1.3	3.3	3.0	2.7	2.3	2.0	1.7	5.0	4.5	4.0	3.5	3.0	2.5
130	1.8	1.6	1.4	1.2	1.0	0.8	2.3	2.0	1.8	1.5	1.3	1.0	3.0	2.7	2.3	2.0	1.7	1.3
120	1.3	1.2	1.0	0.8	0.7	0.5	1.6	1.4	1.2	1.0	0.8	0.6	2.0	1.8	1.5	1.3	1.0	0.8
110	1.0	0.9	0.7	0.6	0.4	0.3	1.2	1.0	0.8	0.7	0.5	0.3	1.4	1.2	1.0	0.8	0.6	0.4
100	0.8	0.6	0.5	0.4	0.3	0.1	0.9	0.7	0.6	0.4	0.3	0.1	1.0	0.8	0.7	0.5	0.3	0.2
	150° Hot Water						140° Hot Water						130° Hot Water					
140	10.0	9.0	8.0	7.0	6.0	5.0	–	–	–	–	–	–	–	–	–	–	–	–
130	4.5	4.0	3.5	3.0	2.5	2.0	9.0	8.0	7.0	6.0	5.0	4.0	–	–	–	–	–	–
120	2.7	2.3	2.0	1.7	1.3	1.0	4.0	3.5	3.0	2.5	2.0	1.5	8.0	7.0	6.0	5.0	4.0	3.0
110	1.8	1.5	1.3	1.0	1.8	0.5	2.3	2.0	1.7	1.3	1.0	0.7	3.5	3.0	2.5	2.0	1.5	1.0
100	1.2	1.0	0.8	0.6	0.4	0.2	1.5	1.3	1.0	0.2	0.5	0.3	2.0	1.7	1.3	1.0	0.7	0.3

$$G_H = \frac{(t_M - t_C)}{(t_H - t_M)} \tag{16-1}$$

where G_H = gallons of hot water required per gallon of cold water
  t_M = temperature of the mixture, °F
  t_C = temperature of the cold water, °F
  t_H = temperature of the hot water, °F

Table 16-5 shows the *percentage of a mixture of hot and cold water that is hot water* for various mixture, hot, and cold water temperatures. For example, if hot water at 200°F and cold water at 40°F are to be mixed to a temperature of 170°F, the table shows that 81% of the water would have to be 200°F hot water. The percentages given are a solution of the formula:

$$P = \frac{(t_M - t_C)}{(t_H - t_C)} \times 100 \tag{16-2}$$

where P = the percentage of the mixture that is hot water
  t_M = temperature of the mixture, °F
  t_C = temperature of the cold water, °F
  t_H = temperature of the hot water, °F

Installation

The storage tank should be installed as high as possible above the boiler. The tank's bottom should be at or above the boiler water line. When this is not possible and boiler water is the heating medium, a circulation pump should be installed to assure positive circulation of the boiler water through

Chapter 16—Hot Water System Design

Table 16-5 Mixing of Hot and Cold Water (Percentage Basis)

Temperature of Mixture, °F	Temperature of Cold Water, °F														
	40	50	60	70	80	90	40	50	60	70	80	90	40	50	60	70	80	90
	210° Hot Water						200° Hot Water						190° Hot Water					
200	94	94	93	93	92	92	–	–	–	–	–	–	–	–	–	–	–	–
190	88	88	87	86	85	83	94	93	93	92	92	91	–	–	–	–	–	–
180	82	81	80	79	76	75	88	87	86	85	83	82	93	93	92	92	91	90
170	76	75	73	71	69	67	81	80	79	76	75	73	87	86	85	83	82	80
160	71	69	67	65	62	58	75	73	71	69	67	64	80	79	78	75	73	70
150	65	62	60	57	54	50	69	67	65	62	58	55	73	71	69	67	64	60
140	59	56	53	50	46	42	62	60	57	54	50	45	67	65	62	58	55	50
130	53	50	46	43	38	33	56	53	50	46	42	36	60	57	54	50	45	40
120	47	44	40	36	31	25	50	46	43	38	33	27	53	50	46	42	36	30
110	41	38	33	29	23	17	44	40	36	31	25	18	46	43	38	33	27	20
100	36	31	26	21	15	10	38	33	29	23	17	9	40	36	31	25	18	10
	180° Hot Water						170° Hot Water						160° Hot Water					
170	93	92	92	91	90	89	–	–	–	–	–	–	–	–	–	–	–	–
160	86	85	83	82	80	78	92	92	91	90	89	88	–	–	–	–	–	–
150	79	76	75	73	70	67	85	83	82	80	78	75	92	91	90	89	88	86
140	71	69	67	64	60	56	76	75	73	70	67	63	83	82	80	78	75	71
130	65	62	58	55	50	44	69	67	64	60	56	50	75	73	70	67	63	57
120	57	54	50	45	40	33	62	58	55	50	44	38	67	64	60	56	50	43
110	50	46	42	36	30	21	45	50	45	40	33	25	58	55	50	44	38	29
100	43	38	33	27	20	11	46	42	36	30	21	12	50	45	40	33	25	14
	150° Hot Water						140° Hot Water						130° Hot Water					
140	91	90	89	88	86	83	–	–	–	–	–	–	–	–	–	–	–	–
130	82	80	78	75	71	67	90	89	88	86	83	80	–	–	–	–	–	–
120	73	70	67	63	57	50	80	78	75	71	67	60	89	88	86	83	80	75
110	64	60	56	50	43	33	70	67	63	57	50	40	78	75	71	67	60	50
100	55	50	44	38	29	17	60	56	50	43	33	20	67	63	57	50	40	25

Figure 16-13 Typical Hot Water Storage Tank Heater

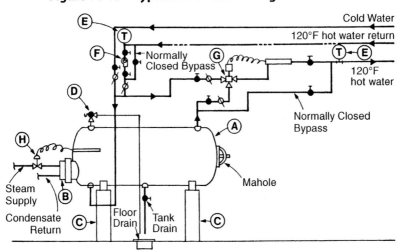

- (A) Hot Water Storage Heater
- (B) Heating Element
- (C) Tank Supports (Steel Saddle, Concrete Piers)
- (D) Combination Pressure and Temperature Relief Valve
- (E) Thermometer (Stem must extend into flow)
- (F) Hot Water Circulating Pump
- (G) Thermostatic Mixing Valve
- (H) Steam Regulating Valve

Figure 16-14 Dual Temperature Hot Water Heating

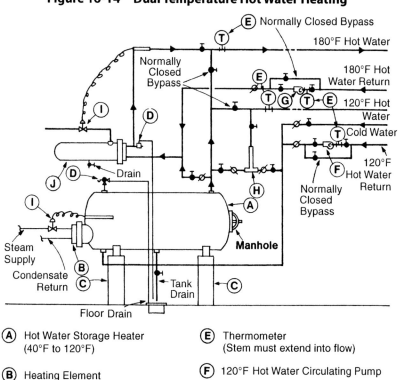

- (A) Hot Water Storage Heater (40°F to 120°F)
- (B) Heating Element
- (C) Tank Supports (Steel Saddle, Concrete Piers)
- (D) Combination Pressure and Temperature Relief Valve
- (E) Thermometer (Stem must extend into flow)
- (F) 120°F Hot Water Circulating Pump
- (G) 180°F Hot Water Circulating Pump
- (H) Thermostatic Mixing Valve
- (I) Steam Regulating Valve
- (J) Instantaneous Booster Heater (120°F to 180°F)

the tank. Horizontal tanks less than 20 in. in diameter are not recommended due to the difficulty of preventing the entering cold water from mixing with the hot water. Figures 16-13 and 16-14 illustrate piping arrangements for one and two-temperature systems.

Hot-Water Temperature

The generally accepted hot-water temperatures for various plumbing fixtures and equipment are given in Table 16-6. Both temperature and pressure should be verified with the client and checked against local codes and the manuals of equipment used.

Safety and Health Concerns

Legionella Pneumophila (Legionaires' Disease)

Legionnaires' disease is a potentially fatal respiratory illness. The disease gained notoriety when a number of American Legionnaires contracted it during a convention. That outbreak was attributed to the water vapor from the

Table 16-6 Typical Hot-Water Temperatures for Plumbing Fixtures and Equipment

Use	Temperature °F	(°C)
Lavatory		
Hand washing	105	(40)
Shaving	115	(45)
Showers and tubs	110	(43)
Therapeutic baths	95	(35)
Surgical scrubbing	110	(43)
Commercial and institutional laundry	140–180	(60–82)
Residential dishwashing and laundry	140	(60)
Commercial, spray-type dishwashing(as required by the NSF):		
Single or multiple-tank hood or rack type:		
Wash	150 min.	(66 min.)
Final rinse:	180–195	(82–91)
Single-tank conveyor type:		
Wash	160 min.	(71 min.)
Final rinse	180–195	(82–91)
Single-tank rack or door type:		
Single-temperature wash and rinse	165 min.	(74 min.)
Chemical sanitizing glassware:		
Wash	140	(60)
Rinse	75 min.	(24 min.)

Note: Be aware that temperatures, as dictated by codes, owners, equipment manufacturers, or regulatory agencies, will occasionally differ from those shown

building's cooling tower(s). The bacteria that cause Legionnaires' disease are widespread in natural sources of water, including rivers, lakes, streams, and ponds. In warm water, the bacteria can grow and multiply to high concentrations. Drinking water containing the Legionella bacteria has no known effects. However, inhalation of the bacteria into the lungs, e.g., while showering, can cause Legionnaires' disease. Much has been published about this problem, and yet there is still controversy over the exact temperatures that foster the growth of the bacteria. Further research is required, for there is still much to be learned. It is incumbent upon designers to familiarize themselves with the latest information on the subject and to take into account when designing their systems. Designers also must be familiar with and abide by the rules of all regulating agencies with jurisdiction.

Scalding[2]. A research project by Moritz and Henriques at Harvard Medical College[3] looked at the relationship between time and water temperature necessary to produce a first-degree burn. A first-degree burn, the least serious type, results in no irreversible damage. The results of the research show that it takes a 3-s exposure to 140°F (60°C) water to produce a first-degree burn. At 130°F (54°C), it takes approximately 20 s, and at 120°F (49°C), it takes 8 min to produce a first-degree burn.

The normal threshold of pain is approximately 118°F (48°C). A person exposed to 120°F (49°C) water would immediately experience discomfort; it is unlikely then that the person would be exposed for the 8 min required to produce a

Table 16-7 Time/Water Temperature Combinations Producing Skin Damage

Water Temperature		
°F	°C	Time (s)
Over 140	Over 60	Less than 1
140	60	2.6
135	58	5.5
130	54	15
125	52	50
120	49	290

Source: Tom Byrley. 1979. 130 degrees F or 140 degrees F. *Contractor Magazine* (September). First published in *American Journal of Pathology*.

Note: The above data indicate conditions producing the first evidence of skin damage in adult males.

---

[2] For more information regarding "Scalding," refer to ASPE Research Foundation, 1989.
[3] Moritz and Henriques, 1947

first-degree burn. People in some occupancies (e.g., hospitals), as well as those over the age of 64 and under the age of 1, may not sense pain or move quickly enough to avoid a burn once pain is sensed. If such a possibility exists, scalding protection should be considered. It is often required by code. (For more information on skin damage caused by exposure to hot water, see Table 16-7.)

Sizing the Hot Water Circulating System

The objective of the design of a hot water circulation system is to have hot water of required temperature readily available at any fixture as needed. Hot water supply piping, whether insulated or not, transmits heat to the surrounding lower-temperature air by convection, radiation, and conduction. If the water remains stationary in the piping for any length of time, its temperature will drop to a level that is unsatisfactory for the proposed use. This cooled water must be withdrawn before hot water at the acceptable temperature appears at the outlet. This results in wastage of water and is inconvenient for the user who wants his/her hot water immediately.

In the past, the additional costs of installing a circulation system and reheating water to compensate for the heat loss in the piping were not considered to be warranted for buildings where the length of run from the hot water heater to the furthest outlet was less than 100 ft and where the building was four stories or less. Today these rules are no longer suitable for design. The new allowable distances for uncirculated, dead-end branches represent a trade-off between the energy utilized by the hot water maintenance system and the cost of energy to heat the excess cold water makeup, the cost of wasted potable water, and the cost of extra sewer surcharges. Chapters 15 and 16 of the *Domestic Water Heating Design Manual*, Second Edition[1] are an excellent resource on this topic. The manual indicates that reasonable delays in obtaining hot water at a fixture are 0–10 sec. A delay of 11–30 sec is considered marginal. Normally, this means that there should be a maximum of approximately 25 ft between the hot water maintenance system and the fixture requiring hot water, the distance dependent on the water flow rate of the plumbing fixture and the size of the line.

System Types

Hot water supply and circulation piping systems can be classified into three general classifications—upfeed, downfeed, and combined upfeed and downfeed. The heater location can be at the bottom or top of the system. Provisions must always be provided at the high points of the system to relieve air that may collect and adversely affect the circulation in the system. (Figures 17-1 through 17-6 illustrate hot water circulation systems.)

In an upfeed supply riser, the circulation riser is always connected below the highest outlet, so that every time the outlet is opened any air that may have collected at the top of the riser is relieved. It is also recommended that the connection be made above the lowest outlet on a downfeed riser, so that

---

[1] American Society of Plumbing Engineers, Chicago, IL, 2003.

Engineered Plumbing Design

Figure 17-1   Upfeed System (Heater Located at Bottom of System)

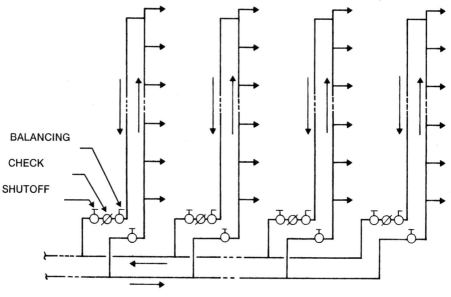

Figure 17-2   Downfeed System (Heater Located at Bottom of System)

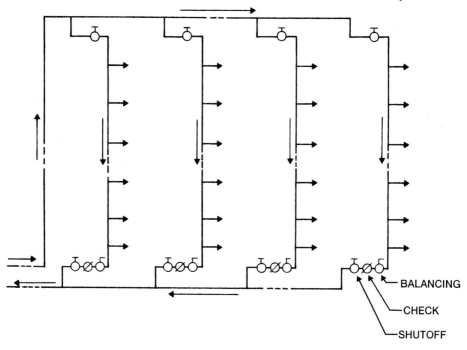

Figure 17-3 Combination Upfeed and Downfeed System (Heater Located at Bottom of System)

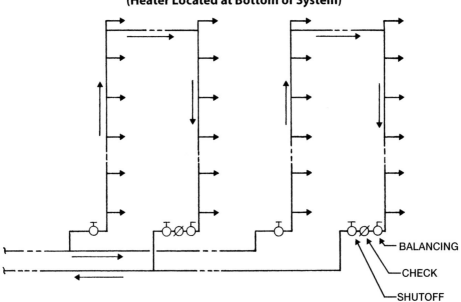

Figure 17-4 Downfeed System (Heater Located at Top of System)

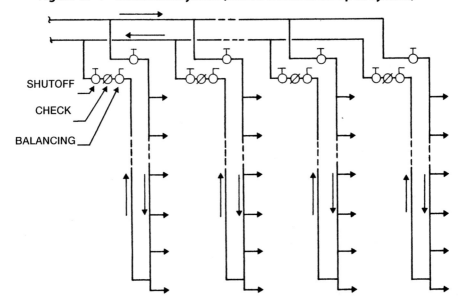

Engineered Plumbing Design

Figure 17-5 Combination Upfeed and Downfeed System (Heater Located at Top of System)

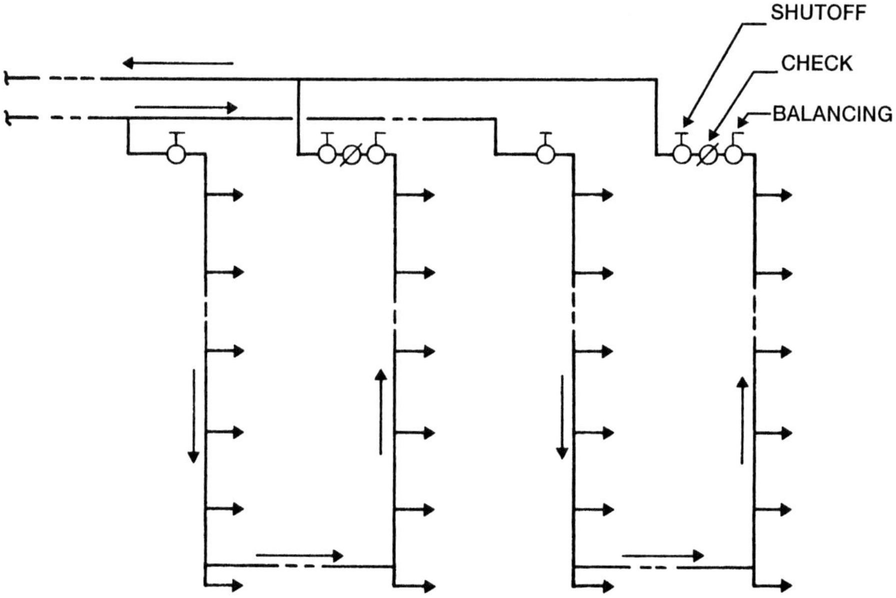

Figure 17-6 Upfeed System (Heater Located at Top of System)

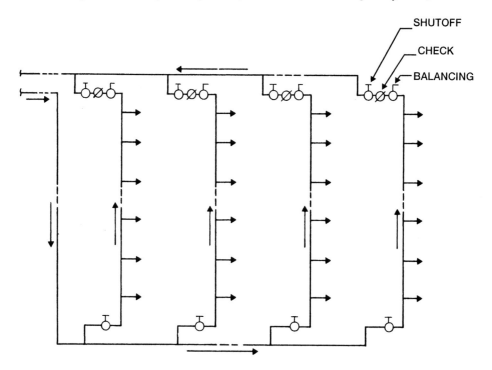

any sediment that might have collected at the base of the riser is washed out when the bottom outlet is opened.

Sizing

Proper sizing of the hot water circulating system is essential for the efficient and economical operation of the hot water system. Oversizing will cause extra expense for materials and create additional heat losses, which in no way contribute to a more efficient operating system. Undersizing will seriously hamper circulation, and thus adequate hot water will not be immediately available at all fixtures.

Empirical methods of sizing circulating piping are usually adequate and satisfactory for the majority of installations, but there are many installations where applying these methods results in a system or parts of a system that are extremely undersized or oversized. For this reason, designers should be thoroughly familiar with the engineering principles involved in the design of hot water circulating systems. The principles are applicable to whatever type of hot water distribution system is employed.

Before any system is designed, the functions of that system and the criteria to be achieved should be thoroughly understood. The hot water circulating system, for example, is designed to maintain the hot water at point of use at required temperatures at all times. To accomplish this accurately and economically, the rate of circulation in the piping and the size of the circulating piping to obtain that rate must be accurately determined. Three basic factors govern this determination:

1. The heat loss rate of the piping
2. The temperature differential at which the system is to operate
3. The allowable friction head loss in the piping.

It is assumed that the hot water supply system, including the mains and risers, has been properly sized. The sizing of the hot water circulating piping can now proceed. The step-by-step procedure is listed below, and then each step is thoroughly discussed.

Procedure

1. Calculate the heat loss rates of the hot water supply piping.
2. Calculate the heat loss rates of the hot water circulating piping.
3. Calculate the circulation rates for all parts of the circulating piping and the total circulation rate required.
4. Determine the allowable uniform friction head loss and the total head required to overcome friction losses in piping when the water is flowing at the required circulation rate.
5. Calculate rates of flow for various pipe sizes, which will give the uniform pressure drop, established in Step 4, and tabulate the results.
6. Size the system based upon the tabulation set up in Step 5.
7. With the sizes established in Step 6, repeat Steps 2 through 6 as a check on the assumptions made.

The above procedure may seem cumbersome and time consuming at first, but once the procedure has been established and applied a few times, and the necessary calculations have been set up in permanent chart form, it will be found to be a simple, rapid method of sizing. It is the only way to be assured of accurate, dependable, and economical results. Each step of the procedure is now thoroughly discussed:

1. Calculate the heat loss rates of the hot water supply piping: Domestic hot water is often distributed at 140°F. The Btu losses per hour per lineal foot at 140°F water temperature and 70°F room temperature for various pipe materials (bare and insulated) are given in Table 17-1. The values for insulated pipe were derived by figuring the insulation to be ½ in. thick fiberglass. Tables for other types and thicknesses of insulation can be compiled. To obtain the heat loss rates of the piping, multiply the length of piping by the appropriate values found in the table.

Table 17-1 Piping Heat Loss (Btu/hr. Per Lineal Ft. For 140°F. Water Temp and 70°F. Room Temp.)

Nominal Pipe Size	Insulated Pipe (½" Fiberglass)	Bare Pipe		
		Sched. 40 Steel	Brass, Copper, T.P.	Type K Copper
½"	15	35	26	19
¾"	17	43	32	26
1"	19	53	38	32
1¼"	21	65	46	39
1½"	25	73	53	46
2"	28	91	65	58
2½"	32	108	75	68
3"	38	129	90	81
4"	46	163	113	103
5"	55	199	138	127
6"	63	233	161	149
8"	80	299	201	188

2. Calculate the heat loss rates of the hot water circulating piping. At this time, sizes for any part of the circulating system are not known. An assumption must be made in order to use the values in Table 17-1. There are two choices:

   A. Assign sizes to the circulating mains equal to one-half the size of the accompanying hot water main, and assign a ¾ in. size to all circulating risers.

   B. Assume the heat losses of the circulating mains and risers to be two-thirds the heat loss of the accompanying hot water mains and risers when both supply and circulating piping are bare or both are insulated. Assume one and one-third the heat loss of the supply piping when the supply is insulated and the circulating piping is bare.

   The latter assumption is favored by the author and seems more compatible with the procedure outlined. By utilizing this assumption, the heat loss rates for the circulating piping can be established for all parts of the system at the same time that the hot water supply heat losses are established. The calculations are thus simplified and expedited.

3. Calculate the circulation rates for all parts of the circulating piping and the total circulation rate required. To find the required circulation rate in any part of the circulating piping, the heat loss rate for that part is divided by 2,500 Btu/hr. This will give gallons per minute of flow required. The 2,500 Btu/hr figure is obtained in the following manner:
   One gal of water weighs 8.3 lb.
   One Btu is equal to the heat necessary to raise 1 lb of water 1°F.
   It is good practice to design a pumped circulation system at a temperature differential of 5°F. Thus, for a circulation rate of 1 gpm:
   8.3 lb/min × 5°F = 41.5 Btu/min and
   41.5 Btu/min × 60 min = 2,490 Btu/hr
   For ease of calculation, 2,500 Btu/hr can be used without introducing a significant error in the results. If the total heat loss of all the piping is 300,000 Btu/hr, then the required circulation rate to overcome this heat loss is 300,000 ÷ 2,500 = 120 gpm.
4. Determine the allowable uniform friction head loss and the total head required to overcome friction losses when water is flowing at the required circulation rate. The uniform friction head loss can be established in two ways:
   A. Take the total circulation rate established in Step 3 and select an efficient pump of that capacity from a manufacturer's catalog. From the curves for the pump selected, note the total head. This will be the total head available. Divide by the equivalent length of run of the longest run of piping to obtain the uniform friction head loss. The longest run of piping is taken from the furthest point of connection to the hot water supply and thence back to the source of the hot water supply. The length of run of the hot water piping is not included because the pressure drop in the hot water piping is insignificant. The size of the hot water piping has been established for extremely high quantities of flow as compared to the circulating flow, so that when only the circulating rate of flow is occurring in the hot water piping the friction loss is extremely low and can be ignored.
   The measured length of run is the developed length. To obtain the equivalent length of run, the resistance of valves, fittings, etc., must be converted to equivalent lengths of pipe. Since sizes are still not known and there are very few fittings in the recirculating line, a reasonable assumption is to add 10% of the developed length to the measured developed length.
   B. Establish the uniform friction head loss empirically. The uniform friction head loss can be selected for a minimum of one foot per hundred feet to a maximum of ten feet per hundred feet. This will permit economical pipe sizes at reasonable velocities of flow and still not create too great a head for the required circulation pump. Select a circulation pump of the capacity established in Step 3 and a head equal to the selected uniform friction head loss times the equivalent length of run divided by 100.

5. Calculate the rates of flow for various pipe sizes that will give the uniform pressure drop selected in Step 4. This can be done from available charts or calculated by means of the following formulas:
For brass or copper pipe (Equation 11-24):

$$q = 40.1 d^{2\frac{1}{2}} \left(\frac{h}{L}\right)^{\frac{1}{2}}$$

For galvanized iron or steel pipe (Equation 11-26):

$$q = 28.3 d^{2\frac{1}{2}} \left(\frac{h}{L}\right)^{\frac{1}{2}}$$

where q = gpm
d = actual internal diameter of pipe, in.
h = total head available for friction loss, ft
L = equivalent length of run, ft

6. Size the system in accordance with the values established in Step 5.

7. With sizes now established, the exact heat losses can be determined and the exact equivalent length of run can be determined. The accuracy of the entire procedure can thus be checked.

Example 17-1

Figure 17-7 shows the riser diagram of an installation for a 24-story building. The lengths of runs are shown as well as the sizes of the hot water piping. The material of the entire system is brass. Both the hot water and circulating piping are insulated with ½" thick fiberglass. The runouts from the mains to the risers are all 10' in developed length. The hot water circulating system is to be sized. (See Table 17-2).

Steps 1, 2 and 3 are done simultaneously by setting up the calculations in tabular form for a grand total of 99,984 + 66,664 = 166,648 Btu per hour. The total circulation rate required is then:

166,648 ÷ 10,000 = 16.7 gallons per minute (using a 20°F temperature difference)

Step 4: Select a uniform friction head loss of 2'/100'. The longest developed length of run is from the 23rd floor ceiling at riser 11, down to the cellar and back to the heater.

Developed length of run =
207' (riser) + 10' (runout) + 275' (main) = 492'
Equivalent length of run = 492 + (.10 × 492) = 541'
The total head required to overcome friction head losses is 2 × 541/100 = 10.82'.
<u>Select a circulation pump to deliver 17 gpm at a discharge head of 11'.</u>

Step 5: Calculate the rates of flow for various pipe sizes which will give a uniform pressure drop of 2'/100'.
$q = 40\ d^{2\frac{1}{2}} \sqrt{h/L} = 40\ d^{2\frac{1}{2}} \sqrt{.02} = 5.66\ d^{2\frac{1}{2}}$
or values can be selected from tables of pressure drops.

½ in. = 1.8 gpm	2 in. = 34.6 gpm
¾ in. = 3.5 gpm	2½ in. = 56 gpm
1 in. = 6.6 gpm	3 in. = 93 gpm
1¼ in. = 12.4 gpm	4 in. = 181 gpm
1½ in. = 18.3 gpm	

Figure 17-7 Riser Diagram for 24-Story Building

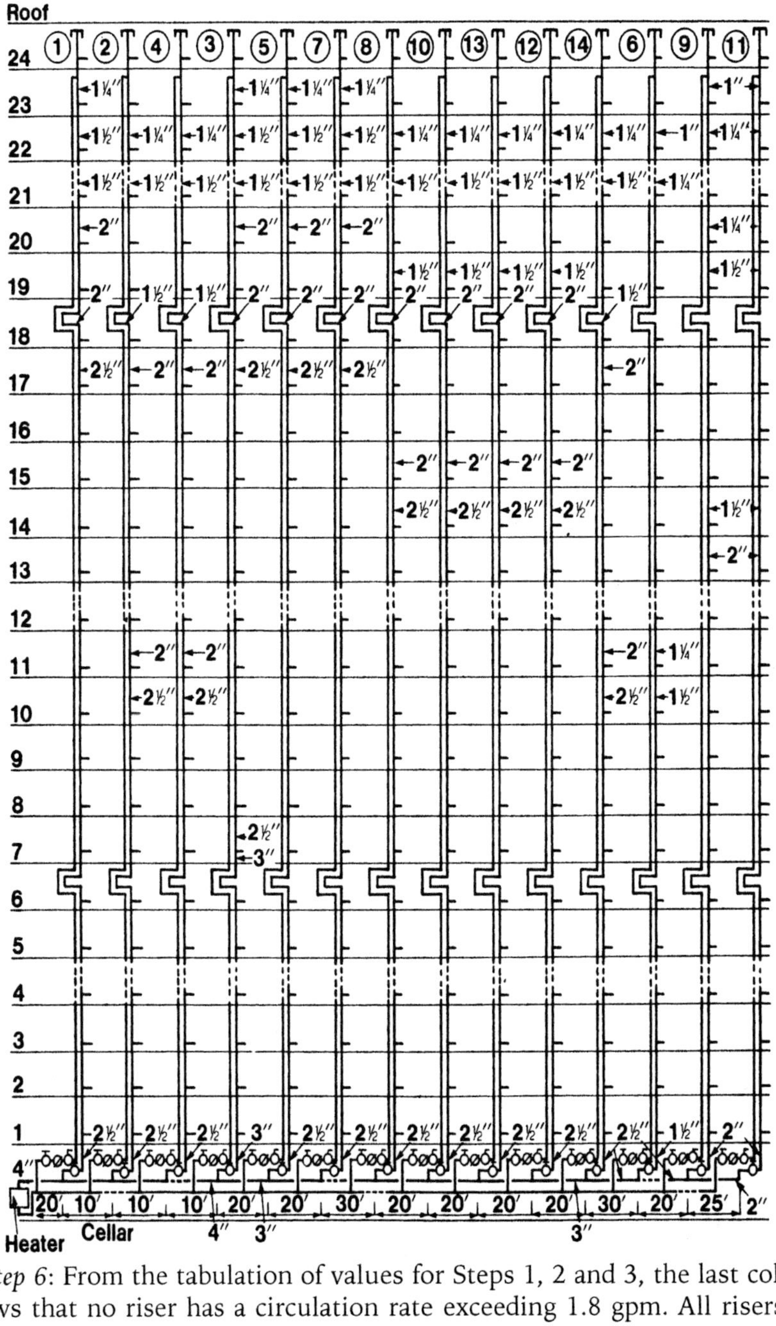

Step 6: From the tabulation of values for Steps 1, 2 and 3, the last column shows that no riser has a circulation rate exceeding 1.8 gpm. All risers can therefore be ½" in size. If galvanized steel piping is used, it is recommended that a minimum size of ¾" be used to accommodate the decrease in pipe area due to the accumulation of deposits with age. Starting at riser 11, add the circulation rate of each riser as it is picked up by the circulating main plus the

Table 17-2 Heat Loss and Circulation Rate

	Hot Water Piping B.T.U. Loss Per Hour	Hot Water Circulating Piping B.T.U. Loss Per Hour	G.P.M. To Overcome Heat Losses (20°F temperature difference)
From Heater to Riser 1	20 × 46 = 920	⅔ (920) = 613	1,533 ÷ 10,000 = 0.1533
From Riser 1 to 2	10 × 46 = 460	⅔ (460) = 307	767 ÷ 10,000 = 0.0767
From Riser 2 to 4	10 × 46 = 460	⅔ (460) = 307	767 ÷ 10,000 = 0.0767
From Riser 4 to 3	10 × 46 = 460	⅔ (460) = 307	767 ÷ 10,000 = 0.0767
From Riser 3 to 5	20 × 30 = 760	⅔ (760) = 507	1,267 ÷ 10,000 = 0.1267
From Riser 5 to 7	20 × 30 = 760	⅔ (760) = 507	1,267 ÷ 10,000 = 0.1267
From Riser 7 to 8	30 × 38 = 1140	⅔ (1140) = 760	1,900 ÷ 10,000 = 0.1900
From Riser 8 to 10	20 × 38 = 760	⅔ (760) = 507	1,267 ÷ 10,000 = 0.1267
From Riser 10 to 13	20 × 38 = 760	⅔ (760) = 507	1,267 ÷ 10,000 = 0.1267
From Riser 13 to 12	20 × 38 = 760	⅔ (760) = 507	1,267 ÷ 10,000 = 0.1267
From Riser 12 to 14	20 × 38 = 760	⅔ (760) = 507	1,267 ÷ 10,000 = 0.1267
From Riser 14 to 6	30 × 32 = 960	⅔ (960) = 640	1,600 ÷ 10,000 = 0.1600
From Riser 6 to 9	20 × 32 = 640	⅔ (640) = 427	1,067 ÷ 10,000 = 0.1067
From Riser 9 to 11	25 × 28 = 700	⅔ (700) = 467	1,167 ÷ 10,000 = 0.1167
	10,300	6,870	
Risers 1, 5, 7	(10 + 160) × 32 = 5440 27 × 28 = 756 18 × 25 = 450 6 × 21 = 126 6772 3 × 6772 = 20,316	⅔ (6772) = 4515 3 × 4515 = 13,545	11,287 ÷ 10,000 = 1.1287
Risers 2, 4, 14	(10 + 97) × 32 = 3424 63 × 28 = 1764 36 × 25 = 900 15 × 21 = 315 6403 3 × 6403 = 19,209	⅔ (6403) = 4269 3 × 4269 = 12,807	10,672 ÷ 10,000 = 1.0672
Riser 3	(10 + 61) × 38 = 2698 99 × 32 = 3168 27 × 28 = 756 18 × 25 = 450 6 × 21 = 126 7,198	⅔ × (7198) = 4,799	11,997 ÷ 10,000 = 1.1997
Riser 6	(10 + 97) × 25 = 2675 99 × 21 = 2079 15 × 19 = 285 5,039	⅔ (5039) = 3,359	8,398 ÷ 10,000 = 0.8398
Risers 8, 10, 12, 13	(10 + 133) × 32 = 4576 36 × 28 = 1008 27 × 25 = 673 15 × 21 = 315 6574 4 × 6574 = 26,296	⅔ (6574) = 4383 4 × 4383 = 17,532	10,957 ÷ 10,000 = 1.0957
Risers 9, 11	(10 + 124) × 28 = 2782 54 × 25 = 1350 27 × 21 = 567 6 × 19 = 114 5813 2 × 5813 = 11,626 99,984	⅔ (5813) = 3876 2 × 3876 = 7,752 66,684	9,689 ÷ 10,000 = 0.9689

circulation rate of that portion of the main. (Table 17-3). All values are obtained directly from the tabulation set up for Steps 1, 2 and 3. Sizes are obtained from tabulation set up for Step 5.

Step 7: With these sizes, Steps 1 through 6 can now be performed without the necessity of any assumptions and the accuracy of the assumptions made can be checked.

Table 17-3 Sizing The Hot Water Circulating System

riser 11 + main (9-11)	0.97 + 0.12 = 1.09	¾"
1.09 + riser 9 + main (6-9) =	1.09 + 0.97 + 0.11 = 2.17	¾"
2.17 + riser 6 + main (14-6) =	2.17 + 1.20 + 0.16 = 3.53	¾"
3.53 + riser 14 + main (12-14) =	3.53 + 1.07 + 0.13 = 4.73	1"
4.73 + riser 12 + main (13-12) =	4.73 + 1.10 + 0.13 = 5.96	1"
5.96 + riser 13 + main (10-13) =	5.96 + 1.10 + 0.13 = 7.19	1¼"
7.19 + riser 10 + main (8-10) =	7.19 + 1.10 + 0.13 = 8.42	1¼"
8.42 + riser 8 + main (7-8) =	8.42 + 1.10 + 0.19 = 9.71	1¼"
9.71 + riser 7 + main (5-7) =	9.71 + 1.13 + 0.13 = 10.97	1¼"
10.97 + riser 5 + main (3-5) =	10.97 + 1.13 + 0.13 = 12.23	1¼"
12.23 + riser 3 + main (4-3) =	12.23 + 1.20 + 0.08 = 13.51	1½"
13.51 + riser 4 + main (2-4) =	13.51 + 1.07 + 0.08 = 14.66	1½"
14.66 + riser 2 + main (1-2) =	14.66 + 1.07 + 0.08 = 15.81	1½"
15.81 + riser 1 + main (heater-1) =	15.81 + 1.13 + 0.15 = 17.09	1½"

Rules of Thumb

As a guide to sizing circulation piping and circulation pumps, the following empirical methods are listed; but they are not recommended in lieu of the more accurate procedures just discussed.

1. An allowance of ½ gpm is assigned for each small hot water riser (¾ in. to 1 in.), 1 gpm for each medium sized hot water riser (1¼ in. to 1½ in.), and 2 gpm for each large size hot water riser (2 in. and larger).
2. An allowance of 1 gpm is assigned for each group of 20 hot water supplied fixtures.

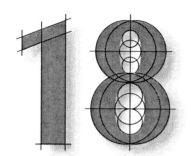

Pipe Expansion and Contraction

All pipelines, which are subject to changes in temperature, expand and contract. Piping increases in length (expands) with an increase in temperature and decreases in length (contracts) with a decrease in temperature. The unit increase in length of a material per 1° F. increase in temperature is called its coefficient of expansion.

The total changes in length may be calculated by

$$L_2 - L_1 = C_E L_1 (T_2 - T_1) (12 \text{ in/ft}) \tag{18-1}$$

where L_1 = original pipe length, ft
L_2 = final pipe length, ft
T_1 = original temperature, °F
T_2 = final temperature, °F
C_E = coefficient of expansion of material, in/ft · °F

The range of temperature change in a hot water piping system is from 40°F entering water to 120°F distribution water, for an 80°F temperature differential. Total linear expansion or contraction for a 100 ft length of run when subjected to an 80°F change in temperature has been calculated for the usual piping materials in a hot water system and is shown in Table 18-1.

Table 18-1 Pipe Expansion For 80° F Temperature Change

Material	Coefficient of Expansion (CE), in/in·°F	Total Expansion of Contraction, in./100 ft
Steel	6.5×10^{-6}	0.62
Copper	9.5×10^{-6}	0.91
Brass	10.4×10^{-6}	1.00

Provisions must be made for the expansion and contraction of all hot water and circulation mains, risers, and branches. If the piping is restrained from moving, it will be subjected to compressive stress on a temperature rise and to tensile stress on a temperature drop. The pipe itself can usually withstand these stresses, but failure frequently occurs at pipe joints and fittings when the piping cannot move freely.

There are two methods commonly employed to absorb pipe expansion or contraction without danger to the piping:
1. Expansion loops and offsets
2. Expansion joints.

It is good engineering practice to limit the total movement to be absorbed by any expansion loop or offset to a maximum of 1½ in. Thus, by anchoring at the points on the length of run that produce 1½-in. movement and placing the expansion loop or joint midway between the anchors, the maximum movement that must be accommodated is ¾ in. The loop or joint can be omitted in the great majority of piping systems by taking advantage of the changes in direction normally required in the layout.

The developed length of piping required for absorbing a given amount of movement by flexure without causing excessive stress in the piping can be calculated by the following formulas:

(18-2)
$$\text{For steel pipe: } L = 6.16\,(de)^{\frac{1}{2}}$$

(18-3)
$$\text{For brass pipe: } L = 6.83\,(de)^{\frac{1}{2}}$$

(18-4)
$$\text{For copper pipe: } L = 7.4\,(de)^{\frac{1}{2}}$$

where L = developed length of piping used to absorb movement, ft.
d = outside diameter of pipe, in.
e = amount of movement to be absorbed, in.

The piping used to absorb the movement can be in the shape of a U-bend, a four-elbow U-bend, a two-elbow offset, or a three-, five-, or six-elbow swing loop. (See Figure 18-1.)

Figure 18-1 Piping to Absorb Movement

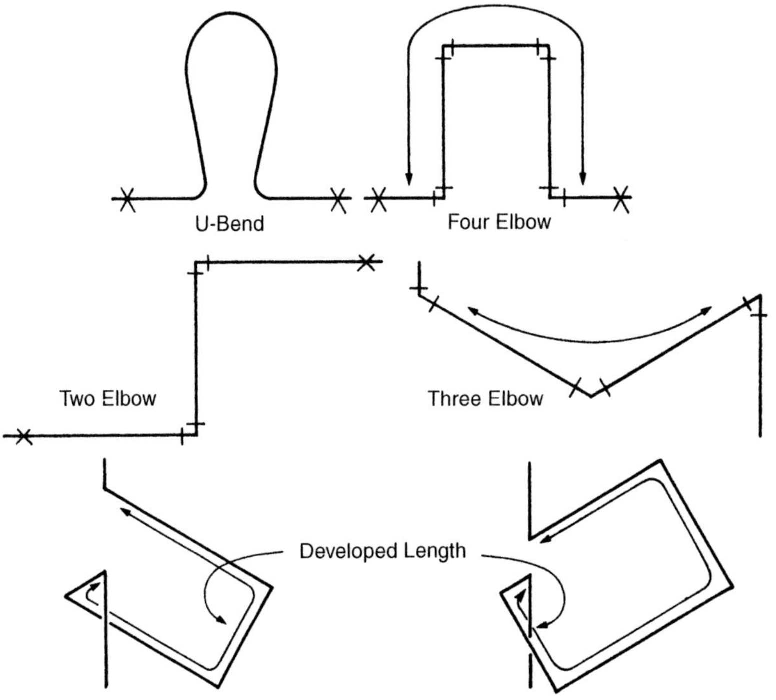

The developed length is measured from the first elbow to the last elbow. Table 18-2 gives the total developed length required to absorb 1½-in. expansion.

Table 18-2 Developed Length of Pipe to Absorb 1½"-In. Movement

Pipe Size, in.	Steel Pipe, ft.	Brass Pipe, ft.	Copper Pipe, ft.
½	6.9	7.7	8.3
¾	7.7	8.6	9.3
1	8.7	9.6	10.4
1¼	9.8	10.8	11.7
1½	10.4	11.5	12.5
2	11.5	12.7	13.8
2½	12.8	14.2	15.4
3	14.2	15.7	17.0
4	16.0	17.7	19.2

One of the most persistent myths in plumbing is that turning motion occurs at the elbows of an expansion loop. This is absolutely false. If any elbow turned in a counterclockwise motion, the joint would be loosened and a leak would occur. A good mechanic will never "back off" a fitting if he/she has overshot the makeup. The mechanic will always make another turn to center the outlet at the correct location, knowing that backing off a fitting will always result in a leak.

Movement is never accommodated by the rotating motion of the fittings: It is always absorbed by the flexure of the pipe. From this it can be seen that the number of elbows in a loop has no bearing whatsoever on its ability to absorb movement. The developed length of pipe available for flexure is the only thing that limits the amount of expansion that can be accommodated.

Figure 18-2 indicates by the dashed lines, in an exaggerated manner for illustrative purposes, how the pipe deforms in a four-elbow loop. The length of leg B should be at least twice the length of A. Wherever possible, B should be larger than 2 × A.

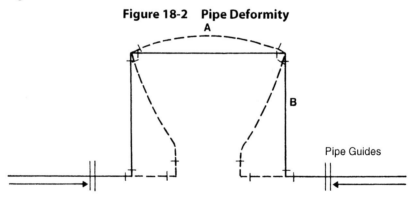

Figure 18-2 Pipe Deformity

There are two types of expansion joints—the slip type and the bellows type. The slip type joint requires packing and lubrication, which dictate that it be placed in an accessible location for maintenance. Guides must be installed in the lines to prevent the pipes from bending and binding in the joint.

The bellows type expansion joint is very satisfactory for the 1½-in. design limitation in movement usually employed in plumbing work. It should be guided or in some other way restrained to prevent collapse.

The main point to remember when applying pipe support, anchors, expansion loops, or joints is that expansion occurs on any temperature rise. The greater the temperature rise, the greater the expansion. The supports, anchors, and guides are installed to restrain the expansion and cause it to move in the

Chapter 18—Pipe Expansion and Contraction

direction desired by the designer, so that trouble does not develop because of negligent or improper installation. If a takeoff connection from mains or risers is located too close to floors, beams, columns, or walls, a change in temperature could cause a break in the takeoff with subsequent flooding damage. As illustrated in Figure 18-3, trouble develops when the riser expands more than dimension "X."

Figure 18-3 Riser Expansion

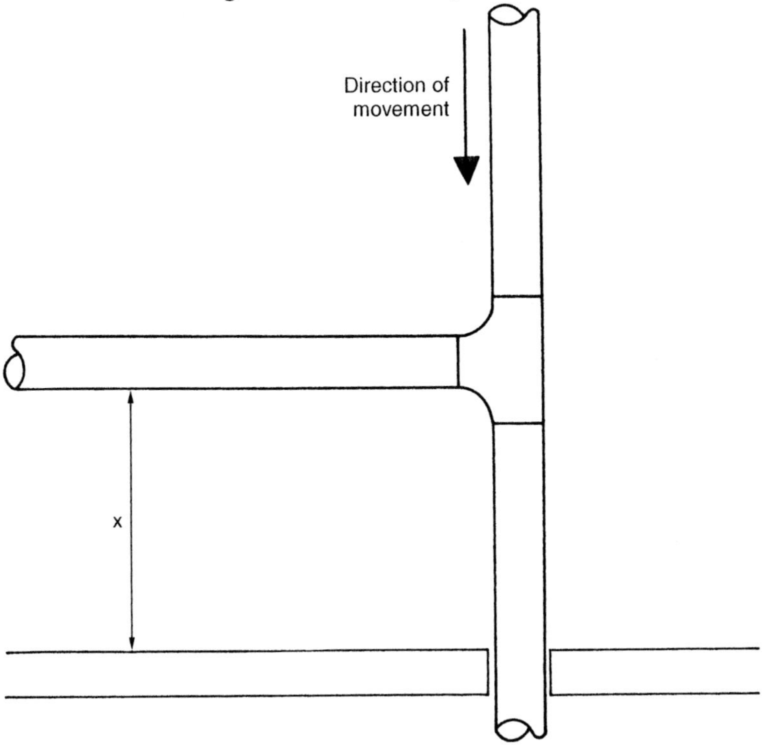

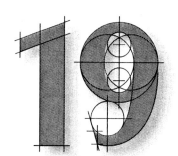

Water Piping Tests

All water piping in a building should be subjected to a water test to assure a watertight job. Any leaks or defects discovered must be corrected. The testing of the piping should be performed before any insulation is applied to the piping and before any part of the system is covered or concealed. Potable water should be used for the test so as not to introduce any possible contaminants into the system. The test should be performed before fixtures, faucets, trim, or final connections are made to equipment.

The rough piping installation should be subject to a hydrostatic pressure of 1½ times the working pressure of the system, but in no case less than 125 psi. The test should extend over a period of at least three hours and demonstrate water tightness without any loss of pressure.

When the system has been completely installed—including all fixtures, faucets, trim, hose connections, and final connections to all equipment—the entire system should be placed in operating condition and thoroughly checked for leaks. All valves, faucets, and trim should be operated and adjusted for maximum performance.

Disinfection

Although the utmost caution is exercised in the installation of piping, there is always the danger that some form of contamination has been introduced into the system. Human life is far too precious to permit the use of any water supply system before it has been thoroughly disinfected and the water has been proven safe for human consumption. The following is written in the form of a specification for the disinfection of water systems.

Disinfection of Water Systems

1. General
    A. Before being placed in service all potable water piping shall be chlorinated as specified herein, in accordance with AWWA (American Water Works Association) Standard C651 and as required by the local Building and Health Department Codes.
    B. Chlorine may be applied by the use of chlorine gas-water mixture, direct chlorine-gas feed or a mixture of calcium hypochlorite and water. If calcium hypochlorite is used, it shall be comparable to commercial products such as Perchloron, HTH, or Maxochlor. The powder shall be mixed with water to form a paste thinned to a slurry and pumped or injected into the lines as specified below.

C. If direct chlorine-gas feed is used, it shall be fed with either a solution-feed chlorinator or by a pressure-feed chlorinator with a diffuser in the pipe.
2. Procedure
    A. Prior to chlorination, all dirt and foreign matter shall be removed by a thorough flushing of the potable water system. The chlorinating agent may be applied to the piping systems at any convenient point. Water shall be fed slowly into the potable water system and the chlorine applied in amounts to produce a dosage of 50 parts per million of available chlorine. Retention shall be for a period of eight hours. During the chlorination process, all valves and accessories shall be operated.
    B. After completion of the above requirements, the system shall be flushed until the water in the system gives chemical and bacteriological tests equal to those of the permanent potable water supply.
    C. Chemical and bacteriological tests shall be conducted by a state laboratory and approved by the local authorities having jurisdiction. Copies of the tests shall be submitted to the architect and all governing authorities.
    D. Warning signs shall be provided at all outlets while chlorinating the system.
    E. If it is impossible to disinfect the potable water-storage tank as provided above, the entire interior of the tank shall be swabbed with a solution containing 200 parts per million of available chlorine and the solution allowed to stand two hours before flushing and returning to service.

Chilled Drinking Water Systems

It is a well-known fact that water that is tepid is not as thirst quenching as water that has been cooled to a temperature of from 40 to 50°F. The 40° water is satisfactory for people in sedentary occupations, but people engaged in physical activities prefer water at 50°. To the latter, water tastes just as cool at 50° as the 40° water does to the former. Although some beverages, such as colas, are more palatable at temperatures below 40°, the 40–50° range has proven to be more than satisfactory for drinking water. A person requires from one-half to two-and-one-half gal of water daily to maintain a level of general good health, depending upon the degree of physical activity in which the individual is engaged.

Generally, public tap water temperatures range well above 50° during the summer months throughout the country. Table 20-1 lists the average summer tap water temperatures of some of the larger cities in the United States. Even water from a deep well warms up in the piping distribution system and is generally higher than 50°. Because of this, it is desirable to cool the drinking water in offices, factories, restaurants, schools, and theaters. It has been determined that drinking water that has been cooled actually improves the efficiency of workers. In restaurants it increases the enjoyment of meals; in schools it improves the alertness of pupils; and in theaters it enhances the pleasure of the show.

Drinking Water Coolers

Drinking fountains should not be confused with drinking water coolers. There is a definite difference between the two. A drinking fountain delivers water at the same temperature as tap water without cooling it in any way. Drinking water coolers are mechanically refrigerated units consisting of a factory assembly in a structure composed of a complete mechanical refrigeration system and having the prime purpose of cooling potable water

Table 20-1 Summer Tap Water Temperatures

City	Temp., °F	City	Temp., °F
Albany, NY	66	Muncie, IN	76
Baltimore, MD	70	Newark, NJ	72
Boston, MA	74	New York, NY	70
Camden, NJ	58	Oakland, CA	64
Chicago, IL	69	Pasadena, CA	74
Cleveland, OH	74	Portland, OR	62
Detroit, MI	75	Rochester, NY	68
Duluth, MN	71	Sacramento, CA	81
Erie, PA	73	St. Louis, MO	85
Fort Wayne, IN	79	Salt Lake City, UT	58
Gary, IN	70	Seattle, WA	62
Hartford, CT	70	Syracuse, NY	65
Indianapolis, IN	82	Toledo, OH	84
Jackson, MS	82	Trenton, NJ	79
Jackson, MI	52	Utica, NY	70
Kansas City, KS	93	Washington, DC	75
Lansing, MI	59	Youngstown, OH	69
Los Angeles, CA	76		

and delivering the chilled water by integral or remote means, or a combination of both.

The capacity of a water cooler is expressed in gallons per hour and is the quantity of water cooled in 1 hr under specific predetermined conditions, i.e., the temperature of the water supply to the unit, the temperature of the delivered chilled water, and the ambient room temperature.

There are three basic types of water coolers:

1. **Bottle Type.** A bottle water cooler is one that uses a 5-gal bottle of water for storing the supply of water to be cooled and uses a faucet or similar device for filling a glass or cup and that also includes a waste water receptacle to catch any drippings. No water or waste connections are required.
2. **Pressure Type.** A pressure type water cooler is supplied with potable water under domestic water pressure. This water is cooled by an integral mechanical refrigeration unit. The waste water is piped to the drainage system. Pressure type water coolers utilize a faucet or similar device for filling cups or may utilize a valve to control the flow to a bubbler, which delivers the water in a trajectory, thus obviating the need of cups.
3. **Self-Contained Remote Type.** A self-contained remote type cooler is a factory assembly in one structure that employs a complete mechanical refrigeration system to cool potable water for delivery to separately installed drinking fountains.

Water coolers are available in water-cooled or air-cooled models.

Specialized water coolers include the following:

1. The explosion-proof type is constructed for safe operation in especially hazardous areas.
2. The cafeteria-type cooler is one that is supplied with water under pressure from the domestic water system and is used primarily in restaurants and cafeterias for rapidly dispensing chilled water into glasses or pitchers. The waste water is piped to the drainage systems.
3. A refrigerated compartment may be provided in a water cooler. The compartment may or may not make ice cubes.
4. Hot water may also be provided from a water cooler. A means of heating water for making instant hot beverages, soups, etc., is provided.

Water cooler installations can be categorized in four basic types:
1. Free standing
2. Flush to wall
3. Wall hung
4. In the wall (semi- or fully recessed).

Refrigeration Components

- Compressors are hermetically sealed and similar to those used in home-type refrigerators.
- Condensers are usually of the fan-cooled type. They can be air cooled or water cooled.
- Refrigerant flow control is generally achieved by capillary tubes in sealed systems.

- Evaporators are generally formed by refrigerant tubing bonded to the outside of a water circuit.
- A precooler is an energy conservation feature of many water coolers. It is used to exchange heat from the supply water to the waste water. A bubbler causes the waste of approximately 60% of the chilled water used. This 60% flows down the waste connection to the drainage system. A precooler pipes the incoming water piping so there is a heat exchange relationship with the waste piping. Another energy conserving scheme is to have the chilled waste water subcool the liquid refrigerant. Coolers equipped with cup fillers in lieu of bubblers are not equipped with precoolers as there is no appreciable waste water flow. Figure 20-1 is a schematic representation of the refrigeration cycle showing the various components of the system.

Figure 20-1   Schematic of Refrigeration Cycle

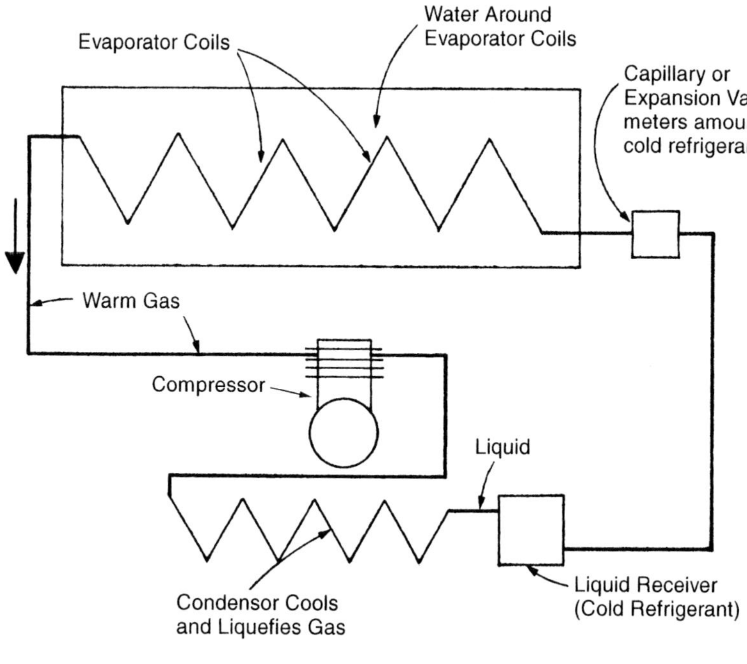

Stream Regulators

A flow rate of ½ gpm from a bubbler produces the best trajectory stream. To achieve this flow, a water cooler should always be equipped with a flow-regulating valve.

Central and Unitary Systems

Although the need and advantages of water coolers are recognized, the layout and quantity cannot be accurately determined until the floor plans are finalized. The floor plans affect the size and type of water cooler to be specified. The plans are also a determining factor in the feasibility of employing a central chilled water system versus the installation of unitary coolers. If the floor layout is such that it is possible to locate the drinking water fixtures one above the other on most floors, utilization of a central system should be analyzed. The economic feasibility increases when the supply can be run behind one bank of fountains and the return (really supply) run behind another bank of fountains. See Figures 20-2 and 20-3 for illustration of this type of loop service. For other types, see Figures 20-4 and 20-5.

Figure 20-2   Downfeed Return Loop

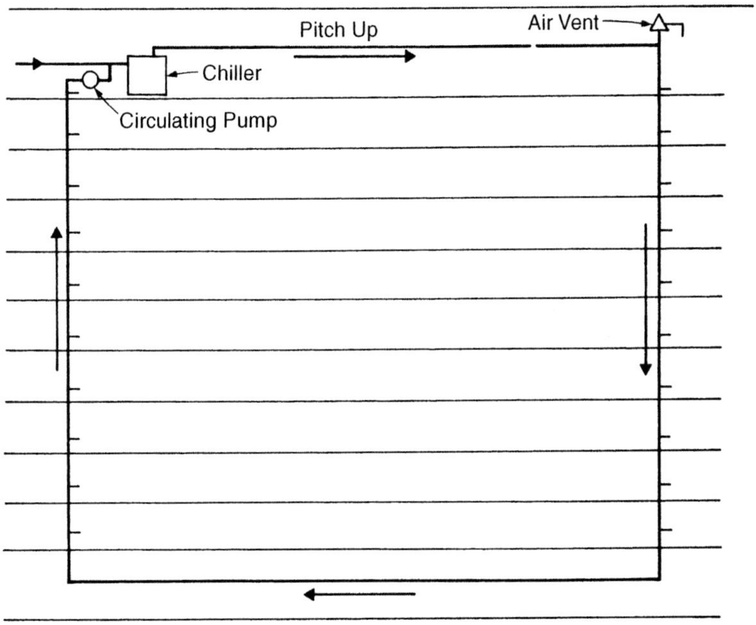

Note: Only major components of system are shown for clarity.

In a central system, maintenance is restricted to only one piece of equipment. However, when repairs are required, chilled water is unavailable for use at any of the fountains. Unless a bypass is provided around the unit, no water is available at all.

The architect has the greatest flexibility in floor layouts when individual water coolers are specified; however, more compressors and chillers are required, and all these units must be maintained. To reduce the number of mechanical refrigeration units, a single unit of adequate capacity can be installed to supply three fountains, one above the other. If the unit is placed on the middle floor, circulation piping can be eliminated. The maximum distance from the unit to a fountain should be limited to 15 ft. If the branch piping is kept to the minimum size of ⅜ in., the quantity of water that must be withdrawn from the dead leg before chilled water of adequate temperature is obtained will be minimal.

In a high-rise building, where pressures can be relatively high, unless precautions are taken to maintain pressures within prescribed limits, the chilled

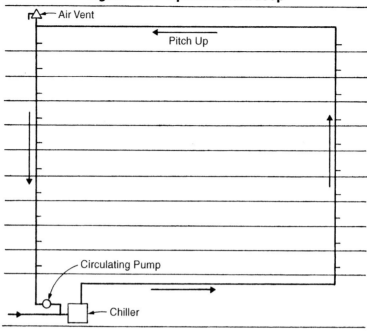

Figure 20-3   Upfeed Return Loop

Note: Only major components of system are shown for clarity.

water can become milky. The milkiness is caused by the entrainment of air. The appearance of the water may cause concern to the user, but it is in no way detrimental to health. If the water is permitted to stand, the milky appearance very rapidly disappears. Proper control of pressure and proper use of air reliefs in the piping will eliminate this problem.

Central Chilled Drinking Water Design

To estimate the chilled drinking water requirements for any type of building, determine the following:
1. Type of building (office, school, hospital, etc.).
2. Total number of people to be served during peak period.
   Determine the cooling load as follows:
1. **Usage Load.** Use 5 gph per outlet for average corridor and office use. For other usages see Table 20-2. Use Table 20-3 for the Btu refrigeration load.
2. **Circulating Line Losses.** See Table 20-4. Include all branch lines in calculating total length.
3. **Circulating Pump Heat Input.** See Table 20-5.
4. **Total Cooling Load.** Use the sum of the Btu obtained in 1, 2, and 3 above and add a 15% factor of safety. This percentage increase is recommended to allow for future expansion or increased population. The total cooling load is used for specifying the condensing unit capacity.

Table 20-2 Drinking Water Requirements

Type of Service	Delivered Water Temp., °F	Gal/Person/Hr	Waste and Consumption/Person/Hr, oz (liquid)	Consumption Only/Person/Hr, oz (liquid)	People Served/Gal
Office (cup)	45–50	0.033	4.2	4.2	30
Office (bubbler)	45–50	0.083	10.5	4.2	12
Light mfg.	45–50	0.143	18.3	7.32	7
Heavy mfg.	50–55	0.20	25.6	10.24	5
Hot heavy mfg.	55–60	0.25	32.0	12.8	4
Restaurant[a]	40–45	0.1 gal/person			
Cafeteria[a]	40–45	0.083 gal/person			
Soda fountain	40–45	0.5 gal/seat			
Theater[a]	45–50	1.0 gal/100 seats continuous capacity			
Schools	45–50	same as office			
Hospitals					
A. Per bed	45–50	0.083 gal			
B. Per attendant	45–50	0.083 gal			
Hotels	45–50	0.08 gal/hr/room			
Public fountains, amusement parks, fairs, etc.	45–50	20–35 gal/hr			
Dept. stores, hotel and office building lobbies	45–50	4–5 gal/hr/fountain			

a Special consideration should be given to peak-load demands for this application.

Table 20-3 Refrigeration Load

	Btu/Hr/Gal Cooled to 45°F				
Water inlet temp., °F	65	70	75	80	85	90
Btu/gal	167	208	250	291	333	374

Notes: Multiply load for 1 gal by total gph. Total Btu/hr is usage load (Table 20-2 × Table 20-3) plus Btu/hr from Tables 20-4 and 20-5 plus 15% safety factor.

Table 20-4 Circulating System Line Loss (Heat Gain) Approx. 1-In. Insulation

Pipe Size (in.)	Btu/Hr/Ft/°F (Temp. Diff.)	Btu/Hr/100 Ft (45°F Circulating Water) Room Temperature, °F		
		70	80	90
½	0.110	280	390	500
¾	0.119	300	420	540
1	0.139	350	490	630
1¼	0.155	390	550	700
1½	0.174	440	610	790
2	0.200	500	700	900
2½	0.228	570	800	1030
3	0.269	680	940	1210

Table 20-5 Circulating Pump Heat Input

Motor H.P.	¼	⅓	½	¾	1	1½	2
Btu Hourly	636	850	1272	1908	2545	3816	5090

Engineered Plumbing Design

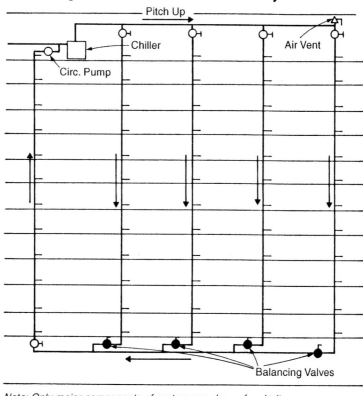

Figure 20-4 Multi-Riser Downfeed System

Note: Only major components of system are shown for clarity.

Circulating Pump Capacity

See Table 20-6. Size the pump to circulate a minimum of 3 gpm per branch circuit or the gpm necessary to limit the temperature rise of the circulating water to a maximum of 5°, whichever is the greater value. The pump should be located in the return line to discharge into the cooler with the makeup water connection between the pump and the cooler.

Makeup Water Mixture

Calculate the temperature of the mixture of makeup water and circulating water. The quantity and temperature of this mixture are used as the specified capacity of the water-cooling unit.

Storage Tank

Size the storage tank for a capacity of 50% of the usage per hour.

Table 20-6 Circulating Pump Capacity

Pipe Size, in.	Room Temperature, °F		
	70	80	90
	GPH/100 ft		
½	8.0	11.1	14.3
¾	8.4	11.8	15.2
1	9.1	12.8	16.5
1¼	10.4	14.6	18.7
1½	11.2	15.7	20.2

Notes:
1. GPH/100 ft of pipe (including all branch lines) circulation rate to limit temperature rise to 5°F (water at 45°F).
2. Divide total gph by 60 to obtain gpm. Add 20% for safety factor.
3. For pump head figure longest branch only.
4. Install pump on the return line to discharge into the cooling unit.
5. Makeup connection should be between the pump and the cooling unit.
6. Btu = gal to be cooled × temp difference × 8.3

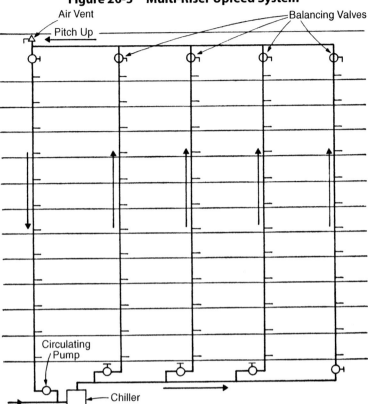

Figure 20-5 Multi-Riser Upfeed System

Note: Only major components of system are shown for clarity.

Piping

The piping should be sized to keep friction losses below 10-ft/100 ft of equivalent length of run.

Example 20-1

Take, for instance, an office building with 60 drinking fountains, room temperature 80°F, inlet water 75°F, chilled water 45°F, 5 branch circuits, 600 ft of 1-in. pipe, 200 ft of 1¼-in. pipe, 100 ft of 1½-in. pipe, ¾ H.P. circulating pump.

Usage Load

(Table 20-3)

60 outlets × 5 gph × 250 Btu/gal = 75,000 Btu/hr

Circulating Loss

(Table 20-4)

600 ft 1-in. pipe @ 490 Btu/hr/100 ft	=	2,940
200 ft 1¼-in. pipe @ 550	=	1,100
100 ft. 1½-in. pipe @ 610	=	610
Circulation Pump Heat Input (¾ H.P.)	=	1,908

(Table 20-5)

Total Btu/hr   81,558
15% factor of safety   + 12,234
                         93,792 Btu/hr

Storage Capacity

.50 × 300 = 150 gal

Circulation Pump Capacity

(Table 20-6 or 3 gpm/circuit)

5 circuits × 3 gpm = 15 gpm

or

600 ft 1-in. line:	12.8 × 6	=	76.8 gph
200 ft 1¼-in. line:	14.6 × 2	=	29.2 gph
100 ft 1½-in. line:	15.7 × 1	=	15.7 gph
			121.7
	Plus 15%		18.3
			140.0 gph

140 gph ÷ 60 min. = 2.3 gpm
Use 15 gpm.

Private Sewage Disposal Systems

With the ever-increasing cost of land located in proximity to urban centers, more and more construction is being implemented in outlying areas. Sanitary sewers are not usually available in these remote locations and it becomes necessary for the plumbing engineer to design private sewage systems to handle the wastes from buildings. Before the rapid escalation of land values, most private sanitary disposal systems were used almost exclusively for private residences. It is estimated that 15 million such systems are presently in use in the United States. Of greater significance, roughly 25% of all new home construction now employs the septic tank–soil absorption sewage disposal system.

Where the concentration of population is not sufficient to economically justify the installation of public sewer systems, installation of a septic tank in conjunction with a subsurface soil absorption field has proven to be an exceptionally satisfactory method of sewage disposal. When properly designed, installed, operated, and maintained, it compares very favorably with the most sophisticated municipal sewage treatment plants.

In 1946, the U.S. Public Health Service, in cooperation with other federal agencies involved in housing, embarked upon a five-year study to establish criteria for the design, installation, and maintenance of the septic tank. Most of the information in this chapter is freely drawn from that study and a later report issued in 1967.

Sewage System Criteria

The proper disposal of sewage is a major factor affecting the health of the public. When improper or inadequate disposal of sewage occurs, many diseases, such as dysentery, infectious hepatitis, typhoid, paratyphoid, and various types of diarrhea are transmitted through contamination of food and water. To avoid such hazards, any system of sewage disposal must meet the following criteria:
- It must not contaminate any drinking water supply.
- It must not be accessible to insects, rodents, or other possible carriers that might come in contact with food or drinking water.
- It must not be accessible to children.
- It must not violate laws or rules and regulations governing water pollution or sewage disposal.
- It must not pollute or contaminate the waters of any bathing beach, shellfish breeding ground, or any stream used for public or private water supply or for recreational purposes.
- It must not become malodorous or unsightly in appearance.

All these criteria are admirably fulfilled by a public sewage disposal system. Every effort should be made to utilize such facilities if at all possible. When public sewers are not available, some other satisfactory method must be employed.

Any method of sewage disposal is merely an attempt to complete the hydrologic cycle, or as it is now popularly called, the ecological cycle. Contaminated water (wastes) of undesirable quality is received and, after processing, returned at an acceptable level of quality. The systems to be discussed are those that return the waste water to the soil and ultimately to the ground water (water table).

There are presently two systems that return waste water to the soil. They are the cesspool and the septic tank–soil absorption systems.

Cesspools

A cesspool is nothing more than a covered pit with an open-jointed or perforated lining into which raw sewage is discharged. The liquid portion of the sewage is disposed by seepage or leaching into the porous soil surrounding the cesspool. The solids (sludge) are retained in the pit.

A cesspool finds its greatest application in receiving the effluent from one-family homes and it is not recommended even for this use. The raw sewage tends to seal the openings in the pit lining as well as the surrounding soil, thus necessitating frequent visits from the "honey dippers" (cesspool cleaning services). Cloggage may become so severe that complete abandonment of the existing cesspool and the construction of a new pit is often necessary. A cesspool should never be recommended as a substitute for a septic tank with a soil absorption field.

A seepage pit (discussed in another portion of this chapter) should never be confused with a cesspool. Although the construction is the same for both, a seepage pit receives the effluent from a septic tank (where the solids have been liquified), whereas a cesspool receives raw sewage.

Septic Tanks

A septic tank is a liquid-tight structure, with inlet and outlet connections, which receives raw sewage. It is basically a sewage settling tank in which raw sewage is retained for a specified period of time, usually 24 hr. The primary purpose of the septic tank is to act as a settling tank and to break up solids so that the resulting effluent will not clog the pores of the soil in the leaching field. Very little purification is accomplished in the tank; the actual treatment and digestion of harmful waste materials takes place in the ground after discharge from the tank.

Three functions are performed by a septic tank to produce an effluent suitable for acceptance by a subsoil absorption system of sewage disposal: (1) removal of solids, (2) biological treatment, and (3) sludge and scum storage.

Removal of Solids

Clogging of the soil varies directly with the amount of suspended solids in the liquid. The rate of flow entering the septic tank is reduced within the tank so that solids sink to the bottom or rise to the surface of the liquid in the tank. These solids are retained and the clarified effluent is discharged.

Solids and liquid in the tank are exposed to bacterial and natural processes, which decompose them. The bacteria present in the wastes are of the anaerobic type, which thrives in the absence of oxygen. Decomposition of the sewage under anaerobic conditions is termed "septic" and it is from this the tank derives its name.

After such biological action, the effluent causes less clogging of the soil than untreated sewage containing the same quantity of suspended solids.

Sludge and Scum Storage

Sludge is an accumulation of solids at the bottom of the tank. Scum is a partially submerged floating mat of solids that forms at the surface of the liquid in the tank. The sludge is digested and compacted into a smaller volume. The same action occurs with the scum but to a lesser degree. Regardless of the efficiency of the operation of the septic tank, a residual of inert solid material will always remain. Adequate space must be provided in the tank to store this residue during the intervals between cleanings. Sludge and scum will flow out of the tank with the effluent and clog the disposal field in a very short period of time if pumping out of the residue is not performed when required.

Septic tanks are eminently effective in performing their purpose when adequately designed, constructed, operated, and maintained. They do not accomplish a high degree of bacteria removal. Although the sewage undergoes some treatment in passing through the tank, infectious agents present in the sewage are not removed. The effluent of a septic tank cannot be considered safe. In many respects, the discharged liquid is more objectionable than the influent because it is septic and malodorous. This should not be construed in any way as detracting from the value of the tank because its primary purpose is simply to condition the raw sewage so that it will not clog the disposal field.

Continued treatment and the removal of pathogens are accomplished by percolation through the soil. Disease-producing bacteria will die out after a time in the unfavorable environment of the soil. Bacteria are also removed by physical forces during filtration through the soil. This combination of factors achieves the eventual purification of the septic tank effluent.

Septic Tank Location

The location of the septic tank should be chosen so as not to cause contamination of any well, spring, or other source of water supply. Underground contamination can travel in any direction for considerable distances unless effectively filtered. Tanks should never be closer than 50 ft to any source of water supply and, where possible, greater distances are preferable. They should be located where the largest possible area will be available for the disposal field and should never be located in swampy areas subject to flooding. Ease of maintenance and accessibility for cleaning are important factors to be considered. When it is anticipated that public sewers will be available in the future, provisions should be made for the eventual connection of the house sewer to such a public source.

Tank Capacity

Studies have proven that liberal tank capacity is not only desirable from a functional viewpoint but is good economical design practice. The liquid capacities recommended in Table 21-1 make allowances for all household appliances including garbage grinders.

Tank Material

Septic tanks must be watertight and constructed of materials not subject to excessive corrosion or decay. Acceptable materials are concrete, coated metal, vitrified clay, heavyweight concrete blocks, or hard-burned bricks. Properly cured precast and cast-in-place, reinforced concrete are believed to be acceptable everywhere. Local codes should be checked as to the acceptability of the other materials. Steel tanks conforming to U.S. Department of Commerce Standard 177-62 are generally acceptable. Precast tanks should have a minimum wall thickness of 3 in. and should be adequately reinforced to facilitate handling. When precast slabs are used as covers, they should be watertight, at least 3 in. thick and adequately reinforced. All concrete surfaces should be coated with a bitumastic paint or similar compound to minimize corrosion.

Table 21-1 Liquid Capacity of Tank (gal)
(provides for use of garbage grinders, automatic clothes washers, and other household appliances)

Number of Bedrooms	Recommended Minimum Tank Capacity	Equivalent Capacity per Bedroom
2 or less	750	375
3	900	300
4[a]	1000	250

a For each additional bedroom, add 250 gal.

Tank Access

Access should be provided to each compartment of the tank for cleaning and inspection by means of a removable cover or a 20-in. minimum size manhole. When the top of the tank is more than 18 in. below grade, manholes and inspection holes should be extended to approximately 8 in. below grade. They can be extended to grade if a seal is provided to prevent the escape of odors.

Tank Inlet

The invert elevation of the inlet should be at least 3 in. above the liquid level in the tank. This will allow for momentary surges during discharge from the house sewer into the tank and also prevent the backup and stranding of solids in the piping entering the tank.

A vented inlet tee or baffle should be provided to direct the influent downward. The outlet of the tee should terminate at least 6 in. below the liquid level but in no case should it be lower than the bottom of the outlet fitting or device.

Tank Outlet

The outlet fitting or device should penetrate the liquid level just far enough to provide a balance between the sludge and scum storage volumes. This will assure usage of the maximum available tank capacity. A properly operating tank divides itself into three distinct layers: scum at the top, a middle layer free of solids (clear space), and sludge at the bottom layer. While the outlet tee or device retains the scum in the tank, it also limits the amount of sludge that can be retained without passing some of the sludge out with effluent.

Data collected from field observation of sludge accumulations indicate that the outlet device should extend to a distance below the liquid level equal to 40% of the liquid depth. For horizontal cylindrical tanks the percentage should be 35. The outlet device or tee should extend up to within 1 in. of the top of the tank for venting purposes. The space between the top of the tank and the baffle permits gas to pass through the tank into the building sanitary system and eventually to atmosphere where it will not cause a nuisance.

Tank Shape

Available data indicate that for tanks of a given capacity and depth, the shape of a septic tank is unimportant and that shallow tanks function equally as well as deep ones. It is recommended, however, that the minimum plan dimension be 2 ft and the liquid depth range from 30 to 60 in.

Scum Storage Space

Space is required above the level of the liquid in the tank for the accumulated scum, which floats on top of the liquid. Although there is some variation, approximately 30% of the total amount of scum will accumulate above the liquid level and 70% will be submerged. In addition to the scum storage space, 1 in. should be provided at the top of the tank for free passage of gas through the tank back to the inlet and building drainage system.

For tanks with vertical walls, the distance between the top of the tank and the liquid level should be approximately 20% of the liquid depth. For horizontal cylindrical tanks, the liquid depth should be 79% of the diameter of the tank. This will provide an open area at the top of the tank equal to 15% of the total cross-sectional area of the tank.

Compartments

Although a number of arrangements are possible, compartments refer to the number of units in series. They can be separate units connected together or sections enclosed in one continuous shell with watertight partitions separating the individual compartments.

A single-compartment tank gives acceptable performance, but available research data indicate that a two-compartment tank with the first compartment equal to ½ to ⅔ of the total volume provides better suspended solids removal. Tanks with three or more equal compartments perform about on an equal basis with a single-compartment tank of the same total capacity. The use of a more than two-compartment tank is therefore not recommended. All the requirements of construction stated previously for a single-compartment tank apply to the two-compartment tank. Each compartment should be provided with an access manhole and venting between compartments for the free passage of gas.

Figure 21-1 illustrates all the salient features of a typical two-compartment septic tank.

Cleaning of Tanks

Before too much sludge or scum is allowed to accumulate, septic tanks should be cleaned to prevent the passage of sludge or scum into the disposal field. Tanks should be inspected at least once a year and cleaned when necessary.

Figure 21-1 Precast Septic Tank

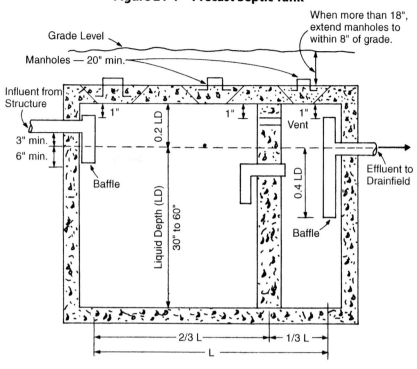

Cleaning is usually accomplished by pumping the contents of the tank into a tank truck. A small residual of sludge should be left in the tank for "seeding" purposes. Tanks should never be washed or disinfected after cleaning.

Chemical Additives

The operation of a septic tank is not improved in any way whatsoever by the addition of chemicals; and such additions are not recommended. Some products that claim to "clean" septic tanks contain sodium hydroxide or potassium hydroxide as the active agent. Such compounds may result in sludge bulking and a sharp increase in alkalinity, which may interfere with digestion. The effluent may severely damage the soil structure of the disposal field and cause accelerated clogging even though some immediate temporary relief may be experienced shortly after application of the product.

On the other hand, ordinary household chemicals in general use around the home will not have a harmful effect on the operation of a septic tank. Small amounts of chorine bleach or small quantities of lye or caustic are not objectionable. If tanks are sized as recommended herein, the dilution of the lye or caustics in the tank will be enough to minimize any harmful effects. Soaps, detergents, bleaches, drain cleaners, etc., will have no appreciable adverse effect on the system. However, since both the soil and the organisms might be susceptible to large doses of chemicals, moderation is recommended.

Toilet paper substitutes, paper towels, newspaper, wrapping paper, rags, and sticks should not be introduced into the septic tank. They may not decompose and are likely to lead to clogging of the disposal field.

Backwash from a household water-softening unit has no adverse effect on the operation of a septic tank, but may cause a slight shortening of life of the disposal field installed in a structured clay type soil.

Septic Tanks for Nonresidential Buildings

Table 21-1 gives the liquid capacity of tanks on the basis of the number of bedrooms. When designing a septic tank for other types of buildings, Table 21-2 may be used to estimate the quantity of sewage flow. The quantities listed are merely the best averages presently available and should be modified in localities or establishments where available information indicates a need to do so.

The retention period of the sewage in a septic tank should be 24 hr. Table 21-2 gives the gallons per person per day (24 hr). The required liquid capacity of the tank can then be determined by multiplying the values given in the table by the estimated population.

Tables 21-3 and 21-4 give daily gallonages

Table 21-2 Quantities of Sewage Flows

Type of Establishment	Gallons Per Person Per Day (unless otherwise noted)
Airports (per passenger)	5
Apartments—multiple family (per resident)	60
Bathhouses and swimming pools	10
Camps:	
Campground with central comfort stations	35
With flush toilets, no showers	25
Construction camps (semi-permanent)	50
Day camps (no meals served)	15
Resort camps (night and day) with limited plumbing	50
Luxury camps	100
Cottages and small dwellings with seasonal occupancy	50
Country clubs (per resident member)	100
Country clubs (per non-resident member present)	25
Dwellings:	
Boarding houses	50
additional for non-resident boarders	10
Luxury residences and estates	150
Multiple-family dwellings (apartments)	60
Rooming houses	40
Single-family dwellings	75
Factories (gallons/person/shift, exclusive of industrial wastes)	35
Hospitals (per bed space)	250
Hotels with private baths (2 persons per room)	60
Hotels without private baths	50
Institutions other than hospitals (per bed space)	125
Laundries, self service (gal/wash, i.e., per customer)	50
Mobile home parks (per space)	250
Motels with bath, toilet, and kitchen wastes (per bed space)	50
Motels (per bed space)	40
Picnic parks (toilet wastes only) (per picnicker)	5
Picnic parks with bathrooms, showers, and flush toilets	10
Restaurants (toilet and kitchen wastes per patron)	10
Restaurants (kitchen wastes per meal served)	3
Restaurants, additional for bars and cocktail lounges	2
Schools:	
Boarding	100
Day, without gyms, cafeterias, or showers	15
Day, with gyms, cafeterias, and showers	25
Day, with cafeterias, but without gyms or showers	20
Service stations (per vehicle served)	10
Swimming pools and bathhouses	10
Theaters:	
Movie (per auditorium seat)	5
Drive-in (per car space)	5
Travel trailer parks	
Without individual water and sewer hookups (per space)	50
With individual water and sewer hookups (per space)	100
Workers:	
Construction (at semi-permanent camps)	50
Day, at schools and offices (per shift)	15

in terms of fixtures for country clubs and public parks, respectively.

Subsurface Soil Absorption System

Criteria for Design

The first step in the design of a subsurface soil absorption sewage disposal system is to determine whether the soil is suitable for the absorption of the septic tank effluent. If it is, the next step is to determine the area required for the disposal field. The soil must have an acceptable percolation rate and should have adequate clearance from ground water. In general, two criteria must be met:

1. The percolation rate should be within the range shown in Table 21-5 or Table 21-6.
2. The maximum elevation of the groundwater table should be at least 4 ft below the bottom of the trench or seepage pit. Rock formation or other impervious strata should be at a depth of more than 4 ft below the bottom of the trench or seepage pit.

If these two primary conditions cannot be met, the site is unsuitable for a soil absorption system and some other seepage disposal system must be employed.

Table 21-3 Sewage Flow from Country Clubs

Type of Fixture	Gallons per Day per Fixture
Showers	500
Baths	300
Lavatories	100
Toilets	150
Urinals	100
Sinks	50

Table 21-4 Sewage Flow at Public Parks
(during hours when park is open)

Type of Fixture	Gallons per Day per Fixture
Flush toilets	36
Urinals	10
Showers	100
Faucets	15

Table 21-5 Absorption Area Requirements for Individual Residences[a]
(provides for garbage grinder and automatic clothes washing machines)

Percolation Rate (time required for water to fall 1 in.), in minutes	Required Absorption Area, in ft²/bedroom[b], standard trench[c], seepage beds[c], and seepage pits[d]
1 or less	70
2	85
3	100
4	115
5	125
10	165
15	190
30[c, e]	250
45[c, e]	300
60[c, e, f]	330

a It is desirable to provide sufficient land area for an entire new absorption system if needed in the future.
b In every case, sufficient land area should be provided for the number of bedrooms (minimum of two) that can be reasonably anticipated, including the unfinished space available for conversion as additional bedrooms.
c Absorption area is figured as trench bottom area and includes a statistical allowance for vertical side wall area.
d Absorption area for seepage pits is figured as effective side wall area beneath the inlet.
e Unsuitable for seepage pits if over 30.
f Unsuitable for absorption systems if over 60

Percolation Tests

Percolation tests help to determine the acceptability of the site and establish the size of the disposal system. The length of time required for percolation tests varies for different types of soil. The safest method is to make tests in holes that have been kept filled with water for at least 4 hrs and preferably overnight. Percolation rates should be determined on the basis of test data obtained after the soil has had the opportunity to become wetted or saturated.

Enough tests should be made in separate holes to

assure the validity of results. The percolation test as developed at the Robert A. Taft Sanitary Engineering Center has proven to be one of the best in the country and is given here in its entirety:

Procedure for Percolation Tests

1. **Number and Location of Tests.** Six or more tests shall be made in separate test holes spaced uniformly over the proposed absorption field site.
2. **Type of Test Hole.** Dig or bore a hole, with horizontal dimensions of from 4 to 12 in. and vertical sides to the depth of the proposed absorption trench. In order to save time, labor, and volume of water required per test, the holes can be bored with a 4-in. auger.
3. **Preparation of a Test Hole.** Carefully scratch the bottom and sides of the hole with a knife blade or sharp-pointed instrument, in order to remove any smeared soil surfaces and to provide a natural soil interface into which water may percolate. Remove all loose material from the hole. Add 2 in. of coarse sand or fine gravel to protect the bottom from scouring and sediment.
4. **Saturation and Swelling of the Soil.** It is important to distinguish between saturation and swelling. Saturation means that the void spaces between soil particles are full of water. This can be accomplished in a short period of time. Swelling is caused by intrusion of water into the individual soil particle. This is a slow process, especially in clay-type soil, and is the reason for requiring a prolonged soaking period.

Table 21-6 Allowable Rate of Sewage Application to a Soil Absorption System

Percolation Rate (time for water to fall 1 in.), in minutes	Maximum Rate of Sewage Application (gal/ft^2/day)[a] for Absorption Trenches[b], Seepage Beds, and Seepage Pits[c]
1 or less	5.0
2	3.5
3	2.9
4	2.5
5	2.2
10	1.6
15	1.3
30[d]	0.9
45[d]	0.8
60[d,e]	0.6

a Not including effluents from septic tanks that receive wastes from garbage grinders and automatic washing machines.
b Absorption area is figured as trench bottom area, and includes a statistical allowance for vertical sidewall area.
c Absorption area for seepage pits is effective sidewall area.
d Over 30 unsuitable for seepage pits.
e Over 60 unsuitable for absorption systems.

In the conduct of the test, carefully fill the hole with clear water to a minimum depth of 12 in. over the gravel. In most soils, it is necessary to refill the hole by supplying a surplus reservoir of water, possibly by means of an automatic syphon, to keep water in the hole for at least 4 hrs and preferably overnight. Determine the percolation rate 24 hrs after water is first added to the hole. This procedure is to ensure that the soil is given ample opportunity to swell and to approach the condition it will be in during the wettest season of the year. Thus, the test will give comparable results in the same soil, whether made in a dry or wet season. In sandy soils, containing little or no clay, the swelling is not essential, and

the test may be made as described under item 5C, after the water from one filling of the hole has completely seeped away.
5. **Percolation Rate Measurement.** With the exception of sandy soils, percolation rate measurements shall be made on the day following the procedure described under item 4, above.
   A. If water remains in the test hole after the overnight swelling period, adjust the depth to approximately 6 in. over the gravel. From a fixed reference point, measure the drop in water level over a 30-min. period. This drop is used to calculate the percolation rate.
   B. If no water remains in the hole after the overnight swelling period, add clear water to bring the depth of water in the hole to approximately 6 in. over the gravel. From a fixed reference point, measure the drop in water level at approximately 30-min intervals for 4 hrs, refilling 6 in. over the gravel as necessary. The drop that occurs during the final 30-min period is used to calculate the percolation rate. The drops during prior periods provide information for possible modification of the procedure to suit local circumstances.
   C. In sandy soils (or other soils in which the first 6 in. of water seeps away in less than 30 min, after the overnight swelling period), the time interval between measurements shall be taken as 10 min and the test run for 1 hr. The drop that occurs during the final 10 min is used to calculate the percolation rate.

Absorption Area

For locations where the percolation rates and soil characteristics prove to be satisfactory, the next step is to determine the required absorption area from Table 21-5 for residences or from Table 21-6 for other types of buildings. As noted in the tables, soil in which the percolation rate is slower than 1 in. in 30 min is not suitable for seepage pits and a rate slower than 1 in. in 60 min is not satisfactory for any type of soil absorption system.

There are three types of soil absorption systems:
1. Absorption trenches
2. Seepage beds
3. Seepage pits.

The selection of the system will be affected by the location of the system in the area under consideration. A safe distance must be maintained between the site and the source of any water supply. No specific distance can be absolutely safe in all localities because of the many variables involved in the underground travel of pollution. Table 21-7 can be used as a guide for establishing minimum distances between various components of a sewage disposal system.

Seepage pits should never be installed in areas of shallow wells or where there are limestone formations and sinkholes with connection to underground channels through which pollution could travel to water sources.

Absorption Trenches

The drain pipe for a soil absorption field may be 12-in. lengths of 4-in. agricultural drain tile, 2–3 ft lengths of open-joint vitrified clay sewer pipe,

Table 21-7 Minimum Distance Between Components of Sewage Disposal System

Component of System	Horizontal Distance (ft)				
	Well or Suction Line	Water Supply Line (pressure)	Stream	Dwelling	Property Line
Building sewer	50	10[a]	50	—	—
Septic tank	50	10	50	5	10
Disposal field and seepage bed	100	25	50	20	5
Seepage pit	100	50	50	20	10
Cesspool[b]	100	50	50	20	15

a Where the water supply line must cross the sewer line, the bottom of the water service within 10 ft of the point of crossing shall be at least 12 in. above the top of the sewer line. The sewer line shall be of cast iron with leaded or mechanical joints at least 10 ft on either side of the crossing.

b Not recommended as a substitute for a septic tank. To be used only when found necessary and approved by the health authority.

or perforated nonmetallic pipe. Individual laterals should not exceed 100 ft in length and the trench bottom and piping should be level. Use of more and shorter laterals is recommended because if a breakdown should occur in any one lateral, most of the field would still be operative. The space between laterals should be at least twice the depth of gravel to prevent overtaxing the percolative capacity of the adjacent soil.

The depth of the absorption trenches should be at least 24 in. to provide the minimum required gravel depth and earth cover. Additional depth may be required for ground contour adjustment, for extra aggregate specified under the pipe, or for other design purposes. The minimum distance of 4 ft between the bottom of the trench and the water table is essential to minimize groundwater contamination. Freezing is an extremely rare occurrence in a well-constructed system that is kept in continuous operation. It is of course extremely important that the pipe be completely surrounded by the gravel to provide for free movement of the waste water.

The required absorption area is based upon the results of the percolation tests and may be selected from Table 21-5 or 21-6.

Example 21-1

For a three-bedroom house and a percolation rate of 1 in. in 15 min, the necessary absorption area will be 3 bedrooms $\times$ 190 ft^2 per bedroom (Table 21-5) = 570 ft^2. For 2-ft-wide trenches with 6 in. of gravel below the drain pipe the total length of trench will be: 570 $\div$ 2 = 285 ft. If this length is divided into three portions (3 laterals), the length of each lateral will be 285 $\div$ 3 = 95 ft. If this length is too long for the site, the number of laterals must be increased. Using 5 laterals, the length of each lateral will be 57 ft. If the trenches are separated by 6 ft, the width of the field will be 2-ft-wide trenches $\times$ 5 trenches = 10 ft plus 6 ft between trenches $\times$ 4 spaces = 24 ft. The total field will then be 57 ft in length by 34 ft. in width for a total area of 1938 ft^2 plus the additional land required to keep the field an acceptable distance from property lines, wells, etc.

Construction

Careful construction is extremely important in achieving a satisfactory soil absorption system. Care must be exercised so as not to seal the surfaces on the bottom and sides of the trenches. Trenches should not be excavated when the soil is wet enough to smear or compact easily. Open trenches should always be protected from surface runoff to prevent entrance of silt and debris. All smeared or compacted surfaces should be raked to a depth of 1 in. and loose material removed before placing gravel in the trench.

The pipe should be completely surrounded by clean, graded gravel ranging in size from ½ to 2½ in. Cinders, broken shells, or similar materials are unsuitable as they are too fine and will lead to premature clogging of the soil. The gravel should extend at least 2 in. above the top of the pipe, at least 6 in. below the bottom of the pipe and fill the entire width of the trench. The top of the gravel should be covered with untreated building paper or a 2-in. layer of hay, straw, or similar pervious material to prevent the earth backfill from clogging the gravel. If an impervious covering is used, it will interfere with evapotranspiration at the surface. This is an important factor in the operation of a disposal field and, although evapotranspiration is not generally taken advantage of in the calculations, it provides an added factor of safety.

If tile pipe is used, the upper half of the joint openings should be covered. Drain tile connectors, collars, clips or other spacers with covers for the upper half of the joints may be used to assure uniform spacing, proper alignment, and protection of the joints. They are available in galvanized iron, copper, and plastic.

The problem of root penetration can be avoided by the use of a liberal quantity of gravel around the pipe. There should be at least 12 in. of gravel beneath the pipe when a trench is within 10 ft of large trees or dense shrubbery.

Backfilling of the trench should be hand tamped and the trench should be overfilled at least 4 to 6 in. This will prevent settlement to a point lower than the surface of the adjacent ground where storm water could collect and cause premature saturation of the absorption field and possible complete washout of the trench. Machine tamping or hydraulic backfilling should never be permitted. Figure 21-2 illustrates a typical absorption trench.

Seepage Beds

The use of seepage beds in lieu of standard trenches has been around for over twenty-five years. Common design practice for soil absorption fields is for trenches with widths varying from 12 to 36 in. When trenches are wider than 3 ft they are called seepage beds. Typically rectangular in shape, seepage beds are compact and used when less land is available for system design. Dry climates prove to be a better environment for use than climates having wet, humid conditions. Keep in mind, seepage beds do not have the sidewall area to provide oxygen to the center of a bed and long-term performance depends on the condition of the sidewall area. Slopes greater than 5% are not suitable for this absorption system application. Care must be taken during construction so as not to destroy soil structure by compacting the soil in the bottom of the bed. Additionally, the Federal Building Administration has sponsored studies indicating that seepage beds are a satisfactory method for disposing of the effluent from septic tanks in soils that are satisfactory for soil absorption systems. The studies have demonstrated that the empirical relationship

Figure 21-2 Section through Typical Absorption Trench

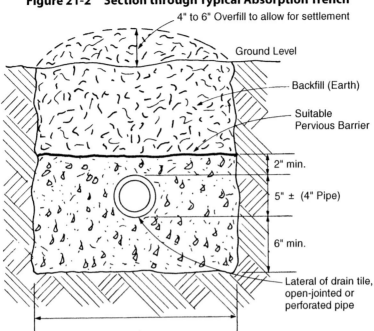

between percolation tests and the bottom area of trenches is applicable for the design of seepage beds.

The three main elements of a seepage bed are the same as those of trenches:
1. The absorption surface
2. The gravel layer
3. The distribution system.

The advantages of seepage beds are (1) a wide bed makes more efficient use of land than a series of long narrow trenches with wasted land between the trenches and (2) efficient use can be made of a variety of modern earth-moving equipment already at the site, which will result in lower costs for the system.

Design Criteria for Seepage Beds

The following criteria should be adhered to in the design of seepage beds:
1. The amount of bottom absorption area shall be the same as for trenches, shown in Table 21-5 or 21-6.
2. Percolation tests should be performed as previously outlined.
3. The bed should have a minimum depth of 24 in. to provide a minimum earth backfill cover of 12 in.
4. The bed should have a minimum of 12 in. depth of gravel extending at least 2 in. above and 6 in. below the pipe.
5. The bottom of the bed and the distribution tile or perforated pipe should be level.

6. The drain lines for distributing the effluent from the septic tank should be spaced no greater than 6 ft apart and no greater than 3 ft from the bed sidewalls.

Distribution Boxes

Although many codes specifically require the use of a distribution box in a soil absorption system, research and field tests have conclusively demonstrated that they offer practically no advantages and can be a source of serious problems in many installations. As a result of its study of distribution boxes, the Public Health Service set forth the following conclusions in the report to the Federal Housing Administration:

1. Distribution boxes can be eliminated from septic tank-soil absorption systems in favor of some other method of distribution without inducing increased failure of disposal fields. In fact, evidence indicates that distribution boxes as presently used may be harmful to the system.
2. Data indicate that on level ground, equal distribution is not necessary if the system is designed so that an overload trench can drain back to the other trenches before failure occurs.
3. On sloping ground a method of distribution is needed to prevent excessive buildup of head and failure of any one trench before the capacity of the entire system is utilized. It is doubtful that distribution boxes as presently used give equal distribution. Rather, they probably act as diversion devices sending most of the liquid to part of the system.

Because of the above findings, it is recommended that distribution boxes be eliminated in all disposal field systems where they are not specifically required by local codes.

Seepage Pits

Where absorption fields are impractical, seepage pits may be applicable. The capacity of a seepage pit should be computed on the basis of percolation tests made in each vertical stratum penetrated. The weighted average of the results should be used to obtain the design figure. Soil strata in which percolation rates are in excess of 30 min/in. should not be included in computing the absorption area.

Effective Area of Seepage Pit

The effective area of a seepage pit is the vertical wall area of the pervious ground below the inlet. The area of the bottom of the pit is not considered in calculating the effective area nor is any impervious vertical areas. Table 21-8 is a compilation of vertical surface area for various pit diameters and depths. The bottom of the pit must always be at least 4 ft above groundwater table.

When more than one pit is required to obtain the necessary absorption area, the distance between the walls of adjacent pits should be equal to three times the diameter of the largest pit. For pits 20 ft or greater in depth the minimum spacing between walls should be 20 ft.

Table 21-8 Vertical Wall Areas of Seepage Effective Strata Depth Below Flow Line (below inlet)

Diameter of seepage pit (feet)	1 foot	2 feet	3 feet	4 feet	5 feet	6 feet	7 feet	8 feet	9 feet	10 feet
3	9.4	19	28	38	47	57	66	75	85	94
4	12.6	26	38	50	63	75	88	101	113	126
5	15.7	31	47	63	79	94	110	126	141	157
6	18.8	38	57	75	94	113	132	151	170	188
7	22.0	44	66	88	110	132	154	176	198	220
8	25.1	50	75	101	126	151	176	201	226	251
9	28.3	57	85	113	141	170	198	226	251	283
10	31.4	63	94	126	157	188	220	251	283	314
11	34.6	69	101	138	173	207	212	276	311	346
12	37.7	75	113	151	188	226	264	302	339	377

Example: A pit of 5-foot diameter and 6-foot depth below the inlet has an effective area of 94 square feet. A pit of 5-foot diameter and 16-foot depth has an area of 94 + 157, or 251 square feet.

Construction of Seepage Pit

All loose material should be removed from the excavated pit. The pit should be backfilled with clean gravel to a depth of 1 ft above the pit bottom to provide a sound foundation for the pit lining. Material for the lining may be clay or concrete brick, block, or rings. Rings should have weepholes or notches to provide for seepage. Brick and block should be laid dry with staggered joints. Brick should be laid flat to form a 4-in. wall. The outside diameter of the lining should be 12 in. less than the diameter of the pit to provide a 6-in. annular space between the lining and pit wall. This annular space should be filled with clean, coarse gravel to the top of the lining.

Flat concrete covers are recommended. They should be supported by undisturbed earth and extend at least 12 in. beyond the excavation. The cover should not bear on the lining for support. A 9-in. capped opening in the pit cover is convenient for pit inspection. All concrete surfaces should be coated with a bitumastic paint or similar product to minimize corrosion.

All connecting piping should be laid on a firm bed of undisturbed soil throughout their length and at a minimum grade of 2% (¼ in./ft). The pit inlet pipe should extend at least 1 ft into the pit with a tee or ell to direct the flow downward to prevent washing and eroding of the sidewalls. When more than one pit is utilized they should be connected in series.

Valves

It would probably be impossible to find an industry anywhere in the world in which valves do not perform an important and vital function. Valves perform their function everywhere; from high-rise buildings to single-story buildings, from space probes to deep-sea explorations, from nuclear submarines to surface vessels, from high-temperature to cryogenic applications, from airplanes to automobiles, from power plants to refineries, and factories of every type imaginable.

What's available in valves today? The answer is basically the same types that were available ten or more years ago. However, some valves are rapidly replacing others in popularity. Two of the most commonly specified valves (gate and globe) are being hard pressed by ball and butterfly valves.

Valves are of primary importance in piping systems because of their basic function: controlling the flow of liquids or gases by on-off service, throttling service, or backflow prevention.

Valve Selection

Selection of the correct type of valve for a specific installation is dictated by the purpose for which it is to be used. For on-off (starting-stopping) service, the selection of a gate, butterfly, ball, or plug valve is indicated. Where the service requirement is flow regulation or throttling, the choice should be globe, butterfly, or ball valves. For prevention of backflow, the selection is check valves.

The function of a valve is to control the flow of liquids or gases. It then becomes obvious that valve selection is dependent upon the characteristics of the fluid to be controlled. The following factors must be evaluated for satisfactory valve selection:
- Whether the fluid is a liquid or a gas
- The viscosity of the fluid (free-flowing characteristics)
- Whether the fluid contains abrasives, granular or fibrous particles
- Fluid temperature ("normal" range, elevated, or cryogenic)
- The fluid pressure
- The degree of leak tightness required
- The maximum pressure drop that can be tolerated through the valve.

Gate Valves

Valve Stems

The sole function of the stem in a gate valve is to raise and lower the disc. The stem should not be subject to corollary stresses and strains of service conditions on the disc. For this reason, a relatively loose disc-stem connection is required. If the disc-stem connection were rigid, side thrust on the disc by pressure and flow would be transmitted to the stem with possible straining and bending of the stem.

The most common stem configurations are
- Rising stem, outside screw and yoke (OS and Y) (see Figure 22-1).
- Rising stem, inside screw (see Figure 22-2).
- Nonrising stem, inside screw (see Figure 22-3).

For the OS and Y construction, the stem threads are outside the valve. When the handwheel is rotated to open the valve, the stem-threading mechanism causes the stem to rise while the handwheel remains in the same location. The OS and Y construction is especially recommended for high temperatures, corrosive liquids, and where the liquid contains solids that might damage stem threads located inside the valve. Lubrication is a simple and easy procedure with external threads, but since the threads are exposed, care must be exercised to protect them from damage.

The rising stem, inside screw construction is generally employed with bronze gate valves (Figures 22-4 and 22-5). The handwheel and stem both rise as the valve is opened; it is therefore important that adequate clearance be provided for valve operation.

Figure 22-1   Rising Stem — Outside Screw and Yoke

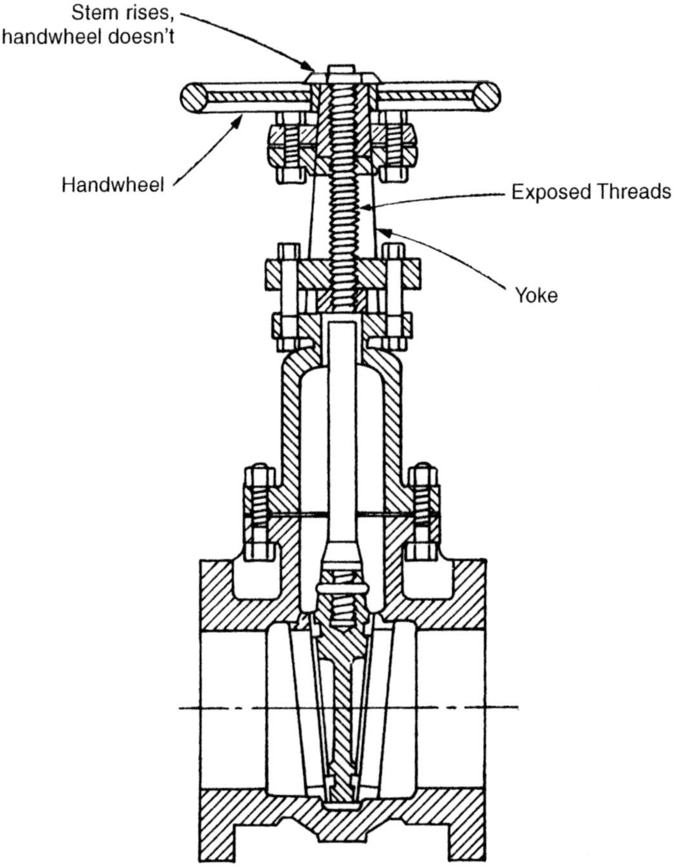

Figure 22-2 Rising Stem — Inside Screw

Handwheel and stem rise together

Figure 22-3 Nonrising Stem — Inside Screw

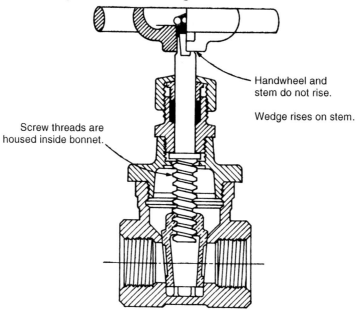

Handwheel and stem do not rise.

Wedge rises on stem.

Screw threads are housed inside bonnet.

The nonrising stem, inside screw construction requires minimum clearance for its operation. The disc moves on the stem as the handwheel is turned. Some disadvantages of this design should be noted. Heat, corrosion, erosion, and solids in the fluid could damage the stem threads due to their constant exposure to the line fluid. In addition, the position of the disc (open or closed)

Figure 22-4 Bronze Gate Valve Rising Stem — Wedge Disc

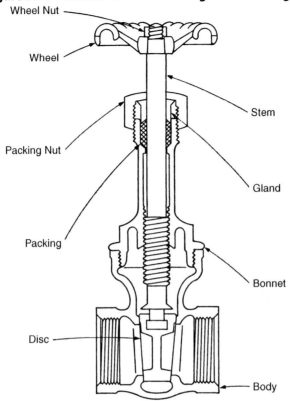

cannot be determined by the position of the handwheel or stem as it is with the rising-stem types (Figures 22-6 and 22-7).

Bonnets

Bonnets are available in various configurations to satisfy various applications:
- Union bonnet
- Screwed bonnet
- Bolted bonnet
- U-bolted bonnet
- Pressure seal bonnet.

The union bonnet configuration is a three-piece construction and is the preferred choice over the screwed bonnet for rugged service and where frequent dismantling for replacement or maintenance is anticipated. It is stronger and safer than the two-piece screwed configuration.

The simplest and cheapest design is the screwed bonnet. It finds its greatest application in low-pressure installations where shock and vibration are not present. It should not be used where frequent disassembly of the valve is required.

Figure 22-5 Bronze Gate Valve Rising Stem — Double Disc

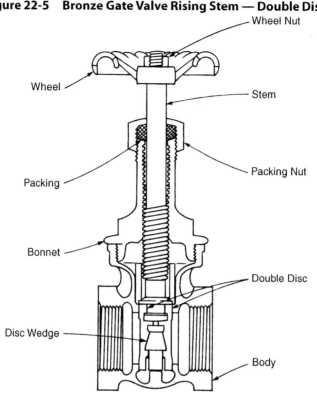

Bolted bonnets are easily disassembled with conventional wrenches. The body and bonnet are generally furnished with flat faces. Bolted bonnets are suitable for rugged service and all pressures and temperatures. They are practical for small and large valves, whereas the screwed and union bonnets are generally found on small valves only.

The U-bolt is a modified version of the bolted bonnet and is used where moderate pressures are encountered. This type of bonnet is usually found in the oil and chemical industries because of its relative ease of disassembly for cleaning and repair, its ruggedness, and its economy.

Pressure-seal bonnets are used in high-pressure and high-temperature applications where bonnet seals that provide more compact, safer body-to-bonnet connections are required. This type of bonnet is found in ANSI Class 600 and higher valves.

Discs

The control mechanism in a gate valve is a sliding disc (wedge) that is moved in and out of the flow passage of the body. The disc is restrained by guides in the valve body. In the fully opened position, the disc is completely out of the flow passage and thus permits a straight-through flow of the fluid through the passageway. The diameter of the passageway is nominally equal to the pipe diameter, which results in a pressure loss through the valve that is

Figure 22-6 Non-Rising Stem Valve Open Position

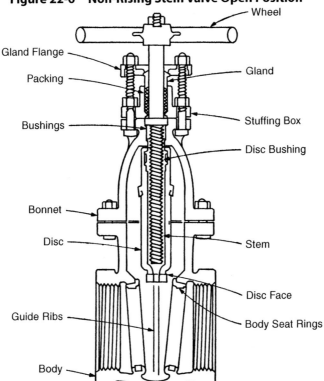

Figure 22-7 Bronze Gate Valve Non-Rising Stem — Wedge Disc

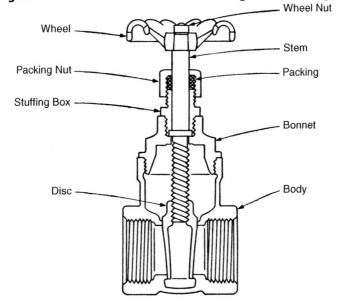

lower than that through a valve that has a restricted flow passage or a design which changes the direction of flow.

Four main types of discs are available in gate valves:
- Solid-wedge disc
- Double disc, parallel faced
- Split-wedge disc
- Flexible-wedge disc

Figure 22-8 Solid Wedge Disc

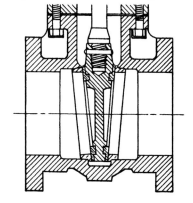

The solid-wedge disc (Figure 22-8) is the most widely used type in gate valves. It is noted for its simplicity of design and versatility. The solid wedge makes closure by descending between two tapered seats in the valve body. Solid-wedge disc seating is available in brass, iron, and steel gate valves.

Selection of the double disc is indicated where the application requires a tight seal to assure a leakproof shutoff. The double disc makes closure by descending between two parallel or tapered seats in the valve body. After parallel-faced double discs are lowered into position, they are seated by being spread against the body seats. A disc spreader makes contact with a stop in the bottom of the valve and forces the discs apart. Valves with double discs are widely used in the waterworks and sewage fields and in the oil and gas industries.

Figure 22-9 Split-Wedge Design

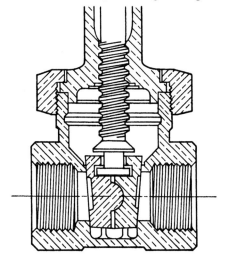

The split wedge (Figure 22-9) is a two-piece disc that seats between matching tapered seats in the body. The spreader device for pressing the discs against the body seats is simple and integral with the disc halves. As the valve is opened, pressure on the disc is relieved before the disc is raised, thus avoiding friction and scoring of the seat. Another type of split wedge is one with a ball-and-socket joint that forces each disc to align itself against the body seat for tight closure.

Flexible discs were developed especially to overcome sticking in high-temperature service with extreme temperature changes. It is solid through the center but not around the outer portion, where it is flexible. This type finds little application in plumbing work.

Materials

There is a wide selection of materials available for gate valves. In the selection procedure, the valve body and bonnet should be considered first, and then the valve trim. Factors to be considered in the selection of materials are
- Pressure and temperature
- Corrosion resistance
- Thermal shock
- Line stresses
- Fire hazard.

The most commonly used materials for the majority of gate valves in plumbing applications are bronze and iron. These materials are the most economical and readily available, and generally satisfy most requirements of pressure, temperature, and corrosion resistance.

Carbon steel and various alloy steels are used when greater strength is required or where high temperature, cryogenic temperature, or special corrosive conditions are encountered.

Valve bodies and bonnets are available in cast, forged or fabricated construction. For greater strength, forged and fabricated steel bodies are the choice over cast steel. The forged steel bodies are available only in a limited size range.

Trim

The term *trim* as applied to valves encompasses the elements of a valve relative to seating. There are many material options available for seat rings, stems, discs, and backseat bushings. An iron body, bronze-mounted (IBBM) gate valve has bronze seat rings, disc rings, stem, backseat bushing and packing gland. When used within their pressure/temperature ratings, IBBM gate valves offer excellent service for most plumbing systems.

The physical properties of a specific metal can often indicate its unsuitability for a particular service. For example, bronze gate valves are available with the body seats machined in the body. Bronze is a relatively soft metal (hardness range of 60–70 BHN [Brinell Hardness Number]) and if grit or dirt is present in the fluid, the valve seats on the body or disc can be easily scored or scratched, with subsequent leakage of the valve. In this case, the selection of bronze valves with seats of stainless steel, copper-nickel, or monel is indicated. Of course, optional trim material increases the initial cost of a valve, but the extended valve life and lower maintenance costs often justify their selection.

Gate valves are now generally available in nonmetallic materials, such as polyvinyl chloride (PVC), chlorinated polyvinyl chloride (CPVC), and polypropylene (PP). When these materials are selected, their temperature limitations should be carefully checked.

Table 22-1 lists the ASTM specifications for the most commonly used materials for gate valve bodies and bonnets. Other standards under which valves are also manufactured are listed in Table 22-2.

Packing

Packing is one of the most important and often overlooked features of a valve. Valve manufacturers furnish gate valves with a general-purpose packing, which may not be satisfactory for a particular service.

Most gate valves employ stuffing boxes with packing glands that can be tightened with open-end or adjustable wrenches. Glands that are not sufficiently tightened will permit leakage. Over tightening can squeeze out the lubricant in the packing.

General-purpose packing materials provided by manufacturers are
- TFE (Teflon) braided with graphite and other synthetic fibers for use up to 450°F in bronze and iron valves
- Highly resilient synthetic-based yarn, inconel wire inserted over a plastic core, graphite lubricated for use up to line temperatures of 1000°F in cast or forged steel valves
- TFE filament, square or chevron rings from pure TFE fibers for use up to 400°F in stainless steel valves

There are many other packings available for specialized applications, including Kevlar, pure graphite, cotton filament, nitrile rubber, and rubber in combination with a cotton jacket.

The proper selection, care, and maintenance of valve stem packing can make a significant difference in the efficiency of piping systems.

It is important to note that many municipalities have outlawed the use of asbestos in any form. Manufacturers no longer produce valves with asbestos in the packing material. More and more materials have become available as a substitute for asbestos and asbestos has disappeared from the market.

Table 22-1 Valve Material Specifications

ANSI Class	Material	ASTM Specification
125	Bronze	B-62
125	Cast iron	A126 Cl. B
150	Bronze	B-62
150	Carbon steel	A-216 Gr. WCB
150	Forged steel	A-105
150	Ductile iron	A-395
200	Bronze	B-61
250	Cast iron	A126 Cl. B
300	Carbon steel	A216 Gr. WCB
300	Forged steel	A105
300	Ductile iron	A395
600	Carbon steel	A216 Gr. WCB
600	Forged steel	A105

Table 22-2 Valve Standards Agencies

American National Standards Institute 25 West 43rd Street, Fourth Floor New York, NY 10036	ANSI
American Petroleum Institute 1220 L Street, NW Washington, DC 20005	API
American Society for Testing and Materials 100 Barr Harbor Drive P.O. Box C700 West Conshohocken, PA 19348	ASTM
American Water Works Association 6666 W. Quincy Avenue Denver, CO 80235	AWWA
Manufacturers Standardization Society of the Valve and Fittings Industry 127 Park Street Vienna, VA 22180	MSS
Underwriters Laboratories 333 Pfingsten Road Northbrook, IL 60062	UL

End Connections

Threaded ends (screw) comply with the requirements of ANSI B2.1 for female taper pipe threads and are generally used for valves 2½ in. and smaller.

The larger the pipe size, the more difficult it is to make up the joint. Threaded ends are suitable for all pressures and are found in brass, iron, steel, and alloy steel valves.

Solder ends are used with copper tubing and are available in the smaller sizes. The ends conform to ANSI B16.8. Use of solder ends is limited under pressure and temperature parameters. Flared ends are also available in small sizes for use with copper tubing.

Flanged ends are generally used for valves 3 in. and larger. Flanged-end valves are installed between adjoining pipe flanges and made up with a gasket between the flanges for a tight joint. Cast-iron valves come with flat-face flanges (ANSI B16.1) and ductile iron and steel valves come with raised-face flanges (ANSI B16.5). There are other modified types of flanges available for special conditions, which are included in the ANSI B16.5 standard.

Weld ends (socket or butt weld) are available with steel valves. They are used for high temperatures and pressures and where leakproof connections must be maintained over long periods of time.

Application

Gate valves are used to start or stop flow. They should be used only in the full-open or full-closed position, *never* in the partially open position for throttling service. A partially raised disc can cause turbulence with resultant vibration and chattering of the disc that can cause damage to the valve part. There is also the danger of wire drawing (erosion) with subsequent seat leakage.

Gate valves should not be used where frequent operation is required. A closed 6-in. gate valve with a 300-psi inlet pressure and atmospheric outlet pressure is subjected to a load of more than 4 tons on the disc. While the valve is tightly seated there is no wear or undue stress on the disc or seats, but at each start of the opening or end of the closing cycle, there is the ever-present danger of wire drawing and erosion of the seating surfaces due to the high velocity of flow. Repeated movement of the disc near the point of closure can cause a drag on the seating surfaces causing *galling* and scoring on the downstream side.

A major market for gate valves is in commercial, industrial, and institutional construction. They are widely used by water utilities and conform to American Water Works Association (AWWA) requirements for this application.

Another major market is in the petroleum, gas, chemical, shipbuilding, pulp, metal, and food and beverage industries. Power generation is another area where gate valves (mainly cast steel and stainless steel) are specified.

Additional types of gate valves are available for specialized service, such as

- Knife gate valves (widely used in the pulp and paper industry)
- Slide valves (for low-pressure liquids and gases where absolute shut-off is not generally required)
- Cryogenic gate valves (liquified gases require a stainless steel extended bonnet and stem to keep the packing out of the freeze area and allow gasification in the bonnet chamber).

Operation and Maintenance

Before the piping system is placed in operation, the system should be thoroughly flushed to remove any chips, dirt, scale, etc., that may have entered the lines during construction. No valve should be operated before this is done, in order to avoid damage to the seating surfaces.

Gate valves should be closed slowly as the disc approaches the seat. The increased velocity of flow, caused by the reduced open area, will tend to flush out any solids that may have been trapped between the disc and seat. After opening the valve, the handwheel should be turned back one-quarter turn to ensure that the valve does not jam in the open position.

The packing nut should be tightened at the first sign of packing leakage. Leaks around the stem could cause corrosion and a loose stem could vibrate, causing damage to the disc or seat.

Exposed valve stem threads should be kept clean and lubricated for ease of operation and the prevention of wear from dirt or rust.

Trends

During the past few years, butterfly and ball valves have been increasingly specified in lieu of gate valves. Some traditional gate valve applications are now given to butterfly, ball, or plug valves. At the present moment, gate valves are still the predominant type of valve for shut-off service. Some gate valve designs are now using resilient materials such as TFE and nitrile to offer resilient soft seating.

Globe Valves

Globe valves derive their name from the globular shape of the body. Flow through a globe valve follows a changing course: the fluid enters the valve parallel to the valve port and after two 90° turns leaves the valve again parallel to the valve port. Globe valves are designed for start-and-stop service and are ideally suited for throttling service.

Globe Valve Seating

Unlike the perpendicular seating in gate valves, globe valve seating is parallel to the line of flow. The flow is controlled by a plug (disc) that moves perpendicular to the axis of flow. The seat of the valve is a machined ring insert fitted in the port opening of the valve. The disc and seat can be quickly and conveniently reseated or replaced, which makes the use of globe valves ideal for applications where frequent maintenance is required.

Globe valves are specified for the following applications:
- Frequent operation
- Throttling (flow regulation)
- Positive shut-off for gases and air
- Where the high pressure drop across the valve can be tolerated.

Globe Valve Structure

The structural elements of a typical globe valve are the same as those of gate valves:
- Body
- Bonnet
- Stem
- Disc
- Seats.

Materials and End Connections

Globe valves are available in a wide range of materials: bronze, all iron, cast iron, cast steel, forged steel, and corrosion-resistant alloys.

Body-end connections are the same as those for gate valves: screwed, soldered, flanged, and welded.

Bonnets
The following bonnet types are generally available:
- Screwed in and screwed on
- Union
- Flanged (bolted)
- Pressure sealed
- Lip sealed
- Breech lock.

Stems
The following types of stem configurations are available:
- Inside screw, rising stem
- Outside screw, rising stem and yoke (OS & Y)
- Sliding stem.

Discs
Globe valves regulate fluid flow by varying the size of the port opening through which the fluid flows. This is achieved by varying the position of the disc. All contact between the seat and the disc ends when flow begins. This is a distinct advantage for the throttling of flow with a minimum of wire drawing and seat erosion.

Tapered Plug Disc
This type of flow-control element has a wide seating contact with the tapered seat. The configuration results in a directly proportionate relation of size of seat opening to the number of turns of the handwheel and permits close flow regulation. Because of this feature it is possible to gauge the rate of flow by the number of turns of the handwheel. If it takes four turns to open fully, then one turn permits 25% flow, two turns 50%, and three turns 75% (see Figure 22-10).

Conventional Disc
This disc is constructed of metal and has a line contact between its tapered or spherical seating surface and a conical seat. This particular flow-control element is recommended for positive shut-off of liquids but is not recommended for throttling service (see Figures 22-11 and 22-12).

Composition Disc
This disc has a flat face that is pressed against a flat, annular metal seating surface. The disc unit consists of a metal disc holder, composition disc, and retaining nut. Composition discs are available in materials suitable for hot and cold water, steam, oil, air, gas, gasoline, and many other fluids. This disc type is highly regarded for dependable, tight seating for hard-to-hold fluids, such as gas and compressed air. They can usually tolerate the embedment of dirt without leaking. A composition disc is not recommended for throttling service (see Figure 22-13).

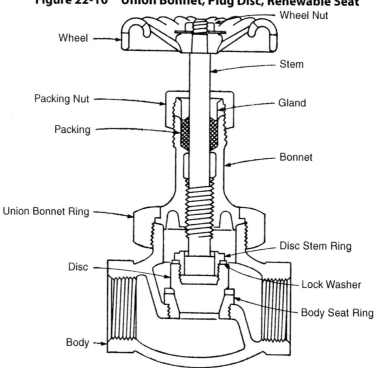

Figure 22-10 Union Bonnet, Plug Disc, Renewable Seat

Seals

There are generally four places in a globe valve where sealing is required. The sealing of three places prevents leakage of the fluid to the outside and the sealing of the fourth place prevents flow of the fluid within the system when the valve is closed. Leakage to the outside is stopped by seals at the valve end connections, the body-to-bonnet joint, and the stem. Valve seats are usually provided as integrally cast or replaceable seat rings that are screwed or pressed into the valve body. Tightness against leakage when the valve is closed is dependent on the fit-up tolerance, material, fluid characteristics, pressure, and temperature. Stem and bonnet-to-body seals are the same as those for gate valves.

Angle Valves

Angle valves are another form of globe valve and have the same operating characteristics as globe valves. They are used when making a 90° turn in the piping to reduce the number of joints and save installation time. An angle valve offers less restriction to flow (less pressure drop) than the globe valve and elbow it replaces.

Installation

Except when pressure under the disc is definitely required, a globe or angle valve will provide more satisfactory service when installed with the pressure (flow) above the disc. One definite exception is a valve that has a renewable composition disc, which should preferably have the pressure below the disc to assure longer disc life. Where continuous flow is a requirement, it is safer to

Chapter 22—Valves

Figure 22-11   Union Bonnet, Conventional Disc

Figure 22-12   Screwed Bonnet, Conventional Disc

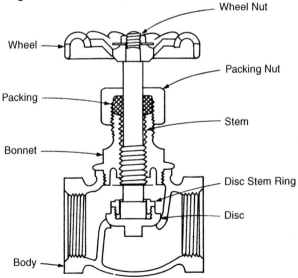

Figure 22-13 Union Bonnet, Composition Disc

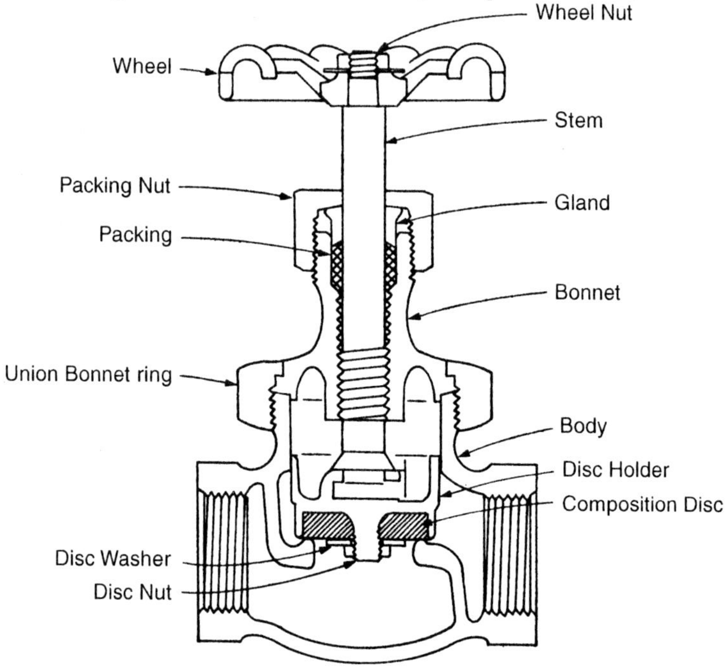

have the pressure below the disc. It is possible for a disc to become separated from the stem and automatically shut off the flow if pressure is above the disc. If this results in a dangerous condition for a particular service, then the pressure should positively be below the disc.

Another factor to be considered is the temperature variation. With pressure under the disc, cooling of the stem when the valve is in the closed position may cause sufficient contraction to unseat the valve and cause leakage. Pressure and temperature above the disc help ensure tight seating.

Check Valves

A check valve is the original truly automatic valve that is actuated by the line fluid. Check valves are designed to perform the single function of preventing the reversal of flow in a piping system. Flow opens these valves (and keeps them open), and reversal of flow plus gravity (or an applied force) causes them to close automatically.

The many varieties of available check valves conform in operating principle to either of the two basic valve types—swing or lift checks. The flow resistance (head loss) through swing checks is less than that through the lift type. The pattern of flow through swing checks is in a straight-through line without restriction at the seat, similar to that of a gate valve.

Swing-Check Valves

Closure of swing checks is dependent upon gravity (weight of the disc) and reversal of flow. The pivot point of the disc is outside the periphery of the disc, increasing the possibility that the fluid will flow back through the valve (backflow) before the disc can seat itself. The disc must travel through an arc of approximately 90° from the open position to the valve seat to effect complete shut-off. Since there is no opposing force to the downward movement of the disc, the speed of the disc, impelled by the force of the reverse flow, results in slamming and possible water hammer upon shut-off (Figure 22-14).

To avoid the dangers of water hammer and to eliminate slamming, swing checks are available with outside lever and weight or spring. By adjustment of the lever arm or spring tension, it is possible to cause valve closure at the moment of zero velocity of flow (just as flow reversal is about to begin) and thus eliminate slamming and water hammer (Figure 22-15).

Double-Disc Check Valves

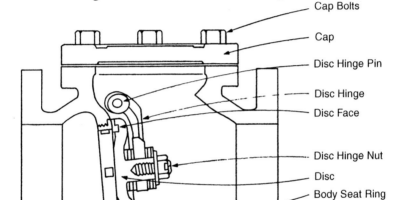

Figure 22-14 Conventional-Swing Check

In the double-disc (double-door) swing check (Figure 22-16), an improved version of the conventional swing check, the disc is split into two separate discs, resulting in a reduction of the mass of the single disc as well as shortening the travel distance from the open to closed position. A further improvement is achieved by the torsion springs, which cause closure of the double discs upon minimal flow reversal. The characteristics of this valve minimize the slam potential as compared to the conventional swing check. The valve also reduces the potential for water hammer but does not eliminate it.

The hinge pin in double-disc check valves is stationary and each disc swings freely when opening or closing. Multiple springs are incorporated in the design of larger-size valves to compensate for the heavier discs and increase the speed of closure.

Slanting-Disc Check Valves

The slanting-disc check valve (Figure 22-17) is a swing-check valve that incorporates features for lower head loss and nonslam operation. The main

Figure 22-15 Swing Check with Outside Lever & Weight

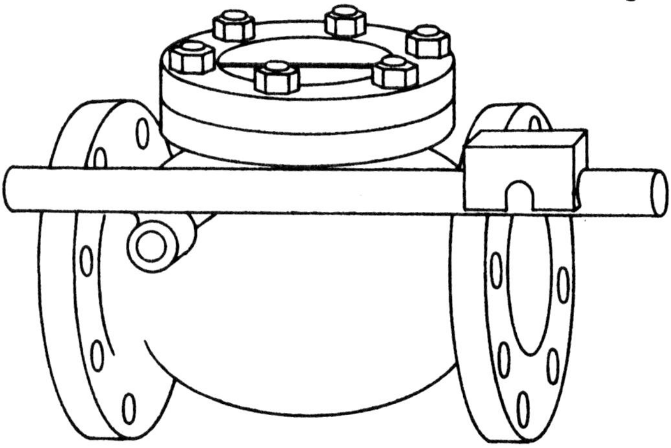

Figure 22-16 Double-Disc Check Valve

body, constructed of two pieces, provides a 50% greater flow area through the disc and seat section. The disc pivots off center with 30% of the disc area above the pivot point to impose resistance against the 70% area below the pivot point. This construction thus has a built-in nonslam characteristic but does not eliminate the possibility of water hammer.

The seat of the valve is placed at a 55° angle. The disc swings from this 55° closed position (rather than the 90° position in a conventional swing check), traveling a short distance to the open position of 15° off the horizontal. The short distance of disc travel permits only a minimal flow reversal before closure and provides a nonslam shut-off.

Lift Check Valves

The conventional lift check valve resembles a globe valve in construction and thus has the same characteristics relative to flow and head loss. The disc is equipped with a short guide, usually above and below, which moves vertically in integral guides in the cap and bridge wall. The disc is seated by backflow, or by gravity when there is no flow. This valve operates in horizontal lines only

Figure 22-17 Slanting-Disc Check Valve

so the disc is free to rise and fall depending upon the pressure under it (Figure 22-18). In addition to the conventional lift check, other available types are the horizontal-ball lift (Figure 22-19), vertical-ball lift, and foot valve.

Silent Check Valves

The design principle of the silent check valve is that it is silent when it closes, has a nonslam characteristic, and there is no water hammer. Silent check valves (sometimes called spring-loaded check valves) are available in the globe or wafer style.

Both styles operate in an identical manner, but the wafer style (Figure 22-20) has a higher head loss than the globe type. The center-guided poppet is spring loaded to be normally closed. The short distance between the poppet and the seat during flow conditions results in silent shut-off. This short distance is

Figure 22-18 Horizontal-Lift Check Valve

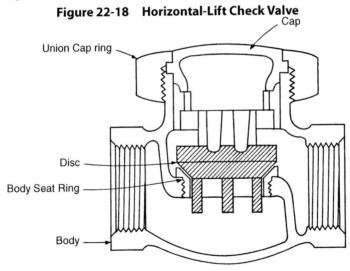

approximately ¼ of the valve size (a 4-in. valve has a 1-in. distance from full open to full closed). It is this short-poppet travel distance coupled with the spring force that accomplishes the silent shut-off. The range of shut-off time is approximately $\frac{1}{10}$ to $\frac{1}{20}$ of a second.

Figure 22-19 Horizontal-Ball Lift Check Valve

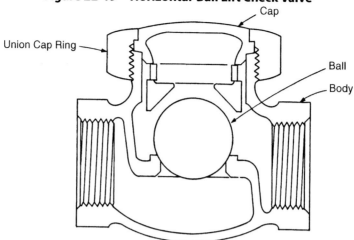

Silent check valves are furnished with helical or conical springs and are available in sizes to suit the specific design-pressure conditions of the piping system. Both types of springs perform equally well. Unlike swing checks that require a flow reversal to cause closure, the silent check is designed to close at the instant of zero-flow velocity before flow reversal occurs and thus eliminates any possibility of water hammer and slam.

It is important to differentiate between a nonslam check valve and a silent check valve. Although nonslam checks, as the name implies, eliminate slam, they do not eliminate the possibility of water hammer. Silent checks eliminate both slam and water hammer.

Figure 22-20 Silent Check Valve

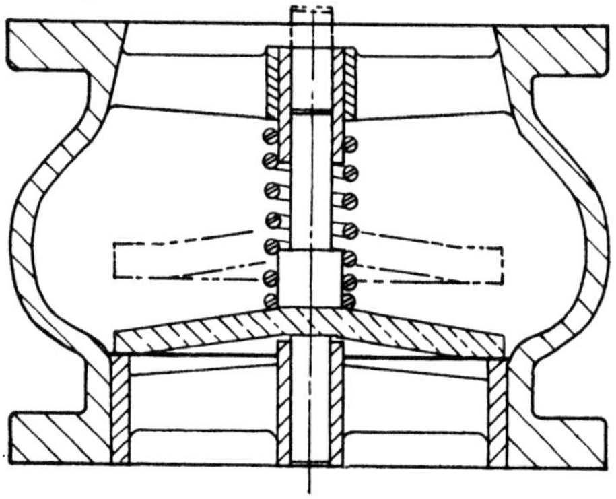

Check valves are available in all the materials, end connections, body closures, seats, etc., gate and globe valves are.

Installation

Swing checks can be installed in either the horizontal or vertical position. When installed in the vertical, the direction of flow must be up. Lift checks (except the vertical type) must be installed in the horizontal position only. Silent check valves can be installed in any position, and, when vertical, the flow can be either up or down.

Sizing

The discs and any associated moving parts of a check valve may be in a state of constant movement if the velocity head is not sufficient to maintain the disc in the wide-open position. The size of the check valve should be selected on the basis of flow conditions to prevent premature wear, noisy operation, and vibration.

The following formulas can be utilized to determine the minimum velocity required to hold the disc in the wide-open and stable position.

For bronze valves
$$V = 35V_s^{1/2} \text{ swing check}$$
$$V = 40V_s^{1/2} \text{ lift check}$$

For cast-iron valves
$$V = 48V_s^{1/2} \text{ swing check}$$
$$V = 30V_s^{1/2} \text{ tilting-disc check}$$
$$V = 25V_s^{1/2} \text{ lift check}$$

where V = Velocity of flow, ft/sec
 V_s = Specific volume of fluid, ft^3/lb

When check valves are sized on this basis it often results in the use of valves that are smaller in size than the pipe in which they are installed. The pressure drop through the valve will be no greater than that of a larger valve that operates partially open.

Pressure-Regulating Valves (PRV)

Excessive pressure in water distribution systems is a major source of trouble. In addition to creating operational difficulties, excessive pressure is a prime contributor to the increased frequency of equipment breakdown and the resultant increase in maintenance costs. Unless a specific high pressure is essential for the proper operation of certain system components, it is economic folly to design a system where the pressure will be in excess of 70 psi. In fact, if it is at all possible, it is desirable to limit the maximum pressure to 60 psi.

When a higher pressure is required for a piece of equipment or an operation, some separate means should be provided to boost the pressure for that specific function. When the pressure is greater than 70 psi it is difficult to maintain the flow velocity (which is a function of pressure) below the critical velocity of 10 fps. Noise begins to occur in a system when the velocity of 10 fps is approached. Even though noise in a water distribution system in certain areas may not be objectionable or damaging, the attendant high pressure and velocity can have the following detrimental effects:
- Accelerated erosion of piping
- Wire drawing of valve seats
- Hydraulic shock (water hammer) with consequent overstressing that can rupture pipe or damage equipment
- Damage to equipment not designed for high pressure or high velocity
- Reduced system life expectancy
- Excessive waste of water due to excessive flow rates at outlets.

These problems can be avoided by maintaining the pressure below the recommended maximum level of 70 psi. Pressure-regulating valves have proven to be the best method for the reduction of available pressure. There are devices other than pressure-regulating valves that can be used to reduce fluid pressure, but none are capable of maintaining the reduced pressure within an acceptable range when the inlet pressure or flow rate varies.

Automatic regulation of pressure in a water distribution system is a relatively simple process. Consider the setup shown in Figure 22-21 which will be used to demonstrate how a pressure-regulating valve functions. A globe valve (A) represents the pressure-regulating valve. Globe valve (B) represents the outlet at the fixture or equipment. Assume valves A and B are closed and a pressure of 80 psi exists at the inlet of valve A. The objective is to maintain a steady, reduced pressure of 40 psi at B. Opening valve A slowly will admit water to the branch and the pressure will rise. If valve A is then closed when the branch pressure reaches 40 psi, that pressure will be maintained as long as valve B remains closed and there are no leaks. As B is gradually opened to permit the water to flow, however, the pressure in the branch will immediately start to drop.

If, at the instant the pressure begins to drop, valve A is opened to a point that admits water into the branch at the same rate it is being discharged at B, the 40 psi pressure will be maintained. Opening valve B wider to simulate an increased demand requires that valve A be opened wider to equalize flow in and out of the branch and keep the pressure from dropping below 40 psi. Similarly, reducing the flow at valve B will cause a pressure rise requiring an immediate throttling adjustment at valve A to reduce the inflow. In practice, a PRV performs the functions of valve A, except that it does it *automatically.*

PRV Characteristics

Pressure-regulating valves fall into four general categories: Single seated, direct or pilot operated; and double seated, direct or pilot operated. Three typi-

Figure 22-21 Operation of a Pressure-Regulating Valve

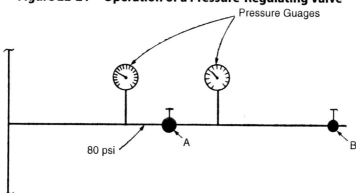

cal single-seated designs are shown in Figures 22-22, 22-23, and 22-24.

Construction of the direct-operated valve is comparatively simple. It has a pressure-adjusting screw that controls valve operation by regulating the action of a spring or weight-loaded pressure diaphragm or piston. The pilot-operated valve is more complex because of the manner in which the pressure is regulated; however, its design enables the valve to function efficiently over an extremely wide range of inlet pressures and flow rates.

In the pilot-operated unit (Figure 22-24), downstream pressure acting on the diaphragm of the control causes flow through the ejector to vary as pressure changes. This action, in turn, alters the pressure in the main valve cover chamber, resulting in a corrective change by the main valve to hold a constant

Figure 22-22   Direct-Operated, Spring Loaded PRV

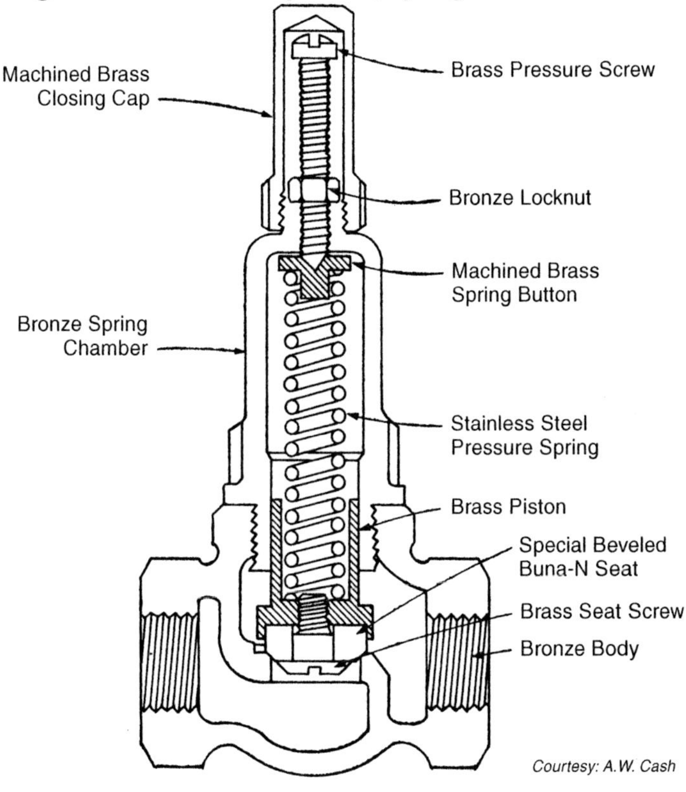

Courtesy: A.W. Cash

downstream pressure. Initial cost and maintenance are generally much less for the direct-operated valve than for the pilot-operated type.

Single-seated PRVs are used for intermittent flow or dead-end service (see Table 22-3 for definition of terms). Double-seated units find application primarily for continuous flow conditions; they are definitely not suitable for dead-end service.

As a general rule, large-capacity, single-seated, direct-operated PRVs should not be used for large flows where a high initial pressure must be reduced to a low system pressure. Under such conditions, these valves are very inefficient due to the extreme unbalance in pressures across the valve seat. For applications of this type the best choice is a pilot-operated PRV, which can handle large pressure reductions. The pilot-operated valve, however, is not well suited for operation with small pressure differentials.

Pressure on the outlet side of a direct-operated valve tends to decrease from the set pressure level as the rate of flow increases. This characteristic is

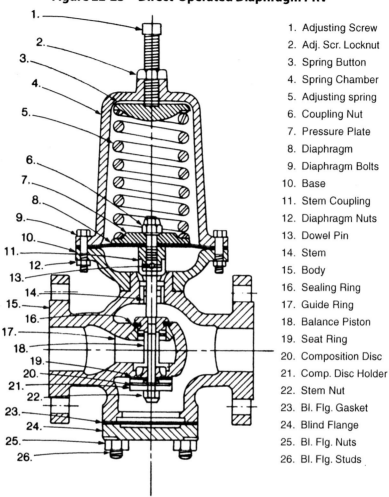

Figure 22-23 Direct-Operated Diaphragm PRV

1. Adjusting Screw
2. Adj. Scr. Locknut
3. Spring Button
4. Spring Chamber
5. Adjusting spring
6. Coupling Nut
7. Pressure Plate
8. Diaphragm
9. Diaphragm Bolts
10. Base
11. Stem Coupling
12. Diaphragm Nuts
13. Dowel Pin
14. Stem
15. Body
16. Sealing Ring
17. Guide Ring
18. Balance Piston
19. Seat Ring
20. Composition Disc
21. Comp. Disc Holder
22. Stem Nut
23. Bl. Flg. Gasket
24. Blind Flange
25. Bl. Flg. Nuts
26. Bl. Flg. Studs

particularly evident with spring-loaded PRVs but is less pronounced with the weight-loaded designs that provide a more constant outlet pressure.

As the valve stem moves in a spring-loaded PRV, the motivating pressure required varies because of the characteristic of the spring action. In a weight-loaded PRV, there is only a very slight change in the loading force, and the motivating pressure required, as the stem moves. Pilot-operated valves, on the other hand, regulate the pressure uniformly throughout their capacity range.

One problem with single-seated valves is that pressure fluctuations are proportional to the seat area divided by the area of the pressure sensing diaphragm or piston. With double-seated valves, variations in inlet pressure have little effect on outlet pressure because the pressure fluctuations act on opposing valve seats and balance each other.

Figure 22-24 Pilot-Operated Pressure-Regulating Valve

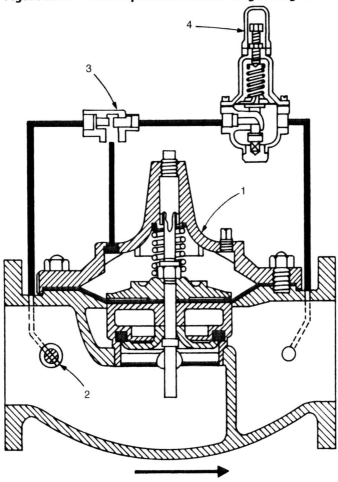

1. Main Valve
2. Strainer

Operating Data:
Downstream pressure acting on diaphragm of control (4) causes flow through ejector (3) to vary as pressure changes. This causes main valve cover chamber pressure to vary, resulting in corrective action of the main valve to hold a constant downstream pressure.

Courtesy: Cla-Val

Outlet Pressure

There have been many field complaints that an installed PRV does not deliver adequate outlet pressure. That is, the reduced pressure was below that required, even after falloff was taken into account, and neither adjustments nor spring replacements could correct the problem.

The common reaction in such cases is to place the blame on a faulty valve. However, because of how a PRV operates, it is much more likely that the ma-

Table 22-3 Pressure-Regulating Valve Glossary

The following terms are commonly used in discussing the operation of a pressure-regulating valve (PRV):

Set pressure. Pressure on the outlet (system) side of the PRV at which the valve begins to open.

Dead end service. Type of service in which the PRV is required to close tight when there is no demand.

Sensitivity. The ability of a PRV to sense a change in pressure. If the valve is oversensitive and reacts too quickly, the result is overcontrol and a hunting effect. Inadequate sensitivity leads to a sluggish operation and excessive variation in the outlet pressure.

Response. The ability of a PRV to react to a change in outlet pressure.

Falloff. The amount of pressure drop from the set pressure to meet flow demand conditions: The greater the flow the greater the falloff. A falloff of 20 psi is considered to be the maximum allowable for proper design.

Accuracy. Falloff from set pressure at full flow. Also used in reference to the capability for reproducing results under repetitive operation with identical conditions of flow.

No-flow pressure. The pressure maintained in the system by the PRV when no water is flowing. When there is no flow the PRV should shut tight to prevent the high-pressure fluid at the valve inlet from entering the system.

Reduced flow pressure. The outlet pressure maintained at the PRV outlet when water is flowing. This flow pressure is always less than the no-flow pressure by the amount of falloff. A PRV that is set to open at 45 psi would deliver a reduced flow pressure of 30 psi at peak demand if the falloff were 15 psi.

In addition to these special terms there are a number of synonymous terms that are used to designate certain operating conditions of a PRV. The following groupings contain terms that are generally considered to be interchangeable:

Inlet pressure, supply pressure, initial pressure, upstream pressure, line pressure.

Outlet pressure, reduced pressure, delivery pressure, downstream pressure, system pressure.

Capacity, peak demand, full flow, design flow rate.

---

jority of these problems are the direct result of improper valve selection and sizing.

When the outlet pressure drops below the acceptable range of falloff, it is a definite indication that demand requirements are greater than the rated capacity of the valve. The valve is simply too small for the job and the outlet pressure drops excessively as the valve attempts to satisfy the flow requirements. Replacement of the valve with one that is properly sized is the only solution for the problem.

On the other hand, a valve that is too large for the application can also create problems. If the valve must accommodate both small and large flows, the oversized unit will exhibit a high noise level during operation. In addition, at minimum flow conditions the valve will operate close to its seat causing wire drawing and erosion.

PRV Sizing

Careful selection and sizing of a PRV is essential for its efficient operation. The valve should *never* be sized solely on the basis of the size of the line in which it is to be installed. In general, a correctly sized valve is usually smaller than the line in which it is installed. Flow and pressure are the basic and all-important PRV selection factors. Furthermore, it is imperative that the outlet piping be sized to accommodate the full volume of flow at the required reduced pressure and suitable low velocity.

Manufacturers' capacity tables should always be checked for guidance in the selection of a properly sized PRV for a specific application. A general rule of practice to obtain good pressure regulation and to keep maintenance to a minimum is to select a valve size just large enough to satisfy the required maximum flow.

The volume of water that most direct-acting pressure-regulating valves will pass is governed by the pressure differential between the valve inlet and the outlet. As the differential increases, the quantity of flow increases. The maximum capacity of the valve is dictated by the differential pressure and the degree of regulation (falloff) that can be tolerated in the system pressure.

Cavitation

On the basis of regulator performance alone, higher inlet pressures tend to improve valve operation in terms of high rates of flow and less falloff of outlet pressure. It can be seen that inlet pressure is an important parameter in valve performance. Inlet pressure should always be a minimum of 70 psi (or higher) for best results.

It should be kept in mind, however, that there is a maximum pressure differential that a particular valve can handle efficiently and chattering begins to occur due to cavitation. The following equation can be used to determine the pressure differential at which cavitation will occur.

$$K = \frac{P_2 + 14.7}{P_1 - P_2}$$

where K = a dimensionless number
P_1 = inlet pressure, psi
P_2 = outlet pressure, psi

Cavitation will occur when K is less than 0.5. This condition is depicted in Figure 22-25.

Series Hookup

When the value of K drops into the cavitation zone it is an indication that the pressure reduction is too great for one valve to handle and the reduction must be accomplished in stages. This staging can be effectively handled by two (or more) valves in series. The primary valve reduces the pressure to a point at which the secondary valve completes the reduction to the required system pressure without causing cavitation.

Parallel Hookup

When one valve is selected to handle a wide range of flow rates it must deliver both minimum and maximum flows at the reduced system pressure. When a small flow volume is required, the valve seat opens slightly and then quickly closes. This action takes place repeatedly, causing undue wear of working parts, wire drawing of the seat, and chattering. The installation of two valves (one small and one large) in parallel can eliminate this problem.

In this arrangement the small valve opens when the demand is minimal. Because this valve is designed for low capacity, the seat will open full, giving

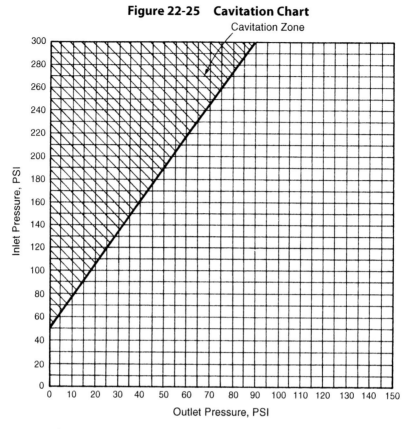

Figure 22-25 Cavitation Chart

After selecting valve size, locate inlet and outlet pressures on cavitation chart. If point located falls in cavitation zone, cavitation can occur.

smooth, quiet, efficient, and automatic operation. When demand increases above the capacity of the small valve the large valve operates to supply the greater flow requirements.

For a parallel hookup with two PRVs of different size, a ratio of 80–20% of total flow is recommended as a basis for sizing the valves. The outlet pressure of the large valve should be set to deliver the required system pressure and the outlet pressure of the small valve should be set for 10 psi higher than the required system pressure. This scheme permits the small valve to operate for light loads up to its capacity while the large valve remains closed. When the small valve reaches its capacity and demand continues to increase, there will be an additional falloff, down to the required system pressure, at which point the large valve will open to take over the load.

For example, if the total demand of a system is 200 gpm at a pressure of 40 psi, the large valve would be sized for a capacity of 160 gpm (80% of 200 gpm) at an outlet pressure of 40 psi; the small valve would be sized for 40 gpm (20% of 200 gpm) with an outlet pressure of 50 psi. In this way, the small valve provides sufficient capacity to satisfy low-flow conditions up to 40 gpm and the demand at the transfer point is sufficiently high that when the large valve begins to operate it will open fairly wide off its seat.

Installation

Pressure-regulating valves should always be installed in straight horizontal runs of piping (Figure 22-26). Spring-loaded valves should be mounted with the stem in the vertical position wherever possible; weight-loaded valves must always be mounted in the vertical position. A strainer should be mandatory at the inlet to every PRV to protect the valve seat from dirt and other foreign matter. A valved bypass should also be provided to permit maintenance, inspection, or replacement of the PRV without interruption of service.

Maintenance problems associated with pressure-regulating valves are often greatly exaggerated. If valves are properly selected and sized and have strainers installed at the inlet, they should cause no more trouble than any other piece of equipment requiring periodic inspection and service.

Quarter-Turn Valves

Plug, ball, and butterfly valves are referred to as quarter-turn valves because they move from the full-open to full-close position with a 90° rotation of the sealing member. They are unique in that quantity of flow is indicated by the position of the operating handle. These valves are applicable to a broad range of services and are available in a wide selection of materials, end connections, seat materials, and sizes.

These valves have achieved greater popularity during the past decade due

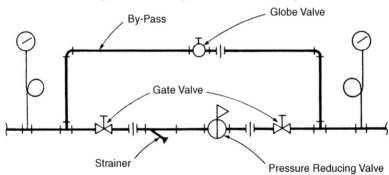

Figure 22-26 Typical PRV Assembly

to their adaptability, relatively small overall size, simple construction, rapid operation, and tight shut-off.

Plug Valves

The plug valve is probably the oldest type of valve in use today. Plug valves were in existence approximately 2,000 years ago and were used in the ancient water systems of Rome. The operation of the valve is extremely simple: a rotary cylindrical or tapered plug, with an opening through it, is fitted into an open body to permit or block the flow of liquids or gases (Figure 22-27). Rotation of the plug one-quarter turn from the closed position permits flow through the opening in the plug.

Plug valves are simple to operate, exhibit fast response, and add relatively little internal disturbance to flow. Because of its straight-through flow pattern, the pressure loss through the valve is low. Plug valves are not normally used for throttling, but some new designs are suitable for some throttling applications.

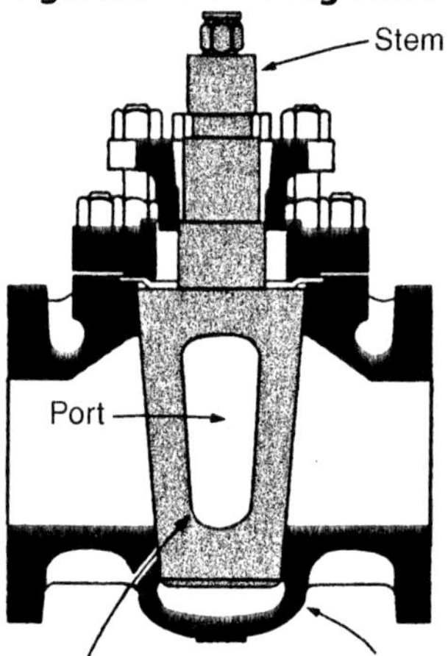

Plug valves come in lubricated or non-lubricated types. The lubricated valve should be specified for hard-to-seal substances. The plug is designed with grooves, which retain a lubricant to seal and lubricate the valve.

The accumulation of sediment and scale in the valve is prevented by the straight passage through the port and the wiping action of the plug as it is rotated 90°.

A disadvantage of the lubricated plug valve is the constant maintenance; lubrication is required at all times to maintain a tight seal between the body and plug.

Figure 22-27 Plug Valve

Ball Valves

Ball valves are similar to plug valves and are considered to be an adaptation of the basic plug valve. The closure member is a ball, instead of a plug, with a hole through it. In the open position the port in the ball connects the inlet and outlet ports in the body (Figure 22-28). Early ball valves had metal-to-metal seating and tight closure was a serious problem. Development of high-temperature elastomeric materials and improved plastics has overcome the problem. Seat materials are available in nylon, neoprene, and Buna-N.

The ball rotates between two resilient seats with concave seating surfaces. Fluid flow is straight through in the open position. A 90° rotation from the open position completely blocks the flow of fluid. Ball valves are available in reduced-port or full-port design (Figure 22-29). The full port design permits full flow through the valve with a negligible pressure loss.

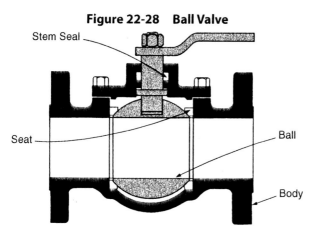

Figure 22-28 Ball Valve

Like plug and gate valves, ball valves are not generally recommended for throttling service.

Ball valves are quick opening, have low pressure drop, require minimum mounting space, require little maintenance, are tight sealing, and operate with the application of low torque.

Butterfly Valves

Butterfly valves are among the most practical devices for throttling control of liquids. The earliest throttling-type butterfly valves did not provide positive shut-off and thus did not find great application. In the early 1900s natural rubber was used for the seating material but achieved only a semipositive low-pressure shut-off capability.

It was not until the early 1940s that the emergence of synthetic rubbers

Figure 22-29   Full and Reduced Port Ball Valves

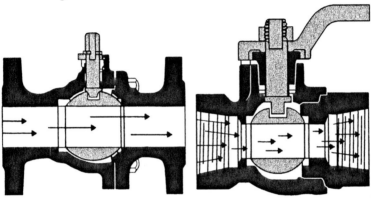

Full Port   Reduced Port

coupled with major advances in valve design led to the leak-free conventional and high-performance butterfly valves available today. The applications of butterfly valves are being constantly expanded through the continuing development of new seat and seal materials as well as new design innovations. At the present time, butterfly valves are being specified in preference to gate and globe valves for most applications.

Butterfly valves control flow by a circular disc with its pivot axis perpendicular to the direction of flow. When the plane of the disc lies parallel with the longitudinal axis of the pipe the valve is fully open. When the disc is rotated 90° the valve is fully closed. As the disc is moved from the fully closed position it offers progressively less resistance to flow until it reaches the position at which the plane of the disc is parallel to the longitudinal axis of the pipe. The force required for actuation does not exhibit a simple linear pattern. In a semi-open position the disc behaves like an airfoil, generating lift and drag forces that attempt to close it. When the disc reaches an open angle of approximately 67° the dynamic forces are at a maximum (Figure 22-30). Throttling is achieved by rotating the disc to an intermediate position and locking the handle in place or using a self-locking actuator. It is essential to have this locking feature because of the dynamic forces acting on the disc (Figure 22-31).

Although a butterfly valve can be used for throttling, its primary function is full-open/full-closed service. The valve is fully opened and fully closed in a quarter turn.

Resilient-lined or rubber-lined styles are the most popular and are available in three basic disc configurations (Figure 22-32). They are
- Vertical disc
- Offset disc
- Angle disc.

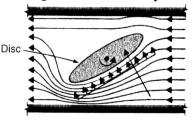

Figure 22-30 Dynamic Forces Acting on Disc of Butterfly Valve

When the disc reaches an open angle of approximately 67 degrees, dynamic forces are at a maximum.

The vertical disc valves generally have a shorter cycle life than the other two types due to the scrubbing and compression set of the elastomer that occurs in the disc boss area. Offset discs eliminate this problem but they have a reduced port diameter, which results in a less efficient valve. The angle disc configuration either minimizes or effectively

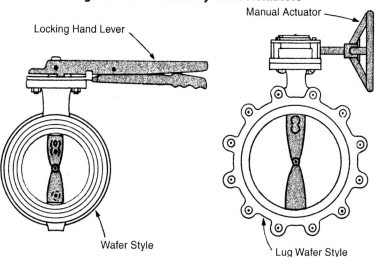

Figure 22-31 Butterfly Valve Actuators

eliminates both of these problems.

The wafer style butterfly valve is used almost exclusively in plumbing work and is available in three configurations (Figure 22-33):
- Span
- Lug (sometimes called "single flange")
- Double flange.

The most popular wafer butterfly valve is the span, followed closely by the lug. The span's popularity is due partly to its ease of installation. It is slipped between properly spaced companion flanges on the piping and is cradled by

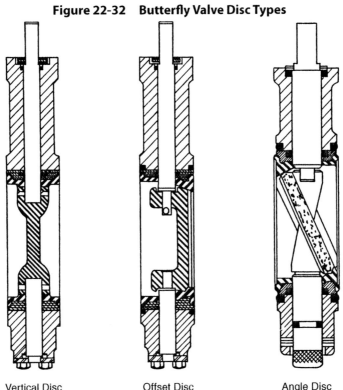

Figure 22-32 Butterfly Valve Disc Types

Vertical Disc
Offset Disc
Angle Disc

Figure 22-33 Butterfly Valve Types

Body
End
Fittings
Span-Type
Lug-Type
Flange-Type

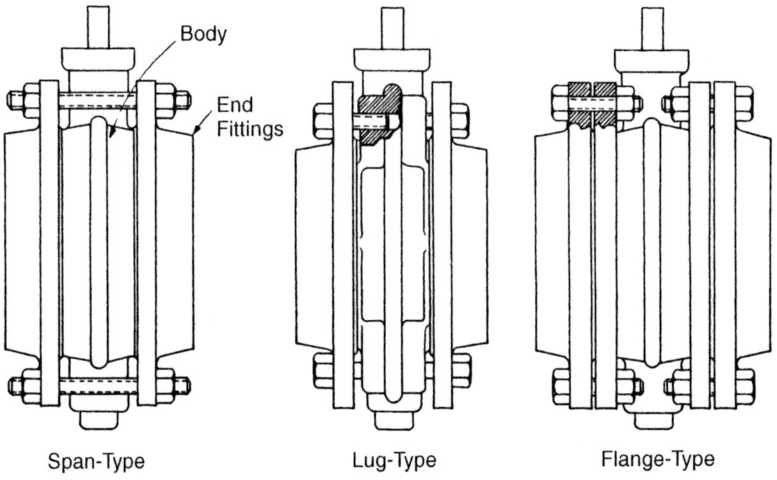

the bolts that join the pipe flanges. The valve does not require any special alignment during the bolting up operation. The lug wafer butterfly valve is also slipped between pipe flanges but the bolts pass through the lugs on the body of the valve. The double-flange wafer valve is installed in the exact same manner as any other flanged valve.

The seats of many resilient-seated wafer valves are designed to extend beyond the valve body face to provide the sealing surface between the body and pipe flanges. Valves with other types of resilient seat design must utilize flange gaskets to provide a leak-free joint.

Butterfly valves generally provide positive shut-off up to 250 psi and there are special designs capable of handling pressures as high as 720 psi.

Particular care must be exercised in closing butterfly valves so as to avoid water hammer. When they are closed too quickly (which can easily happen because of the quarter-turn feature), hydraulic shock will occur. When they are closed quickly the disc is subjected to the full force of the high-pressure wave.

The bodies of butterfly valves are very short compared to bodies of other valve types. They take up very little longitudinal space and this compactness results in a valve that is much lower in weight for a given capacity than other types of valves. A typical iron-body 6-in. wafer butterfly valve measures $2\frac{1}{8}$ in. face to face and weighs approximately 28 lb, whereas a 6-in. flanged gate valve is $10\frac{1}{2}$ in. long and weighs approximately 175 lb.

The advantages of butterfly valves include light weight, economy, simplicity of design, compact configuration, ease of installation and maintenance, quick operation, reliability, and versatility.

Index

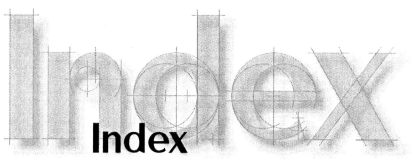

1-compartment sinks, 17
1-piece tank water closets, 4, 5. *See also* water closets
2-circuit systems in water supply, 128–129
2-compartment sinks, 17, 18
2-elbow offsets, 193
2-hand hole traps, 41, 63
2-level drinking fountains, 19
2-piece tank water closets, 4. *See also* water closets
3-compartment sinks, 17
3-elbow swing loops, 193
4-elbow U-bends, 193
5-elbow swing loops, 193
6-elbow swing loops, 193
25-year storms, rate of flow and, 69

A

abrasion, 124, 222
abrasion resistance, 2, 17
absolute pressure, 21
absorption areas, 216
absorption rates for septic systems, 214
absorption trenches. *See also* seepage beds; seepage pits
construction, 218
functions, 216–217
illustrated, 219
accessibility
accessible showers, 18
accessible water closets and toilets, 8, 10
"easy access" defined, 2
to septic tanks, 210
accuracy in pressure-regulating valves, 246
acidity, water system materials and, 128
acid-resistant fixtures, 2
acoustics in plumbing systems
hydraulic shock noise, 120
noise and design issues, 125
velocity and, 124, 139
water closets and, 4
actual velocity, vs. ideal, 115
AGA (American Gas Association), 161, 181
air
cores of, in stacks, 77

flow in stacks, 77
globe valves and, 232
gravity circulation of, 78–79
outlet flow rates, 75–76
properties of, 74
suds pressure and, 88
vent extensions and, 79
vortexes in fixture traps, 23
air chambers
hydraulic shock and, 122
tests, 123
air circulation
in ejector basins, 106
in vents, 78–79
air compression, 74
air compressors, 101, 103
air-conditioning
equipment in drainage system calculations, 58–59
Legionnaires' disease and, 179
makeup water, continuous demand and, 131
water supply in summer, 130
air gaps, overflows and, 152
air inlets
house traps and, 39–41, 42
storm traps and, 63
airports, 213
air pressure, 74
air circulation and, 78
in air columns, 76
in sanitary systems, 75
air-testing drainage systems, 92
alarms in ejector systems, 107
alkalinity, water system materials and, 128
alloy steel gate valves, 229
aluminum fixtures, 2
American Gas Association (AGA), 161
American National Standards Institute (ANSI)
A112.6.1 standard, 10
A112.19.6 standard, 8, 13
B2.1 standard, 230
list of standards, 3
valve material specifications, 230

American Petroleum Institute (API), 230
American Society for Testing and Materials (ASTM), 3, 229, 230
American Society of Heating, Refrigeration, & Air Conditioning Engineers (ASHRAE), 3, 171
American Society of Mechanical Engineers (ASME), 8, 10, 13, 167–168
American Society of Plumbing Engineers (ASPE), 171
American Society of Sanitary Engineering (ASSE), 3, 10
American Water Works Association (AWWA), 196, 230
ammonia in condensers, 153
anaerobic conditions in septic tanks, 209
anchoring water supply systems, 158
ancient water systems, 250
angle-disc butterfly valves, 252–254
angle valves, 234–236
annual costs. *See* economic concerns
ANSI. *See* American National Standards Institute (ANSI)
anti-siphon ballcocks, 10
apartment buildings
fixture-unit studies, 133
hot-water demand and recovery estimates, 167–168, 169, 172–173, 175
recirculating hot-water systems, 188–191
sewage flows, 213
API (American Petroleum Institute), 230
appliances. *See* fixtures and fixture outlets
aquastats, 162
architects, fixture selection and, 2
arm baths, 175
asbestos cement storm-drainage pipes, 67
asbestos in valve packing, 230
ASHRAE (American Society of Heating, Refrigeration, & Air Conditioning Engineers), 3, 171
ASME (American Society of Mechanical Engineers), 8, 10, 13, 167–168
ASPE (American Society of Plumbing Engineers), 171
asphaltum paint, 158
aspirators, 152, 153
ASSE (American Society of Sanitary Engineering), 3, 10
ASTM (American Society for Testing and Materials), 3, 229, 230
athletic establishments, 175
atmospheric pressure
standard air and, 74
in total pressure calculations, 22
trap seals and, 21
attachments, pipe, 45–46. *See also* joints
auditoriums, 134–135
automatic flow-control orifices, 126–127
automatic flushometer valves, 11
automatic pressure regulation, 242
automatic storage water heaters, 162–163
availability. *See* demand
average flow rates
calculating for fixtures, 50
defined, 130
in design loads, 131
average velocity, 110
AWWA (American Water Works Association), 196, 230

B

backfilling
absorption trenches, 218
seepage beds, 219
underground drainage pipes, 43
backflow. *See also* backflow preventers; back-siphonage
back-pressure tests, 8, 9
below-rim connections and, 152
in drainage system branches, 50
replenishing trap seals, 24
in sewer drainage systems, 39, 40
swing check valves and, 237
backflow preventers, 152, 222
backing off fittings, 194
back-outlet water closets, 4, 6–7. *See also* water closets
backseat bushings, 229
back-siphonage, 12. *See also* backflow; siphonage
back-spud bowls, 5
back-spud (flushometer) water closets, 5. *See also* water closets
back-to-back water closets, 10
backup of water
in drainage system branches, 50
pressure fluctuations and, 52
back vents, 83
backwash from water softeners, 213
backwater valves
drainage systems, 39, 40
subsoil drainage, 70
bacteria
Legionnaires' disease, 179
septic tanks, 209
testing for, 197
baffles in septic tanks, 210
baking-sink grease interceptors, 25
ballcocks, 10, 11–12, 152–153
ball-removal tests, 8
ball valves
functions, 222
quarter-turn valves, 249, 250–251
water supply systems, 159
barber shops, 26, 51, 52

barium traps, 25
barrier-free access. *See* accessibility
basins
ejectors and sumps, 100
materials, 100
sizing, 106–107, 108
bathrooms and bathroom groups
fixture units, 52, 133
sizing sanitary systems and, 60
typical demand, 131
bathtubs
demand, 131, 175
fixture units, 133
minimum required pipe sizes, 140
trap size, 51
types and requirements, 18
water storage needs for, 174
beauty salons, 26, 51, 52
bedpan washers, 90, 153
Bell, John, 28
bellows
bellows expansion joints, 194
in shock absorbers, 123
below-rim connections, securing, 152–154
Bernoulli's theorem, 30–31, 114–115
BHN hardness range, 229
bidets
defined, 19
fixture units, 52
trap size, 51
bi-level drinking fountains, 19
bilge pumps. *See* sump pumps
biological and biomedical laboratories. *See* laboratory facilities
bitumastic paint, 100, 210
bituminous fiber, 67
blackouts, sump-pump systems and, 108
bleach in septic tanks, 212
Bloodgood, Don Evans, 28
blow-off valves, 159
blowout fixtures
blowout fixture traps, 79
blowout urinals, 12, 13
blowout water closets, 4, 5, 87
boilers
boiler blowoffs, 39, 47
water heater installation, 176
boiling point of water, 109
bolted bonnets, 225, 226, 233
bolts for water closets, 9
bonnets
gate valves, 225–226

globe valves, 232–233, 234, 235, 236
booster-pump systems
increasing water pressure, 130
pressure in, 138
sizing systems, 141, 148–150
booster-water heaters, 164
bottle-type water coolers, 199
brackets, types of, 46
branches
in drainage systems, 50, 54
in stack systems, 57, 85
branch intervals, 54, 85
branch vents
defined, 83
relief vents, 83
research, 96–97
sizing, 96–97, 98
suds pressure, 89
brass downspouts, 67
brass piping
flow rate equation, 188
friction head loss, 139, 140
hydraulic tables, 142–147
pressure waves, 121
roughness, 118
breech-lock bonnets, 233
bricks
in seepage pits, 221
in septic tanks, 210
Brinell Hardness Number (BHN), 229
British Thermal Units (Btu)
heat loss, 186
water heaters, 167
bronze valves, 159
gate valves, 223, 225, 227, 229
globe valves, 233
hardness range, 229
materials, 241
Btu per hour
heat loss, 186
water heaters, 167
bubblers, 199, 200, 203
bubbles. *See also* detergents; soaps
in cavitation, 124
in fixture traps, 23
near hydraulic jumps, 37
suds pressure, 88–90
building codes. *See* codes and standards
building drains (house drains)
backwater valves, 39, 40
cleanouts, 44
defined, 38

runout to, 58
sanitary. *See* sanitary drainage systems
sizing, 58–62
stack offsets and, 55, 56
building occupancies. *See* occupancies
Building Research Center, 81
buildings. *See also* large buildings; multiple-family buildings; residential systems; specific types of buildings
hot-water demands, 167–168
terminal velocity lengths, 34–35
building sewers. *See also* sewer systems
defined, 38
distance to sewage disposal systems, 217
built-in showers, 18
bulk-media tests, 9
Buna-N seats in ball valves, 251
buried piping. *See* underground piping
burns from hot-water (scalding), 161, 164, 179, 180
bursting tanks, 161
butterfly valves
functions, 222
quarter-turn valves, 249, 251–254
in water supply systems, 159
butt-weld valve-end connections, 231
bypasses
hot-water systems, 161
pressure-regulating valves, 249

C

cafeterias
cafeteria-style water coolers, 199
estimating hot-water demand and recovery, 171
calcium hypochlorite, 196
calculations. *See* equations
Cameron Hydraulic Data, 139
camps, 213
Canada, frost vent-closure prevention in, 92
Canadian Standards Association (CSA), 9, 161
capacity. *See* flow rates; peak demand
cap flashing, frost vent-closure prevention and, 92
carbon-steel gate valves, 229
cast-iron bathtubs, 18
cast-iron downspouts, 67
cast-iron ejector or sump basins, 100
cast-iron fixtures, 2
cast-iron piping
flow rate equation, 188
friction head loss in pipes, 139
pressure waves in pipes, 121
cast-iron roof drains, 67
cast-iron soil pipe

drainage pipe supports, 43–44
storm drainage, 67
cast-iron valves, 159
check valves, 241
gate valves, 231
globe valves, 233
caulking, 158
caustic soda in septic tanks, 212
cavitation
defined, 124
pressure-regulating valves and, 247, 248
velocity and, 139
CDA (Copper Development Association), 124
centersets for faucets, 17
central water-cooling systems, 200–203
centrifugal pumps
lift devices, 101–105
sizing for ejectors, 103–106
cesspool systems, 208, 217
chambers, air
hydraulic shock and, 122
tests, 123
chattering. *See* cavitation
check valves
below-rim connections and, 152
bronze valves, 241
in compound water meters, 156–157
in ejector pumps, 108
in ejectors, 101
functions, 222
hydraulic shock and, 120
installing, 241
overview, 236–237
poppets in, 240
pressure loss and, 236–237
pumps and, 108, 122, 153
sizing, 241
submerged inlets and, 152
sump and well pumps, 153
types
differential, 156–157
double-disc, 237–238
lift, 236–237, 239, 240, 241
non-slamming, 238–240
silent (spring-loaded), 122, 239–240, 241
slanting-disc, 238–239
swing. *See* swing check valves
weight-loaded, 108
vacuum breakers and, 153
in water-supply systems, 159
chemical solutions in septic tanks, 212–213
chemical-solution storage tanks, 153–154

chemical tests for water systems, 197
children's water closets, 8
chilled drinking-water systems
  bubblers and stream regulators, 200
  compared to drinking fountains, 198
  continuous demand and, 131
  defined, 198
  milkiness in systems, 201
  pipes and piping, 205–206
  sizing systems, 202–204
  storage tanks, 204
  water coolers, 19, 198, 200
chlorinated piping, 196
chlorinated polyvinyl-chloride gate valves, 229
chlorine bleach in septic tanks, 212
circuit and loop venting
  defined, 87–89
  fixture traps, 80
  sizing, 97–98
circular lavatories, 15, 175
circular sewers, 32
circulating pumps
  heat gain, 203
  sizing chilled drinking-water systems, 204
circulation, air
  in ejector basins, 106
  in vents, 78–79
circulation systems (fluid)
  circulating pumps in chilled drinking-water systems, 204
  circulation rates, 187
  heat gain from circulating pumps, 203
  types of, 181–185
clamps, types of, 45–46
classifications of occupancy. *See* occupancies
cleanouts
  designing, 48
  in drainage pipes, 44, 47
  in house traps, 41, 42
  in indirect-waste systems, 47
clear floor space
  around urinals, 14
  around water closets and toilets, 10
clevises, types of, 45–46
climate, seepage beds and, 218
clips, types of, 46
clogging
  cleanouts and, 44
  in septic tanks and cesspools, 208, 209
close-coupled water closets, 5
closed-conduit flow, vs. open channel, 30
closure of valves
  hydraulic shock and, 120

non-slamming check valves, 238–240, 239–240
pressure waves and, 122
clothes washers. *See* laundry systems and washers
clubs, 175
coated-metal septic tanks, 210
codes and standards
  ASTM valve material specifications, 229, 230
  controlled-flow roof drainage, 70
  fixture-unit recommendations, 135
  lists, 2, 230
  valve standards, 230
  vent sizing, 97–98
  water-supply systems, 151
coefficent of expansion, 192
coefficent of friction. *See* roughness of pipes
cold-water systems
  chilled drinking-water systems. *See* chilled drinking-water systems
  mixing with hot, 176, 177
  sizing calculations, 148
combination sinks and trays
  fixture units, 52
  trap size, 51
combination waste and vent venting
  defined, 86–87
  fixture traps, 81
combination wye and 1/8 bend, 33
combined pressure and temperature relief valves, 161–162
combined sewers
  connections, 64
  drainage into, 38
  problems with, 38–39
  storm drainage, 38, 63, 70–73
combined upfeed and downfeed systems, 181, 183–184
commercial dishwasher grease interceptors, 25
commercial kitchen sink grease interceptors, 26
commercial laundries, 213. *See also* laundry systems and washers
common vents, 83, 88
compartments in septic tanks, 211
composition discs in globe valves, 233–234, 236
compound meters, 155, 156–157
compression
  air, 74, 101, 103
  compressive stress on pipes, 192
  suds, 88
  water's properties, 109
compressors
  air, 101, 103
  water coolers, 199
concrete and concrete block septic tanks, 210
concrete covers on seepage pits, 221

concrete ejector or sump basins, 100
concrete storm drainage pipes, 67
condensation
drainage for condensates, 39, 47
horizontal drainage pipes and, 42
storm-drainage systems, 64
condensers
in water coolers, 199
water supply for, 153
conductors. *See* downspouts and leaders
conservation of energy, 30–31, 114–115
conserving water
changes in demand, 134
Energy Policy Act of 1992, 132
urinals, 12
water closet fixtures, 5
constant flow in water supply system design, 126–128
consumption. *See* demand
contamination
disinfecting systems, 196–197
septic tanks and, 209
water closets and, 4
in water-supply systems, 125
continuous demand
continuous-flow baths, 175
defined, 131
fixture units in, 136
water meters and continuous flow, 156
continuous venting, 84
contraction
high-temperature wastes and, 39
pipe calculations, 192–195
storm-drainage systems, 64
water-supply pipes, 158
control boards, 157
controlled-flow roof drainage
drain-down time, 69
principles, 68–70
storm intensities, 69
suggested code for, 70
conventional discs in globe valves, 233, 235
cooling jacket drainage, 47
cooling tank drainage, 47
cooling-tower water. *See also* air-conditioning
Legionnaires' disease, 179
water supply in summer, 130
cooling water, water supply systems and, 152
Copper Development Association, 124
copper downspouts, 67
copper roof drains, 67
copper tubing or piping
drainage pipe supports, 44

flow rate equation, 188
friction head loss, 139, 140
hydraulic tables, 142–147
pressure waves in, 121
recommended velocity in, 124
roughness, 118
solder end connections, 231
water supply tubing, 158
corner strainers, 67
corner-tank water closets, 4. *See also* water closets
corrosion and corrosion resistance
ejector or sump basins, 100
fixture standards, 2
globe valves, 233
valve selection and, 223, 224, 229
velocity and, 139
water-supply systems and, 125, 158
costs. *See* economic concerns
cotton filament in valve packing, 230
counterflashing, frost vent-closure prevention, 92
counter-mounted lavatories, 17
counter self-rimming lavatories, 17
country clubs, 213, 214
CPVC gate valves, 229
cross-connections, 151
crown-vented traps, 20
crown weirs, 21
cryogenic gate valves, 231
cryogenic temperatures, 229
CSA (Canadian Standards Association), 9, 161
cubic feet per second flow rates, 29
cup drinking-water systems, 203
current meters, 156, 157
cushions, types of, 46
cuspidors, 51, 52
cutting burrs, 158

D

damage. *See* corrosion and corrosion resistance; hazards; scaling
Darcy formula
friction head loss, 117
maximum length of pipe and, 78
pressure losses in vent systems, 76–77
Dawson, F. M., 34, 35
dead-end service, 246
deck-type drinking fountains, 19
defects in fixture standards, 2
delay of hot water, 242
delivery pressure. *See* outlet pressure
demand

demand factor in equations, 174
estimating, 131
fixture units, 132–137
instantaneous water heaters and, 163
lavatories and, 15
types of, 131
urinals and, 12
water closets and, 5
water consumption tests, 8, 13, 14
demographics. *See* occupancies
density
defined, 74
friction head loss and, 117
water, 109
dental offices
fixture units, 52
plaster traps, 25
sink trap size, 51
deposits. *See* scaling; slime; sludge
design flow rates. *See* peak demand
destructive forces in pipes. *See* hydraulic shock; water hammer
detergents. *See also* bubbles; soaps
in septic tanks, 212
suds, 88–90
developed length
compared to equivalent length, 94
defined, 119
expansion and contraction, 193–194
in recirculating systems, 188
of vent stacks, 95
diaphragming actions
stack loading and, 35
water flow in pipes, 33
diaphragms
in exhaust valves, 102
in pressure-regulating valves, 245, 246
dielectric isolating joints, 158
differential check valves, 156–157
direct chlorine-gas feed, 196, 197
directly-heated automatic-storage water heaters, 162–163
direct-operated pressure-regulating valves, 243–245, 247
disabled individuals, water closets for, 8
discharge
discharging air from outlets, 75
drainage system branches, 50
fixture units, 135
hydrostatic head and, 53
loads in ejector systems, 107
from outlets, 75, 117
pneumatic ejectors and, 101
self-siphonage and, 81

storm drainage rates, 64–65
disc meters, 154, 155, 156–157
discs
butterfly valves, 252
disc-stem connections, 222–223
double-disc check valves, 237–238
gate valves, 222–223, 226–228, 229
globe valves, 233
turbulence and, 231
diseases, improper sewage treatment and, 207
dishwashers
fixture units, 52
hot-water demand, 175
minimum required pipe sizes, 140
suds, 88–90
trap sizes, 51
vacuum breakers and check valves, 153
water heaters, 163
disinfection. *See* sterilization and disinfection
displacement meters, 156
disposal fields (sewage). *See* subsurface soil-absorption systems
disposers, 214
dissection tables, 151, 153
distribution boxes, 220
Dole value, 127
domed strainers, 66–67
domestic systems. *See* cold-water systems; hot-water systems
Domestic Water Heating Design Manual, 160, 181
dormitories
hot-water demand and recovery estimates, 167, 170, 171
shower peak demand, 174
double. *See also* entries beginning with dual-, two-, etc.
double-compartment sinks, 17, 18
double-disc (double-door) check valves, 237–238
double-disc gate valves, 226, 228
double sanitary tees, in drainage pipes, 44
double-seated pressure-regulating valves, 243–244
double tees, 10, 44
douches, 19
downfeed chilled-water systems, 201
downfeed hot-water systems, 181, 182, 183
downspouts and leaders. *See also* stacks
connections, 63
materials, 67
rectangular, 66
storm water systems, 63
downstream pressure. *See* outlet pressure
drain, waste, and vent copper pipes, 44
drainage fixture units, 135
drainage stacks

relief vents, 83
suds pressure in, 90
wet venting, 84–87
drain bodies. *See* sump pumps; sumps
drain-down time, 69
drain fields. *See* subsurface soil-absorption systems
drain line carry tests, 9
drains and drainage systems. *See also* building drains (house drains); horizontal drainage; sanitary drainage systems; specific types of drains or drainage systems
  backwater valves, 39, 40
  below sewer level, 39
  branch connections, 41–42
  cleanouts, 44, 47
  combined systems, 38–39
  drainage fixture units, 135
  drain-down time, 69
  ejectors and sumps, 100–109
  fixture-drain sizing, 50–51
  high-temperature wastes, 39
  house-drain sizing, 58–62
  house traps and air inlets, 39–41
  indirect wastes, 47
  piping installation, 42–44
  sanitary house drains, 41
  special wastes, 47
  stack sizing, 52–58
  storm water, 38
  suds pressure, 88–90
  system sizing, 50
  vent stacks, 79
  water-supply systems and, 151, 157
drain valves, 9
drinking fountains. *See also* chilled drinking-water systems
  compared to water coolers, 198
  fixture units, 52
  minimum required pipe sizes, 140
  trap sizes, 51
  types, 19
drinking water
  chilled drinking water. *See* chilled drinking-water systems
  criteria, 151
  protecting, 151–152
drinking-water coolers. *See* water coolers
drip pan drainage, 47
dry pit pumps, 101, 105
dual. *See also* entries beginning with double-, two-, etc.
dual-flush water closets, 4. *See also* water closets
dual-temperature hot water heating, 178
dual vents (common vents), 83, 88
ductile iron gate valves, 231
ducts. *See* vents and venting systems

duplex. *See also* entries beginning with double-, dual-, two-, etc.
duplex ejector systems, 107
duplex pumps in ejectors, 105
duplex sump pump systems, 108
durability, in fixture standards, 2
duration of peak demand, 168
dwellings, 34–35, 167–168. *See also* large buildings; multiple-family buildings; residential systems
DWV copper pipes, 44
dye tests, 8, 13, 14

E

"easy access," defined, 2
Eaton, Herbert. N., 34, 52–53, 81–82
ecological cycle in sewage treatment, 208
economic concerns
  booster-pump systems, 148
  booster-water heaters, 164
  chilled drinking-water systems, 201
  hot-water systems, 160
  storage water heaters, 165–166
  water-supply system design, 128
Edison Electric Institute, 167–168
educational facilities. *See* schools
effluent, 208, 209, 211
egg-shaped sewers, 31, 32
ejector basins
  defined, 100
  sizing, 106–107
ejector pumps, 100, 103–106
ejectors
  basins, 100, 106–107
  controls, 107
  defined, 100
  in drainage-system calculations, 58, 59
  drainage systems below sewer level, 39
  installing, 107–108
  operation, 101–103
  pneumatic cycles, 107
  pumps, 100, 103–106
  venting, 90–91
elasticity
  air, 74
  water, 109
elastomeric seals or gaskets, 9
elastomeric seats in ball valves, 251
elbows, 193–194
electrical equipment
  horizontal drainage pipes and, 42
  water pipes and, 157

electric water coolers, 19
electric water heaters, 162
electrodes in ejector systems, 101, 107
electronic equipment, 42, 157
elevation of septic tanks, 210
elongated-bowl water closets, 4, 7. *See also* water closets
ELR. *See* equivalent length of run (ELR)
embalming tables, 151, 153
emergency power, storm drainage and, 108
empirical method, 132, 185
enameled cast-iron bathtubs, 18
enameled cast-iron fixtures, 2
enameled pressed-steel fixtures, 2
enameled steel bathtubs, 18
end connections
  gate valves, 230–231
  globe valves, 233
energy conservation (Bernoulli's theorem), 30–31, 114–115
energy-grade line, 31
Energy Policy Act of 1992, 5, 12, 15, 132
entrained air in fixture traps, 23
equations
  absorption trenches, 217
  air circulation, 79
  air column pressure, 76
  air discharge maximum rates, 75
  average fixture flow rates, 50
  Bernoulli's theorem, 30–31, 114–115
  Btu per hour for water heaters, 167
  cavitation pressure, 247
  check valve sizing, 241
  chilled drinking-water systems, 202–206
  combined sewer systems, 71–73
  Darcy formula, 76–77, 117
  developed length and flexure, 193
  drainage stack flow rates, 52
  drainage stack sizing, 55–58
  ejector basin sizing, 106
  ejector pump sizing, 105
  ejector vent sizing, 91
  equivalent length of run, 139
  fixture unit values, 136
  friction head, 115, 117–119
  friction head loss, 105–106
  grease interceptor sizing, 26
  Hazen and Williams formula, 141
  hot-water demand and recovery, 171–173
  hot-water peak demand, 168
  house drain sizing, 59–62
  Hunter's curve, 132–135
  hydraulic shock, 122
  kinetic energy, 111–112
  Manning's formula, 27–28, 97
  mixed-temperature demand for showers, 176
  outlet air discharge, 75
  outlet flow, 115, 117, 126
  parallel-pipe system flow, 130
  pipe expansion and contraction, 192–195
  pipe flow rates, 188
  pipe maximum lengths, 78
  pipe roughness and friction, 118
  potential energy, 111
  pressure waves, 121
  rainwater rates, 64–65
  recirculating hot-water rates, 187
  Reynolds formula, 110
  static head, 112–113
  terminal velocity, 34–35
  Toricelli's, 115
  turbulent flow, 110
  uniform friction head loss, 138, 187–188
  velocity head, 113–114
  velocity of flow, 110–111
  vent stack maximum length, 94
  weight of water, 112–113
  equipment, friction head loss and, 119
equivalent length of run (ELR)
  defined, 118–119
  recirculating systems, 188
  uniform friction head loss, 138–139
  vent stacks, 94
erosion
  discs in globe valves and, 233
  high pressure and, 241
  oversized valves and, 247
  valve selection and, 224
  velocity and, 124, 139
estimating
  costs. *See* economic concerns
  demand, 131
  demand with fixture units, 132–137
  equivalent length of run, 139
  hot-water demand and recovery estimates, 167–168, 169, 170, 171, 172–173, 175
  public utility water pressure, 138
Estimating Loads in Plumbing Systems, 132
evaporation in trap seals, 20
evaporative coolers, 130, 179. *See also* air-conditioning
evaporators in water coolers, 200
expansion
  high-temperature wastes and, 39
  of hot water, 161
  pipe calculations, 192–195

storm-drainage systems, 64
water-supply pipes, 158
expansion joints, 192–193, 194
expansion loops and offsets, 192–193
expenses. *See* economic concerns
exploding tanks, 161
explosion-proof water coolers, 199
eye nuts, 45–46

F

factories. *See* industrial facilities
failures, 20, 125
falloff in pressure-regulating valves, 246, 247
fats in kitchens. *See* grease, fats, and oils
faucets. *See also* fixtures and fixture outlets
  centersets, 17
  demand, 131, 133
  flow rates, 15
  lavatory openings, 17
  residential kitchen sinks, 17
fecal matter, 208, 209, 211
federal agencies. *See* specific agencies under "US"
feet of head, 112–113
feet of surcharge, 30
female urinals, 13. *See also* urinals
ferrous alloy storm drainage components, 67
fiberglass fixtures
  bathtubs, 18
  ejector or sump basins, 100
fibers in packing, 230
fibrous particles in valves, 222
finances. *See* economic concerns
fire hazards, valve selection and, 229
fire protection and suppression systems
  compound water meters, 157
  vacuum breakers and, 152
  valve selection and, 229
fish traps. *See* strainers
fittings. *See also* specific types of fittings
  backing off, 194
  equivalent length of run (ELR), 94, 118–119, 138–139, 188
  friction head loss and, 94
  septic tanks, 210–211
  uniform friction head loss, 138–139
five-elbow swing loops, 193
fixtures and fixture outlets. *See also* specific types of fixtures (water closets, showers, etc.)
  drainage system sizing and, 50–51
  flow rates, 50
  hot-water demand, 175
  indirect wastes, 47

manufacturer's pressure specifications, 125
plumbing engineers as selector, 2
quality standards, 2–3
standardization of hydraulic characteristics, 82
trap-seal loss, 22–24
types of fixtures, 4
fixture traps
  continuous venting, 84
  defined, 20
  distance from vents, 81
  drainage sizing and, 50–51
  house traps and, 40
  installation, 24–25
  interceptors, 25–26
  prohibited types, 20
  siphonage, 21–24
  trap seals, 21
  urinals, 13
  vents, 79–81, 83–88
fixture units
  in combined sewer system sizing, 70–73
  converting gpm to, 73, 137
  drainage stack sizing and, 55, 56
  ejector pump sizing and, 103
  estimating demand with, 132–137
  house drains and, 58
  loads in ejector systems, 107
  maximum permissible loads, 53–54
  sanitary drainage-system sizing, 51–52
  using in equations, 136
  vent sizing and, 99
  water-supply system sizing and, 140–141
  water supply vs. drainage, 135
flanged (bolted) bonnets, 225, 226, 244
flanged-end connections, 233
flanged-end valves, 231
flanged joints, 158
flange-type butterfly valves, 253, 254
flared-end connections, 231
flashing
  frost vent-closure prevention, 92
  roof drainage, 70
flashing rings in roof drains, 66–67
flat-bottomed fixtures, 81
flat-face flanged gate valve connections, 231
flexible-wedge disc gate valves, 228
flexure in piping systems, 192–195
floating-head water heaters, 163
floats in ejector systems, 107
floor drains
  combination waste-and-vent venting, 86
  fixture units, 52

265

grease interceptors, 25
special wastes, 47
floor-mounted urinals, 14. *See also* urinals
floor-mounted water closets, 4, 5, 6–7, 9. *See also* water closets
flow and flow behavior
Bernoulli's theorem, 114–115
check valves and, 236–237
constant flow demand, 126–128
definitions and terminology, 130–131
equivalent length of run (ELR), 94, 118–119, 138–139, 188
flow capacities in stacks, 35, 36, 77–78
friction head, 115
kinetic energy, 111–112
outlet flow, 115, 116
parallel-pipe systems and, 129
in pipes, 109, 115–117
potential energy, 111
static head, 112–113
types of, 110
velocity, 110–111
velocity head, 113–114
water-supply system design, 125
flow-control devices
defined, 126–127
regulating valves in water coolers, 200
showerheads, 174
valves for flow regulation, 222
flow pressure. *See also* pressure
defined, 75, 116, 125–126
minimum available pressure, 138
flow rates
controlled-flow roof drainage, 68–70
drainage pipes, 27–29, 44
fixture outlets, 50
fixture units and, 51
friction head loss and, 187
half-full flow, 28
horizontal drainage, 27–29
hot-water circulation systems, 185
instantaneous water heaters, 163
lavatories, 15
pressure-regulating valves and, 247
septic tank sewage, 213–214
showers, 174
stacks, 35, 52
suds, 89
types of piping and, 188
urinals, 15
water closets, 10–11
water meters, 156
Flowserve publications, 139

flushing
cleaning action, 4
government mandates, 5
performance testing, 8–9
urinal tests, 13–14
water closet system types, 10–11
water conservation and toilets, 5
flushing-rim sinks
fixture trap sizes, 51
fixture units, 52
minimum required pipe sizes, 140
flushometer tanks, 5–6, 10–11
flushometer-tank water closets, 4, 5–6. *See also* water closets
flushometer valves. *See* flush-valve (flushometer valve) water closets
flushometer water closets, 5–6. *See also* water closets
flush tanks, 133, 134
flush-valve (flushometer valve) urinals, 15. *See also* urinals
flush-valve (flushometer valve) water closets, 4, 5, 6, 11. *See also* water closets
flush valves
demand and, 133, 134
vacuum breakers and, 152
FOG. *See* grease, fats, and oils
food-processing areas and kitchens
food-scrap sink grease interceptors, 25
food-service establishments, 170
food-warming tables, 154
food-waste grinders, 214
sinks. *See* kitchen sinks
food-scrap sink grease interceptors, 25
food-service establishments, 26, 167–168, 170, 203, 213
food-warming tables, 154
food-waste grinders, 214
foot basins, 175
foot valves, 239
foreign matter
fixture traps and, 20
septic tanks and, 213
forged-steel globe valves, 233
formulas. *See* equations
fouling surfaces on fixtures, 2
fountains, drinking. *See* drinking fountains
four-elbow U-bends, 193
free-standing drinking fountains, 19
freezing
fixture traps and, 24–25
plugs in cleanouts, 44
vent terminal closures, 91–92
water meters and, 156
water's behavior, 109

French, John L., 81–82
fresh-air inlets, house traps and, 41, 42
friction
  air in stacks, 77
  air pressure and, 75
  Bernoulli's theorem and, 114
  coeffecient of friction. *See* roughness of pipes
  equivalent length of run (ELR), 94, 118–119, 138–139, 188
  friction head, 115
  loss. *See* friction head loss
  water's behavior and, 115
friction head, defined, 115
friction head loss
  ejector pump sizing, 105
  equations, 76–77, 117–119
  fittings and, 94
  flow in piping and, 115
  Hazen and Williams formula, 141
  hot-water circulation systems, 185
  hydraulic tables, 139, 142–147
  outlet air flow, 75
  pipe materials and, 118
  suds and, 89
  uniform friction head loss, 118, 138–139, 142–147, 185, 187–188
  velocity and, 124
  water-supply system sizing, 138–139, 141, 148
frost. *See* freezing
FU. *See* fixture units
full-flow rates. *See* peak demand
full-port ball valves, 251
fully-recessed drinking fountains, 19
fungi, 78
funnel drain trap sizes, 51

G

galling, 231
gallons per hour, drinking-water coolers, 199
gallons per minute
  air discharge, 76
  converting to fixture units, 51, 73, 133, 136, 137, 141
  flow equations, 111
  flow in hydraulic tables, 142–147
  loads in ejector systems, 107
  outlet flow, 126
  in recirculating hot-water systems, 191
  uniform flow rates, 29
galvanized ferrous alloy downspouts, 67
galvanized iron
  flow rate equation, 188
  friction head loss in pipes, 139
  pipe roughness, 118
galvanized-steel downspouts, 67
galvanized-steel piping
  flow rate equation, 188
  friction head loss in pipes, 139
  pipe roughness, 118
garbage grinders, 214
gases
  in septic tanks, 211
  valves and, 222, 232
gas-fired water heaters, 162
gaskets, 158
gate valves
  applications, 231
  backwater valves, 39
  bonnets, 225–226
  discs, 226–228
  end connections, 230–231
  functions, 222
  materials, 229
  operation and maintenance, 232
  packing, 230
  trends in use, 232
  trim, 229
  valve stems, 222–225
  in water-supply systems, 159
globe valves
  angle valves, 234–236
  components, 232–234
  defined, 232
  functions, 222
  installing, 236
  materials and connections, 233
  seals, 234
  seating, 232
  in water-supply systems, 159
glossary of valve terminology, 246
gph (gallons per hour), drinking-water coolers, 199
gpm. *See* gallons per minute
grades. *See* slopes
granular particles in valves, 222
granule tests, 8
graphite in valve packing, 230
gravel
  in absorption trenches, 218
  depth for absorption systems, 217
  in seepage pits, 221
gravity
  gravity flow, 27
  terminal velocity lengths and, 34
gravity circulation of air, 79
gravity drainage systems

elevating to, 39
subdrains and, 108
gravity flushes, 5
gravity-tank systems
increasing water pressure, 130
pressure in, 138
water closets, 4
grease, fats, and oils
grease interceptors, 25–26
grease traps, 25
velocity of flow for scouring action, 30
grease interceptors, 25–26
grease traps, 25
groundwater. *See also* storm water
in sewage treatment, 208
subsurface soil-absorption systems and, 214
gutters, 66
gymnasiums, 175

H

hair and lint traps, 26
half-full flow equations, 28
hammer. *See* water hammer
handicapped-design water closets, 4. *See also* water closets
handicapped individuals, water closets for, 8
hangers. *See* supports and hangers
hardness numbers for metal, 229
hardness of water, 128
Harvard Medical College, 179
hazardous wet wells, 101
hazards
improper sewage treatment and, 207
water-supply systems, 125
Hazen and Williams formula, 141, 142–147
head. *See* pressure
Heald, C. C., 139
health-care facilities
chilled drinking-water systems, 203
hospital hot-water demand, 175
hospital sewage flows, 213
nursing home hot-water demand, 167, 169
health hazards, 125, 207
heated water. *See* hot-water systems
heaters. *See* water heaters
heat gain in chilled drinking-water systems, 203
heating, ventilation, and cooling systems, 152
heat losses
examples, 188–191
hot-water systems, 171, 181, 185
recirculating systems, 185, 190
table, 186

Henriques, F. C., 179
high-pressure plumbing systems, 241
high-sudsing detergents, 88–90
high-temperature energy cut-offs, 162
high-temperature waste drainage, 39
holes for percolation tests, 215
honey dippers, 208
horizontal-ball lift check valves, 239, 240
horizontal branch connections
house drains, 41
sanitary tees and, 53
sizing for drainage systems, 55, 57
stack offsets, 41–42
stacks, 33
venting, 96, 98
horizontal drainage
air flow in, 78
branch connections, 55
flow patterns, 27
scouring action, 30
sewer shapes, 31–32
slope and, 28, 30
surcharging, 30–31
horizontal pipes
hydraulic jumps in, 35–37
installing drainage pipes, 42–44
hose bibbs
continuous demand and, 131
minimum required pipe sizes, 140
typical demand, 131
hospitals
chilled drinking-water systems, 203
hot-water demand, 175
sewage flows, 213
hotels and motels
chilled drinking-water systems, 203
fixture-unit studies, 133
grease traps, 26
hot-water demand, 167, 169, 175
selecting water-heater systems, 166–167
sewage flows, 213
shower peak demand, 174
hot-water systems
boiling points, 109–110
circulation systems, 181, 185–191
combined hot- and cold-water coolers, 199
delays, 181
demand and, 134, 167–168, 175
dual-temperature systems, 178
fixture hot-water demand, 175
heat loss, 171, 186, 190
mixing with cold, 176, 177

objectives, 160–161
occupancies and, 169–170
pipe expansion and contraction, 192
recommended hot-water temperature, 178, 179
safety devices, 161–162
scalding, 161, 164, 179–180
showers and, 174–176
sizing calculations, 149, 185–191
storage systems, 176–178
types of, 181–185
valves, 159
water heaters. *See* water heaters
house drains. *See* building drains (house drains)
houses, 34–35, 167–168. *See also* buildings; large buildings; multiple-family buildings; residential systems; specific types of buildings
house sewers, 38
house traps
clean outs, 44
issues, 39–41
HTH product, 196
Hubbard baths, 175
humidity, frost vent-closure prevention and, 92
Hunter, Roy B., 33, 35, 51, 132–135
HVAC pipes and systems, 152
hydraulic grade line, 31
Hydraulic Institute Engineering Data Book, 139
hydraulic jumps
air flow and, 77
branch connections and, 41–42
defined, 35–37
friction and, 75
hydraulic pressures, house drains and, 58
hydraulic radius
defined, 28
sewer shape and, 32
hydraulics, 109
hydraulic shock. *See also* water hammer
air chambers, 122
defined, 120
erosion, noise, and cavitation, 124
high pressure and, 242
pressure wave traveling time, 120–121
tests, 123
velocity and, 139
hydraulic tables, 139, 141, 142–147
hydrokinetics, 109
hydropneumatic tank systems
pressure in, 130, 138
sizing systems, 141, 148–150
hydrostatic head. *See* static head
hydrostatic pressure. *See* static head

hydrostatics, 109
hydro-therapeutic showers, 175

I

IBBM gate valves, 229
ice. *See* freezing; ice cubes
ice cubes, 199
ideal velocity, 115
impaired individuals, water closets for, 8
inconel wire packing, 230
indirect waste pipes, 47
individual vents
defined, 83
sizing, 96–97, 98
suds pressure, 89
induced siphonage, 21–22
industrial facilities. *See also* specific types of facilities
chilled drinking-water systems, 203
constant flow demand, 126
hot-water demand, 168, 175
sewage flows, 213
inferential water meters, 156
initial pressure (inlet pressure), 246, 248–249
ink tests, 8, 13
inlet pressure
parallel valve hookups, 248–249
synonyms, 246
inlets. *See also* outlets
cavitation and, 247
ejectors, 101
septic tanks, 210
submerged inlets, 152
inserts, types of, 45–46
inside dimensions (ID, internal diameters), 128, 139
inside screw valve configurations, 223, 224, 233
inspecting. *See* cleanouts; maintenance
installing
angle valves, 236
drainage pipes, 42–44
fixture traps, 24–25
gate valves, 232
hot-water systems, 160
pressure-regulating valves, 249
relief valves, 162
subdrains, 107–108
swing and lift check valves, 241
urinals, 14
water closets, 9–10
water coolers, 199
water heaters, 176–178
instantaneous water heaters, 163–166

insulation
  heat loss and, 186
  storm drainage systems, 65
  water heaters, 161
interceptors. *See* specific kinds of interceptors
intermittent demand, 131
internal diameters (ID, inside dimensions), 128, 139
invisible water seals, 4
iron body bronze-mounted gate valves, 229
iron gate valves, 229
iron globe valves, 233
irrigation, 130
isolating water meters, 154

J

joints
  in absorption pipe systems, 218
  end connections and, 231
  expansion joints, 192–193, 194
  high-temperature wastes and, 39
  in water-supply pipes, 158

K

Kalinske, A. A., 34
kettle trench-drain grease interceptors, 25
Kevlar in valve packing, 230
kinetic energy
  in Bernoulli's theorem, 30, 31
  equations, 111–112
  flow in piping and, 115, 116
  hydraulic shock and, 120
  pressure and, 125
  velocity head and, 113
kitchens. *See* food-processing areas and kitchens
kitchen sinks
  fixture units, 52, 133
  hot-water demand, 175
  minimum required pipe sizes for, 140
  size of, 17
  sizing sanitary systems and, 60
  suds, 88–90
  typical demand, 131
  water storage needs for, 174
knife gate valves, 231

L

laboratory facilities
  constant flow demands, 126
  laboratory-sink trap size, 51
  laboratory-table venting, 86

laminar flow, 110
landscaping irrigation, 130
lap baths, 175
large buildings
  maximum permissible fixture unit loads, 53–54
  relief vents in, 83
  sizing stacks, 55–58
  stacks and flow rates, 52
  terminal velocity lengths, 34–35
laundries, self-service, 213
laundry systems and washers
  fixture units, 133
  soil-absorption system rates, 214
  suds, 88–90
  typical demand, 131
  vacuum breakers and check valves, 153
laundry trays
  fixture-trap sizes, 51
  minimum required pipe sizes, 140
  suds and, 88–90
  typical demand, 131
laundry tubs
  hot-water demand, 175
  water storage needs for, 174
lavatories. *See also* sinks and wash basins
  Energy Policy Act of 1992 criteria, 132
  faucets and overflows, 17
  fixture units, 52, 133
  flow rates, 15
  hot-water demand, 175
  minimum required pipe sizes, 140
  self-siphonage and, 22–24, 82
  shapes and sizes, 15–17
  sizing sanitary systems and, 60
  trap sizes, 51
  typical demand, 131
  water-storage needs for, 174
lawn sprinklers
  continuous demand and, 131
  summer months and, 130
  vacuum breakers, 152
leaching trenches. *See* absorption trenches; subsurface soil-absorption systems
leaders. *See* downspouts and leaders; stacks
lead roof drains, 67
leakage
  butterfly valves and, 251
  gate valves and, 228, 231
  globe valves and, 234
  in horizontal drainage pipes, 42
  hydraulic shock and, 120
  testing for, 196–197

valve selection and, 222
ledge-back lavatories, 16, 17
Legionella Pneumophila, 178–179
length of pipes
  equivalent length of run (ELR), 94, 118–119, 138–139, 188
  expansion and contraction, 158, 192–195
  friction head loss and, 117, 138
  hot water at tap and, 181
  level sensors in ejector systems, 107
lift check valves
  installing, 241
  operation, 239, 240
  pressure loss, 236–237
  sizing, 241
lift stations, 100, 101–105
limestone, 216
line pressure
  parallel valve hookups, 248–249
  synonyms, 246
lint traps, 26
lip-sealed bonnets on globe valves, 233
liquids, valve selection and, 222
loads
  controlled-flow roof drainage, 68–70
  ejector systems, 107
  fixture units, 51, 132–137
  stacks, 58
  storm-drainage piping, 66
  water coolers, 202
  water-supply systems, 131–132
local vents, 90
long-turn stack fittings, 82
long-turn tee wyes, 53
loop venting
  defined, 87–89
  fixture traps, 80
  sizing, 97–98
lubrication
  gate valves, 223
  graphite-lubricated packing, 230
  plug valves, 250
lug-type butterfly valves, 253, 254
lye in septic tanks, 212

M

maintenance
  gate valves, 232
  hot-water systems, 160–161
  hydraulic shock and repairs, 120
  plug valves, 250
  pressure-induced equipment problems, 241

pressure-regulating valves, 249
makeup water
  air-conditioning, 131
  chilled drinking-water systems, 204
malls or retail areas, 203
manholes
  ejector venting, 101
  septic tanks, 210
  water in, 30
manifold systems (parallel pipe), 128–130
Manning, Robert, 27, 28
Manning's formula
  half-full flow, 97
  horizontal drainage flow, 27–28
  hydraulic radius and velocity, 31–32
manual flushometer valves, 11
manufacturers' pressure specifications for fixtures, 125
Manufacturers Standardization Society of the Valve and Fittings Industry (MSS), 230
mass equations, 111–112
materials. See specific materials or system fixtures
maximum flow (maximum possible flow)
  defined, 130
  in design loads, 131
maximum probable flow. See peak demand
Maxochlor product, 196
mechanical failures
  fixture traps, 20
  water supply system design, 125
medical and health clinics. See health-care facilities; hospitals
mercantile establishments, 203
mercury switch floats in ejector systems, 107
metals. See specific metals
method of probability, 132
microorganisms. See bacteria
milkiness in chilled-water systems, 201
minimum available pressure, 138
mixing hot and cold water, 174–176, 177
mixing valves, demand and, 133
mobile home parks, 213
mop basins, 17
Moritz, A. R., 179
mortuary tables, 151, 153
motels. See hotels and motels
motors
  removal systems for, 101
  water pipes and, 157
mounting
  drinking fountains, 19
  lavatories, 17
  top-mounted pumps, 101, 103, 104

urinals, 14
water closets, 4, 5, 6–7, 9–10
movement in pipes, 64, 192–195. *See also* contraction; expansion
(MSS) Manufacturers Standardization Society of the Valve and Fittings Industry, 230
multifamily dwellings. *See* multiple-family buildings
multiple. *See also* entries beginning with double-, dual-, etc.
multiple-family buildings
fixture-unit studies, 133
hot-water demand and recovery estimates, 167–168, 169, 172–173, 175
recirculating hot-water systems, 188–191
sewage flows, 213
multiple pumps in ejectors, 105
multistory buildings. *See* large buildings
municipal water system pressure in, 130
mushroom strainers, 66–67

N

National Bureau of Standards (National Institute of Technology & Science)
branch venting, 96–97
distance between vents and traps, 81
Estimating Loads in Plumbing Systems, 132
frost closure, 91–92
Report BMS 126, 23
Self-Siphonage of Fixture Traps, 81–82
National Fire Protection Association, Inc. (NFPA), 157
NBS (National Bureau of Standards). *See* National Bureau of Standards
Neal, Maynard, 139
negative pressure. *See also* vacuum
air flow in stacks, 77
fixture traps and, 22
neoprene seal plug cleanouts, 44
neoprene seats in ball valves, 251
NFPA (National Fire Protection Association, Inc.), 157
nightclubs and bars, 175
nitrile rubber in valve packing, 230
NITS (National Institute of Technology & Science). *See* National Bureau of Standards
no-flow pressure in pressure-regulating valves, 246
noise. *See* acoustics in plumbing systems
nominal flow rates, 127
non-potable water systems, 151
non-residential building septic tanks, 213
non-rising stem valve configurations, 223, 224, 227
non-slamming check valves, 238–240
nursing homes. *See* health-care facilities

nylon seats in ball valves, 251

O

oakum, 158
occupancies
chilled drinking-water systems, 203
fixture-unit studies, 133
hot-water demand and, 167–170, 175
private-occupancy buildings, 133
public-occupancy buildings, 133
recovery and, 169–170
odors, air circulation and, 78
office buildings
chilled drinking-water systems, 203
hot-water demand and recovery estimates, 167, 169, 171–172, 175
offset disc butterfly valves, 252–254
offsets
expansion and contraction control, 193
relief vents and, 83
storm drainage systems, 64, 65
venting and, 86
oil interceptors, 100
one-compartment sinks, 17
one-piece tank water closets, 4, 5. *See also* water closets
open-channel flow
defined, 31
vs. closed conduit, 30
openings for tool access. *See* cleanouts
operating tables, 151, 153
organisms in water. *See* bacteria
orthopedic areas, 25
OS and Y stem configurations, 223, 233
outdoor air pressure, 78
outlet pressure
parallel valve hookups, 248–249
pressure-regulating valves and, 246–247
synonyms for, 246
outlets. *See also* inlets
discharge from, 117
ejectors, 101
flow from, 115, 116, 126
pressure-regulating valves and, 244, 246
self-siphonage and, 81
septic tanks, 210–211
outside-lever swing check valves, 237
outside screw and yoke configurations, 223
oval-shaped sewers, 31, 32
overflow
lavatories, 17
scuppers, 68–69

water closets, 11
overshooting, 164
overtightening packing glands, 230
oxidation, 158

P

P (pressure). *See* pressure
P-traps, 82
packing gate valves, 230
packing glands, 229, 230
packing nuts, 232
painting pipes, 158
panel boards, 157
pans
  for horizontal drainage pipes, 42
  for water-supply pipes, 157
pantry sinks, 175
parallel-faced double-disc gate valves, 228
parallel-pipe systems, 128–130
parallel-valve hookups, 248–249
park sewage flows, 214
paved area drainage, 108
peak demand
  in design loads, 131
  equations, 136
  fixture units in, 136
  flushometer valves and, 11
  full-flow rate equations, 28
  hot-water systems, 168
  maximum probable flow, 130–131
  maximum velocity, 139–140
  minimum available pressure, 138
  pressure-regulating valves and, 247
  sewers, 31
  showers, 174–176
  storage water heaters and, 165
  synonyms for, 246
  urinals, 15
  velocity and, 139–140
peak flow. *See* peak demand
pedestal drinking fountains, 19
peppermint vapor tests, 93
Perchloron product, 196
percolation rates and tests, 214–216, 219, 220
performance standards for fixtures, 2
performance tests. *See* testing
pH, water system materials and, 128
physically-challenged individuals, water closets and toilets for, 8
picnic park sewage flows, 213
pilot-operated pressure-regulating valves, 243–245

pipe compound, 158
pipe hangers. *See* supports and hangers
pipes and piping. *See also* sizing; specific kinds of piping or piping functions; specific materials (copper, plastic, etc.)
applications
absorption trenches, 216–217
chilled drinking-water systems, 205–206
controlled-flow roof drainage, 68–70
drainage installation, 42–44
ejector systems, 106
hot-water systems, 161
seepage pits, 221
storm systems, 64, 67
vent pipe maximums, 78
water-supply systems, 157–158
burrs and edges, 158
expansion and contractions, 39, 64, 158, 192–195
flow behavior, 109, 115–117, 129
flow rates, 27–29, 44, 188
friction head loss, 115, 117, 118, 138–139, 140
friction in, 117–119
frost vent-closure prevention, 92
heat loss, 171, 181, 185, 186, 188–191
high-temperature wastes and, 39
hydraulic tables, 139, 142–147
inside diameters, 128, 139
minimum required pipe sizes, 140
movement in, 192–195
recommended velocity, 124
roughness, 27, 36, 110, 117, 118, 124, 128
sizing, 128, 138–139, 140
supports and hangers, 43–44, 158, 194–195
testing, 196–197
pitch of horizontal drainage pipes, 42
planter drain sand traps, 26
plaster traps, 25
plastic downspouts, 67
plastic fixtures, 2
plastic water closets, 4. *See also* water closets
plastic wraps on toilet seats, 8
plug cocks, 159
plug discs in globe valves, 234
plugs of water in fixture traps, 24
plug valves
  functions, 222
  quarter-turn valves, 249–250
Plumbing Engineering Design Handbook, 171
plumbing engineers
  fixture selection and, 2
  systems designed by, 1–2
plumbing fixtures. *See* fixtures and fixture outlets
plumbing systems, 1–2

plunger duplex meters, 156
pneumatic-assist flush water closets, 4
pneumatic ejectors, 101, 107
pneumatic pressure fluctuations, 58
polypropylene gate valves, 229
polyvinyl chloride (PVC) gate valves, 229
poppets in check valves, 240
positive pressure of air flow in stacks, 77
possible maximum demand, 174
potable water. *See* drinking water; private water utilities; wells
potassium hydroxide, 212
potential energy
Bernoulli's theorem, 30, 31
equations, 111
flow in piping, 115
flow pressure, 125
velocity head, 113
pot sinks
fixture trap sizes, 51
fixture units, 52
grease interceptors, 25
pounding forces in water. *See* hydraulic shock; water hammer
pounds-force per square inch, 75
pounds per square inch (psi)
converting feet of head to, 112–113
maximum in plumbing systems, 241
trap seals and, 21
uniform friction head loss, 138
power outages, sump-pump systems and, 108
PP (polypropylene) gate valves, 229
precipitation. *See* rainwater and precipitation
precoolers for water coolers, 200
pre-fabricated septic tanks, 210
pre-fabricated shower bases, 18
pre-fabricated shower enclosures, 18
pressure. *See also* pressure drops
air pressure, 74, 75, 76, 78
ejector pump sizing, 105–106
fixture-trap siphonage and, 21
flow-control devices, 126–127
flow pressure, 116
fluctuations in, and pressure-regulating valves, 245–246
friction head, 115, 119
in hot-water systems, 161
inadequate, in water-supply systems, 130
losses. *See* pressure drops
methods of increasing, 130
minimum available, 138
pressure-regulating valves. *See* pressure-regulating or reducing valves (PRV)

pressure waves. *See* hydraulic shock; water hammer
relief vents, 83
static head, 75, 112–113
in storm-water systems, 63
suds pressure, 88–90
sump pumps and, 108
surcharging, 30–31
valve selection and, 222, 229
variations in stacks, 33
velocity head, 113–114
water backup and, 52
water-supply system design, 125, 130
water-system tests, 196
pressure and temperature-relief valves. *See* pressure-regulating or reducing valves (PRV)
pressure drops
check valves and, 236–237
fixture traps and, 23
friction head loss, 76–77, 124
ejector pump sizing, 105
equations, 76–77, 117–119
fittings and, 94
flow in piping and, 115
Hazen and Williams formula, 141
hot-water circulation systems, 185
hydraulic tables, 139, 142–147
outlet air flow, 75
pipe materials and, 118
suds and, 89
uniform friction head loss, 118, 138–139, 142–147, 185, 187–188
velocity and, 124
water-supply system sizing, 138–139, 141, 148
globe valves and, 232
length of pipes and, 78
in parallel-pipe systems, 129
valves and, 222, 226–228, 236–237
water meters and, 155
pressure-feed chlorinators, 197
pressure loss. *See* pressure drops
pressure-regulating or reducing valves (PRV)
cavitation and, 247
compared to flow-control devices, 128
glossary, 246
in hot-water systems, 161–162
installing, 249
operation and advantages of, 241–243
outlet pressure, 246–247
in parallel hookups, 248–249
in series hookups, 247–248
sizing, 247
types of, 243–246

in water-supply systems, 159
pressure regulators, 10–11. *See also* pressure-regulating or reducing valves (PRV)
pressure-relief valves. *See* pressure-regulating or reducing valves (PRV)
pressure-seal bonnets, 225, 226, 233
pressure surges, relief valves and, 162
pressure-type water coolers, 199
pressure waves. *See* hydraulic shock; water hammer
priming pumps, 151
prison water closets, 4
privacy, urinals and, 13, 14
private-occupancy buildings, 133
private sewage-disposal systems
  cesspools, 208
  criteria, 207–208
  drainage into, 38
  septic tanks, 208–214
  significance of, 207
  subsurface soil absorption, 214–221
private-use lavatories, 15
private water utilities, 151
probable maximum demand, 174
propane-fired water heaters, 162
proportional meters, 156
protective coatings on tanks, 210
PRVs. *See* pressure-regulating or reducing valves (PRV)
PSI. *See* pounds per square inch (PSI)
public drinking-water fountains, 203
Public Health Service, 220
public-occupancy buildings, 133
public park sewage flows, 214
public sewage-disposal systems
  drainage into, 38
  vs. private systems, 208
public storm sewers
  drainage, 38
  sanitary waste and, 38
public-use lavatories, 15, 175
public water-supply systems, 138
pumps
  alternating in ejector systems, 107
  check valves and, 122
  ejectors and sumps, 100
  hydraulic shock and, 120
  illustrated, 102, 103, 104, 105
  lift stations, 101–105
  priming, 151
  removal systems for, 101
  required pressures for, 150
  sizing
    chilled drinking-water system, 204

  ejectors, 103–106
  recirculating systems, 188
  water-supply systems, 141, 148–150
  submersible, 101, 102
  suction intakes, 106
  sump pumps, 108
  in water coolers, 202
  purity, water properties and, 109
PVC gate valves, 229

Q

quality standards for fixtures, 2–3
quarter-turn valves, 249–254
quietness. *See* acoustics in plumbing systems

R

rainwater and precipitation. *See also* storm water
  angle of fall, 65–66
  rainwater drains. *See* storm-drainage systems
  rates of, 64–65
  storm intensities, 69
raised-face flanged gate-valve connections, 231
rapid closure of valves, 120
rate of flow. *See* flow rates
rats, 40
receiving tanks for subdrain wastes, 100
receptacles for indirect wastes, 47
recirculating systems
  circulation rates and heat loss, 190
  heat losses, 171, 186
  hot water, 181
  sizing, 185–191
  types of, 181–185
recovery
  hot water heaters, 168
  occupancies and, 169–170
  system heat losses and, 171
recreational establishments, 175
rectangular gutters, 66
reduced-flow pressure, 246. *See also* outlet pressure
reduced-port ball valves, 251
refrigeration
  drinking-water coolers, 198–200, 203
  relief valves for units, 153
regulations. *See* codes and standards
regulators. *See* specific types of regulators
relief-valve discharges
  drainage, 47
  pipes, 162
relief valves

installation, 162
refrigeration units, 153
relief vents
  components and properties, 83–85
  sizing, 97
  suds pressure, 89
renewable discs, 236
renewable seats in globe valves, 234
repairs. *See* maintenance
replacement of water-supply system parts, 128
residential systems. *See also* cold-water systems; hot-water systems
  boiling points in hot water, 109–110
  hot-water demand, 175
  hot-water systems. *See* hot-water systems
  kitchen sinks, 17
  lavatory flow rates, 15
  private sewage-disposal systems. *See* private sewage-disposal systems
  sewage flows, 213
  water meters, 154
resilient-seated butterfly valves, 252, 254
resistance to flow, house traps and, 41
response in pressure-regulating valves, 246
restaurants. *See also* food-processing areas and kitchens
  chilled drinking-water systems, 203
  grease traps, 26
  hot-water demand, 167–168
  sewage flows, 213
restrooms. *See* lavatories; water closets
retail stores and shops, 203
retention period in septic tanks, 213
reverse-trap water closets, 4, 5. *See also* water closets
Reynolds formula, 110
rims on water closets, 8, 9
rings, types of, 45–46
risers
  control valves, 159
  expansion, 195
  riser diagrams, 140, 189
  rising stem-valve configurations, 223, 233
Robert A. Taft Sanitary Engineering Center, 215
Roman water systems, 250
roof drains
  components, 66–67
  controlled-flow drainage, 68–70
  design configuration, 69
  drain-down time, 69
  materials, 67
  roof loading, 68–69
  suggested code, 70
  temperature and, 63–64

roofs, extending stacks through, 77, 79
root penetration, 218
rotary-piston meters, 156
rough-in dimensions, 7
roughness of pipes
  comparisons, 118
  friction-head loss and, 117
  horizontal flow rates and, 27
  hydraulic jumps and, 36
  laminar flow and, 110
  velocity and, 124
  in water-supply system design, 128
round bowls on water closets, 7
round fixtures, self-siphonage and, 81
round-front bowl water closets, 4, 7. *See also* water closets
round gutters and downspouts, 66
rubber butterfly valves, 251
rubber in valve packing, 230
runoff. *See* storm water
RV. *See* pressure-regulating or reducing valves (PRV)

S

saddles, types of, 46
safety. *See also* hazards
  hot-water system devices, 161–162
  hot-water systems, 160
  Legionnaires' disease, 178–179
  scalding, 161, 164, 179–180
sand traps, 25
sanitary building drains. *See also* sanitary drainage systems
  connections, 41
  uniform flow rates, 29
sanitary drainage systems
  ejectors, 100–108
  fixture units, 51–52
  maximum permissible loads, 53–54
  testing, 92–93
sanitary sewer systems. *See also* private-sewage disposal systems
  ejectors, 100–108
  pneumatic effects in, 75
  pressure variations in, 24
  public systems, 38
  sewage treatment plants, 38–39
  sewer gases, 40–41, 42
  sizing house drains for, 59–62
  storm-water systems and, 39, 63, 70–73
sanitary tees, 10, 33, 44, 53
sanitation. *See* cleanouts; sterilization and disinfection
saturation in percolation tests, 215
scalding, 161, 164, 179–180

276

Index

scaling
water heaters and, 164
water-supply system design, 125
Schedule 40 steel hydraulic tables, 142–147
Schedule 80 steel hydraulic tables, 142–147
school laboratories. *See* laboratory facilities
schools
  chilled drinking-water systems, 203
  hot-water demand and recovery estimates, 167–168, 170, 173, 175
  sewage flows, 213
  water-heater selection example, 165
scouring action
  flow rates for, 30
  house drains and, 58
  storm-drainage systems, 68
screening urinals, 14
screwed bonnets on valves, 225, 233, 235
screwed ends on valves, 230–231, 233
screwed joints, 158
screwed pipes, 44
scullery sinks
  fixture trap sizes, 51
  fixture units, 52
  grease interceptors, 25
scum in septic tanks, 209, 210, 211
scuppers, 68–69
sealed-pressure actuators, 107
sealing globe valves, 234
seals
  elastomeric seals or gaskets, 9
  globe valves, 234
  trap-seal tests, 8, 9
  water closets, 9
  wax-ring seals, 9
seasonal residence sewage flows, 213
seating in globe valves, 232, 233
seat rings in gate valves, 229
seats (water closets), 8, 9
secondary storm-drainage scuppers, 68–69
secondary valves, 247–248
seepage beds, 218–220
seepage pits
  compared to cesspools, 208
  design, 214
  distance between components, 217
  installing, 220–221
  locations for, 216
self-cleaning fixture traps, 20
self-contained remote-type water coolers, 199
self-siphonage
  defined, 22–24

distance from vents and, 81
tee wyes and, 33
trap vents and, 80
Self-Siphonage of Fixture Traps, 81–82
semi-circular gutters, 66
semi-circular lavatories, 175
semi-instantaneous water heaters, 164–165
semi-recessed drinking fountains, 19
sensitivity of pressure-regulating valves, 246
septic tanks, 208–214
  access, inlets and outlets, 210–211
  chemical additives, 212–213
  cleaning, 211–212
  compartments, 211
  distance between components, 217
  functions of, 208–214
  location, 209
  non-residential buildings, 213
  prevalence of, 207
  shape, 211
  sizing and materials, 210
service sinks
  defined, 17
  fixture-trap sizes, 51
  fixture units, 52, 133
  minimum required pipe sizes, 140
service stations, 213
set pressure, 246
settling basins, 108
sewage-disposal systems. *See* private sewage-disposal systems; public sewage-disposal systems; sanitary sewer systems
sewage effluent, 208, 209, 211
sewage lift stations, 100, 101–105
sewage treatment plants, 38–39
sewer gases, 40–41, 42
sewer systems. *See also* building sewers; cesspool systems; private sewage-disposal systems; public sewage-disposal systems; septic tanks; storm-drainage systems; subsurface soil-absorption systems; specific types of sewers
  combined systems. *See* combined sewers
  drainage systems below sewer level, 39
  indirect and special wastes, 47
  shapes and drainage, 31–32
sharp edges on pipes, 158
shelf-back lavatories, 16, 17
shell and coil water heaters, 165
shields, types of, 46
shock absorbers
  hydraulic shock and, 122–123
  illustrated, 123
  tests, 123

in water supply systems, 159
shock-wave traveling time, 120–121
shopping malls, 203
shops, 203
short-turn fittings
   in horizontal pipes, 44
   trap seal loss and, 82
shower pans, 18
showers
   Energy Policy Act criteria, 132
   fixture-trap sizes, 51
   fixture units, 52, 133
   hot-water demand, 175
   minimum required pipe sizes, 140
   peak demands and, 174–176
   requirements, 18
   shower pans, 18
   types, 18
   typical demand, 131
   water storage needs for, 174
shut-off valves
   ejector pumps, 108
   water-supply systems, 159
silent check valves, 239–240
   installing, 241
spring-loaded check valves, 122
sill cocks, 131
simultaneous use of fixtures, 52
single-compartment sinks, 17
single-flange butterfly valves (lug), 254
single-hand hole traps, 63
single-seated pressure-regulating valves, 243–246
sink-disposal units, 214
sinkholes, 216
sinks and wash basins. *See also* lavatories
   Energy Policy Act criteria, 132
   fixture units, 52
   grease interceptors, 25
   kitchen sinks, 17
   minimum required pipe sizes, 140
   service sinks, 17
   trap size, 51
   typical demand, 131
siphonage, 12, 21–24, 79
siphon-jet reverse-trap water closets, 4, 5. *See also* water closets
siphon-jet urinals, 13. *See also* urinals
siphon-vortex water closets, 4. *See also* water closets
siphon-wash water closets, 4. *See also* water closets
sitz baths, 175
six-elbow swing loops, 193
sizing

absorption areas, 216
absorption trenches, 217
check valves, 241
chilled drinking-water systems, 202–204, 204
circuit and loop venting, 97–98
combined sewer systems, 71–73
controlled-flow roof drainage, 68–70
drainage-system branches, 50
drainage systems, 50
drainage-system stacks, 55–58
ejector basins, 106–107
ejector pumps, 103–106
fixture traps, 25
grease interceptors, 26
gutters, 66
hot-water circulation systems, 185–191, 191
house drains, 58–62
individual and branch vents, 96–97, 98
nonresidential septic tanks, 213
pipes, 128
pressure-regulating valves, 246–247, 247
pumps for water supply, 148
relief vents, 97
sanitary branches, 53–54
seepage pits, 220
septic tanks, 210
storage water heaters, 167–176
storm-drainage systems, 66
sump basins, 108
sump pumps, 108
vent systems, 94–99
   extensions and terminals, 94
   headers, 94, 99
   stacks, 97
water heaters, 162, 165–166
water-supply systems
   friction head loss, 138–139
   hydropnuematic or booster pumps systems, 141, 148–150
   maximum velocity, 139–140
   minimum required pipe sizes, 140
   procedures, 140–141
   skin damage from hot water, 179–180
slab-type lavatories, 15, 16
slanting-disc check valves, 238–239
slide valves, 231
sliding stem-valve configurations, 233
slime, 78
slip-expansion joints, 194
slopes
   air flow in drains and, 78
   horizontal drainage flow and, 27–29, 30
   house drains and, 58

hydraulic jumps and, 36
self-siphonage and, 81, 82
slop sinks
  hot-water demand, 175
  minimum required pipe sizes, 140
sludge
  in cesspools, 208
  chemicals and, 212
  in septic tanks, 209, 210–211
slugs of water
  in fixture traps, 24
  in stacks, 33
smells, 78
smoke tests, 93
smoothness of pipes. *See* roughness of pipes
snow
  frost vent-closure prevention, 92
  roof loads, 69
soaps
  in septic tanks, 212
  soap dispensers, 17
  suds, 88–90
socket-weld valve-end connections, 231
sodium hydroxide, 212
soil-absorption systems. *See* subsurface soil-absorption systems
soils
  percolation rates, 214–216
  subsurface soil-absorption systems, 214–221
soil stacks
  branch intervals, 54
  defined, 33
  extending through roof, 77
  suds pressure in, 90
soil vents. *See* stack venting
solder-end connections, 231, 233
solids in septic tanks and cesspools, 208, 211
solid toilet seats, 8
solid-wedge disc gate valves, 228
solution-feed chlorinators, 197
sounds. *See* acoustics in plumbing systems
spacing
  around urinals, 14
  around water closets, 10
span-type butterfly valves, 253, 254
specific weight, 74
splashback lavatories, 15, 16
split-rim toilet seats, 8
split rings, 45–46
split-wedge disc gate valves, 228
sports facilities, 134–135
spring-loaded check valves. *See* silent check valves

spring-loaded pressure-regulating valves, 245, 249
sprinkler system drainage, 47
spud flushometer water closets, 5. *See also* water closets
stack offsets
  branch connections and, 41–42
  sizing stacks and, 55, 56, 57
  venting, 86
stacks
  air flow in, 77
  capacities, 35, 36
  connections, 33
  defined, 33
  flow in, 33–34, 53, 77
  hydraulic jumps, 35–37
  interference with flow in, 53
  sizing in drainage systems, 52–58
  terminal velocity length, 34–35
  types of, 33
stack venting, 85–88
  branch interval venting, 85
  common vent headers, 94
  fixture traps, 80
  stack offset venting, 86
  vent extensions, 79, 94
  vent stacks. *See* vent stacks
  vent terminals, 79
stainless steel fixtures, 2
stall urinals, 13, 14. *See also* urinals
standard air, 74
standards. *See* codes and standards
standard water closets, 8. *See also* water closets
static head
  air in vent systems, 76
  Bernoulli's theorem, 30, 31, 115
  defined, 75
  ejector pump sizing, 105
  flow in piping, 115–116
  flow pressure and, 125
  hydrostatic pressure in drainage system branches, 50, 53
  in sizing procedures, 140, 148
  velocity head and, 113–114
static pressure. *See* static head
steady flow. *See* continuous demand
steam exhaust, 39
steam expansion tank drains, 47
steam tables, 154
steam water-heating systems, 165
steel bathtubs, 18
steel downspouts, 67
steel ejector or sump basins, 100
steel fixtures, 2
steel piping. *See also* galvanized-steel piping

279

flow rate equation, 188
friction head loss in pipes, 139
pipe hydraulic tables, 142–147
pipe roughness, 118
pressure waves in pipes, 121
steel septic tanks, 210
steel valves
gate valves, 229, 231
globe valves, 233
stems
gate valves, 222–225, 229
glove valves, 233
sterilization and disinfection
equipment vapor vents, 90
sterilizers, 153
sterilizing equipment, 90, 152
water-supply systems, 196–197
stone-composition fixtures, 2
stoppages
cleanouts and, 44
in septic tanks and cesspools, 208, 209
storage tanks. See tanks
storage water heaters, 165–178
components, 166
economic concerns, 165–166
sizing, 167–176
stores, 203
storm building drains. See storm-drainage systems
storm-drainage systems
collection areas, 64–65
combined systems, 70–73
controlled-flow roof drainage, 68–70
drain-down time, 69
flow velocity, 68
gutters, 66
history of house traps, 40
materials, 67
power outages and, 108
provisions for, 38
roof drains, 66–67
roof loading, 68–69
sanitary waste and, 38, 63
sizing, 66
storm-drain flow rates, 29
storm intensities, 69
subsoil systems, 70
sump systems, 108
testing, 92–93
vertical walls and, 65–66
storm water
defined, 63
sanitary systems and, 63

sump basins, 108
storm-water lift stations, 100, 101–105
straight-tee fittings, 82
straight-tube water heaters, 164
strainers
pressure-regulating valves and, 249
in roof drains, 66–67
in water meters, 157
in water-supply systems, 159
streamline flow, 110
stream regulators in water coolers, 200
strength of fixtures, standards, 2
stresses
tensile stress on pipes, 192
valve selection and, 229
stuffing boxes, 230
subdrains
defined, 100
gravity drainage systems and, 108
installation, 107–108
submerged inlets, 152
submersible pumps, 101, 102
subsoil drainage
design, 70
sand traps, 26
subsoil drains compared to subdrains, 100
subsurface soil-absorption systems, 214–221
clogging, 208, 209
construction, 218
disposal fields (sewage), 208, 209
distance between components, 217
distribution boxes, 220
seepage beds, 218–220
seepage pits, 220–221
subsurface waste-disposal systems. See subsurface soil-absorption systems
subsurface water, 208, 214
suction intakes on pumps, 106
suds. See bubbles
summer months
tap-water temperatures, 198
water pressure, 130
sump basins, 100, 108
sump pumps, 58, 100, 153
sumps
defined, 100
drainage systems below sewer level, 39
roof drains, 66–67
subsoil drainage, 70
venting, 90–91
supply pressure
parallel valve hookups, 248–249

synonyms, 246
supports and hangers
  drainage pipes, 43–44
  illustrated, 45–46
  pipe movement and, 194–195
  water-supply systems, 158
surcharging, 30–31, 50
surface-mounted drinking fountains, 19
surface runoff. *See* storm water
surgeon's sinks, 18, 51, 52
surge pressure. *See* water hammer
surging flows
  after hydraulic jumps, 37
  in horizontal drainage, 29
suspended solids in septic tanks, 208
swelling in percolation tests, 215
swimming pools
  sewage flows, 213
  water heaters, 163
swing check valves
  backwater valves, 39
  double-disc (double-door) check valves, 237–238
  ejector pumps, 108
  hot-water systems, 159
  installing, 241
  operation, 237, 238
  pressure loss, 236–237
  pumps and, 122
  sizing, 241
  slanting-disc check valves, 238–239
swing loops, 193
system pressure. *See* outlet pressure

T

T&P devices, 161–162, 166
tall buildings. *See* large buildings
tank-emptying line drainage, 47
tank overflow drainage, 47
tanks. *See also* septic tanks
  chilled drinking-water systems, 204
  disinfecting, 197
  ejector systems, 106–107
  emptying-line drainage, 47
  heat losses, 171
  overflow drainage, 47
  septic tanks, 210
  sizing, 106–107, 174
  water heaters, 176, 177
tapered-plug discs, 233
tap-water average temperatures, 198
target areas in water closets, 7

tee wyes in horizontal branch connections, 33
Teflon packing, 230
telephone equipment, 157
temperature
  air density and, 74
  air pressure and, 78
  chilled-drinking water, 198
  freezing vent terminals, 91–92
  friction head and, 115
  friction head loss and, 117, 141
  frost vent-closure prevention, 92
  globe valves and, 236
  high-temperature wastes, 39
  hot-water systems, 160
  instantaneous water heaters, 164
  pipe expansion and contraction, 39, 192–195
  recommended hot-water temperature, 178, 179
  storm-drainage systems and, 63–64
  valve selection and, 222, 223, 224, 229
  water properties and, 109
temperature and pressure relief devices
temperature regulators in storage water heaters, 166
temperature-relief valves, 161–162
tempering valves, 159, 164
tensile stress on pipes, 192
terminal length, 34
terminal velocity lengths, 34–35
terminal vents. *See* vent terminals
testing
  air chambers, 123
  percolation rates, 214–216
  shock absorbers, 123
  storm and sanitary systems, 92–93
  urinal tests, 13–14
  water-closet flushing, 8–9
  water-piping systems, 196–197
test tees, 92
TFE packing, 230
theaters
  chilled drinking-water systems, 203
  sewage flows, 213
thermal shock, valve selection and, 229
thermoplastic piping, 124
thermostatic mixing valves, 133
thermostats
  in relief valves, 162
  in water heaters, 166
threaded ends on gate valves, 230–231
three-compartment sinks, 17
three-elbow swing loops, 193
throttling-service valves, 232, 251
tilting-disc check valves, 238–239

time intervals for pressure waves, 120
time-to-tap and wastage, 242
TMVs (mixing valves), 133
toilet paper in septic tanks, 213
toilet rooms. *See* bathrooms and bathroom groups
toilets. *See* bathrooms and bathroom groups; water closets
top-mounted pumps
   illustrated, 103, 104
   lift stations, 101
top-spud bowls, 5
top-spud water closets, 5. *See also* water closets
Toricelli
   flow from outlets, 115
   illustration of theorem, 116
total discharge, 55
total load, 52
T&P devices, 161–162, 166
TP (toilet paper), 213
trail flow
   lavatories, 23
   replenishing trap seals, 24
   self-siphonage and, 81
   transformers, 157
traps. *See* fixture traps; house traps
trap-seal loss
   air pressure and, 74
   backup of water and, 52
   limiting, 24
   pressure and, 22
   self-siphonage, 22–24, 81
   stack flow and, 34
   suds, 89
   types of fittings and, 82
trap seals
   defined, 21
   depth of, 82
   evaporation of, 20
   fixture-trap vents, 79
   house traps, 39–41
   limiting loss, 24
   minimum in fixture traps, 20
   replenishing, 24
   storm-water drainage and, 63
   tests, 8, 9
   water closets, 11
trap vents
   fixture-trap vents, 79
   methods, 83–88
trenches. *See* absorption trenches
trim, 229
turbine meters, 156–157
turbulent flow, 110

turnbuckles, 45–46
twenty-five-year storms, rate of flow and, 69
two. *See also* entries beginning with double-, dual-
two-circuit water supply systems, 128–129
two-compartment sinks, 17
two-elbow offsets, 193
two-hand hole traps, 41, 63
two-level drinking fountains, 19
two-piece tank water closets, 4. *See also* water closets
Type K copper
   friction head loss in pipes, 140
   pipe hydraulic tables, 142–147
   sizing calculations, 149
Type L copper
   friction head loss in pipes, 140
   pipe hydraulic tables, 142–147
   sizing calculations, 149
Type M copper
   pipe hydraulic tables, 142–147

U

U. S. agencies and departments. *See* US agencies and departments
U-tube water heaters, 163
U-bends, 193
U-bolted bonnets, 225, 226
U-covers, 158
UL (Underwriters' Laboratories, Inc.), 230
under-counter mounted lavatories and sinks, 17
underground piping
   drainage pipes, 42–43, 67
   water-supply pipes, 158
undershooting, 164
Underwriters' Laboratories, Inc., 230
uniform flow
   branches in drainage systems, 50
   horizontal drainage piping, 27–29
   velocity equations, 29
uniform friction head loss
   calculating, 187–188
   defined, 118, 138
   hydraulic tables, 139, 142–147
   sizing, 185
union bonnets, 225, 233, 234, 235, 236
unitary water-cooler systems, 200–203
United States agencies and departments. *See* US agencies and departments
United States Testing Company, 123
universities. *See* laboratory facilities; schools
unventilated water closets, 4
upfeed chilled-water systems, 205

Index

upfeed hot-water systems, 181, 182, 184
upstream pressure
parallel valve hookups, 248–249
synonyms, 246
urinals. *See also* water closets
Energy Policy Act criteria, 132
female, 13
fixture-trap sizes, 51
fixture units, 52, 133
flushing requirements, 15
installation requirements, 14
minimum required pipe sizes, 140
screening, 14
sizing sanitary systems and, 60
testing, 13–14
traps, 13
types, 12
typical demand, 131
waterless, 15
usable hot-water storage, 168
US Federal Building Administration, 218–219
US Federal Housing Administration, 220
US Public Health Service, 207, 220
utility quotes on water pressure, 130

V

vacuum, 21
vacuum breakers (backflow preventers), 162, 222
valves. *See also* specific types of valves
Ancient Roman, 250
equivalent length of run (ELR), 94, 118–119, 138–139, 188
glossary, 246
overview, 222
in parallel hookups, 248–249
rapid closure of, 120, 122
selecting, 222
in series hookups, 247–248
types of, 159. *See also* names of specific valve types
uniform friction-head loss, 138–139
valve stems. *See* stems
vapor vents, 90
velocity
branch discharge, 53
equations, 34–35, 110–111
erosion, noise, and cavitation, 124
flow from outlets, 115
flow in piping, 115–117
friction-head loss and, 117
half- and full-flow rates, 28
horizontal drainage pipes, 42
horizontal flow rates, 27

hydraulic jumps, 35–37
hydraulic shocks, 120
hydraulic tables, 139, 142–147
kinetic energy and, 111–112
maximum rates of air discharge, 75
plugs of water in fixture traps, 24
Reynolds number, 110
sewer shape and, 31
in sizing water-supply systems, 139–140
storm-drainage systems, 68
surges and, 29
terminal velocity lengths, 34–35
velocity head, 113–114, 116–117
velocity head
defined, 113–114
in piping, 116–117
velocity meters, 156
vent extensions
defined, 79
sizing, 94
vent headers, 94, 99
venting tables, 77
vents and venting systems. *See also* vent stacks; vent terminals
air flow
in drains, 78
from outlets, 75–76
in stacks, 77
air properties, 74
circuit and loop venting, 87–89, 97–98
combination waste and vent venting, 86–87
common vents, 88
continuous venting, 84
distances between vents and traps, 81
ejector and sump vents, 90–91, 107
fixture-trap vents, 79–81, 83–88
friction head loss, 76–77
frost closure, 91–92
gravity circulation of air, 78–79
maximum length of pipes, 78
pneumatic effects in sanitary systems, 75
relief vents, 83–85, 97
sizing
individual and branch vents, 96–97, 98
relief vents, 97
vent systems, 94–99
stack venting, 79, 80, 85–88, 94
static head, 75, 76
suds pressure, 88–90
tables, 77
testing pipe systems, 92–93
vapor or local vents, 90

283

vent extensions, 79, 94
venting tables, 77
vent stacks. *See* vent stacks
vent terminals and headers, 79, 91–92, 94, 99
wet venting, 84–87
vent stacks
  common vent headers, 94
  connections, 79, 80
  defined, 79
  developed length of run vs. equivalent length, 94, 95
  maximum length, 94
  relief vents, 83
  sizing, 97
  suds pressure in, 90
  venting tables, 77
  vent terminals, 79, 91–92, 94
venturi meters, 156
vermin, 40
vertical-ball life check valves, 239
vertical-disc butterfly valves, 252–254
vertical pipes, flow in, 33
vertical stacks. *See* stacks
vertical walls, storm drainage and, 65–66
vibration and vibration isolation
  hydraulic shock and, 120
  in piping. *See* hydraulic shock; water hammer
  velocity of flow and, 124
viscosity
  defined, 109
  turbulent flow and, 110
  valve selection and, 222
viscous flow, 110
vitreous-china fixtures, 2
vitreous-china water closets, 4. *See also* water closets
vitrified-clay septic tanks, 210
vitrified-clay storm pipes, 67
vitrified-earthenware fixtures, 2
vortexes in fixture traps, 23

W

wafer-style butterfly valves, 254
wafer-style silent check valves, 240
waiting for hot water, 242
wall areas of seepage systems, 221
wall carriers, 9–10
wall-hung tank water closets, 4, 5, 6. *See also* water closets
wall-hung urinals, 14. *See also* urinals
wall-hung water closets, 4, 5, 7, 9–10. *See also* water closets
wall hydrants, 140
walls, vertical, 65–66

warming tables, 154
wash basins. *See* sinks and wash basins
washdown urinals, 13. *See also* urinals
wash down water closets, 4, 5. *See also* water closets
washing machines. *See* laundry systems and washers
washout urinals, 13. *See also* urinals
washout water closets, 5. *See also* water closets
wash rooms. *See* bathrooms and bathroom groups; lavatories
wastage. *See* leakage; time-to-tap
waste-disposal units, 214
wasted water. *See* leakage; time-to-tap
waste or soil vents. *See* stack venting
waste stacks
  branch intervals, 54
  defined, 33
  suds pressure in, 90
  wet venting, 84–87
waste systems. *See also* sewer systems
  combination waste and vent venting, 86–87
  drainage, waste, and vent copper pipes, 44
  special wastes, 47
waste-water management. *See* private sewage-disposal systems; sewer systems
waste-water scouring action, 30
wasting water. *See* conserving water; leakage; time-to-tap
water. *See also* water analysis
  Bernoulli's theorem, 114–115
  flow in piping, 109, 115–117
  friction, 115, 117–119
  kinetic energy, 111–112
  minimum daily human needs, 198
  physical properties, 109–110
  potential energy, 111
  static head, 112–113
  types of flow, 110
  velocity, 110–111
  velocity head, 113–114
  wastage. *See* conserving water; leakage; time-to-tap
  weight of, 109, 112–113
water analysis
  private water systems and wells, 151
  in water system design, 128
water-closet compartments, 10
water closets. *See also* urinals
  bolts, 9
  categories, 5–7
  Energy Policy Act criteria, 132
  fixture units, 52, 133
  flushing-performance testing, 8–9
  flushing standards, 5
  flushing systems, 10–11

installation requirements, 9–10
minimum required pipe sizes, 140
mounting, 6–7, 7–8
rough-in dimensions, 7
seats, 8
self-siphonage, 82
shapes and sizes, 7–8
standards, 10
types of, 4–7
typical demand, 131
water closet flanges, 9
wet venting and, 84–85
water conservation. *See* conserving water
water-cooled equipment, 59
water coolers. *See also* chilled drinking-water systems
bubblers and stream regulators, 200
compared to drinking fountains, 198
continuous demand and, 131
defined, 19
milkiness in, 201
pipes and piping, 205–206
sizing systems, 202–204, 204
storage tanks, 204
water-distribution pipes and systems. *See* cold-water systems; hot-water systems; water-supply systems
water hammer
butterfly valves and, 254
high pressure and, 242
pressure-wave traveling time, 120–121
pump shut-off and, 108
relief vents and surge pressure, 84
silent check valves, 239, 240
swing check valves and, 237
as symptom of hydraulic shock, 120
valve closure and surge pressure, 120
velocity and, 139
water hardness, 128
water heaters. *See also* hot-water systems
booster heaters, 164
directly-heated automatic storage, 162–163
dual-temperature systems, 178
installation, 176–178
instantaneous water heaters, 163–164
Legionella Pneumophila and, 178–179
recommended hot-water temperature, 178, 179
recovery, 168
scalding and, 164, 179–180
selection, 161
semi-instantaneous water heaters, 164–165
showers and, 174–176
sizing, 167–176
storage water heaters, 165–178

types of circulation systems and, 181
usable capacity, 168
water impurities. *See* contamination
waterless urinals, 15. *See also* urinals
water losses. *See* leakage
water meters
compound meters, 155, 156–157
current or velocity meters, 156, 157
disc meters, 154, 155, 157
displacement meters, 155–156
plunger-duplex meters, 156
proportional meters, 156
rotary-piston meters, 156
rules for, 157
turbine meters, 156
types and installation, 154–157
venturi meters, 156
water-rise tests, 8, 9
water seals, 4
water siphons, 153
water softeners, 213
water sprinklers
continuous demand and, 131
summer month demand, 130
vacuum breakers, 152
water-storage tanks. *See* tanks
water-supply fixture units, 135
water-supply systems. *See also* cold-water systems; hot-water systems; private water utilities; wells
below-rim connections, 152–154
components, 151
pipes, 157–158
water meters, 154–157
constant flow, 126–128
demand, 131, 175
design loads, 131–132
design objects, 125
drainage systems and, 151
fixture units, 132–137, 135
flow at outlets, 126
flow definitions, 130–131
flow pressure, 125–126
hot-water systems. *See* hot-water systems
inadequate pressure, 130
materials, 128
parallel circuits, 128–130
sizing
factors, 138
friction head loss and, 138–139
hydropneumatic or booster pumps systems, 141, 148–150
maximum velocity and, 139–140
minimum required pipe sizes and, 140

procedures, 140–141
testing, 196–197
valves. See valves
water meters. See water meters
water wells. See wells
wax-ring seals, 9
wedge disc gate valves, 224, 226–228
weighting fixtures, 132–137
weight-loaded check valves, 108
weight-loaded pressure-regulating valves, 245, 249
weight of water, 109, 112–113
weirs
in fixture traps, 81
in roof drains, 69, 70
weld-end connections for gate valves, 231
weld-end connections for globe valves, 233
well pumps, 153
wells
water analysis, 151
well pumps, 153
wet wells, 101
wet venting
defined, 84–87
fixture traps, 80
wet wells, 101
wheelchair accessibility, 8
wind, frost vent-closure prevention and, 92
wire drawing. See also erosion
high pressure and, 241
oversized valves and, 247
women's urinals, 13
work-table venting, 86
wyes, 10, 33, 53
Wyly, R. S., 34, 52–53

Y

yarn packing, 230
Y.M.C.A. hot-water demand, 175